Trajectory Optimization and Guidance Methods for Mars Entry

Shuang Li · Xu Liu · Xiu-qiang Jiang ·
Yu-ming Peng

Trajectory Optimization and Guidance Methods for Mars Entry

北京理工大学出版社
BEIJING INSTITUTE OF TECHNOLOGY PRESS

Springer

Shuang Li
College of Astronautics
Nanjing University of Aeronautics
and Astronautics
Nanjing, Jiangsu, China

Xiu-qiang Jiang
College of Astronautics
Nanjing University of Aeronautics
and Astronautics
Nanjing, Jiangsu, China

School of Aeronautics and Astronautics
Sichuan University
Chengdu, Sichuan, China

Xu Liu
College of Astronautics
Nanjing University of Aeronautics
and Astronautics
Nanjing, Jiangsu, China

School of Information Engineering
Nanchang University
Nanchang, Jiangxi, China

Yu-ming Peng
College of Astronautics
Nanjing University of Aeronautics
and Astronautics
Nanjing, Jiangsu, China

Shanghai Institute of Satellite Engineering
Shanghai, China

ISBN 978-981-99-6281-5   ISBN 978-981-99-6282-2 (eBook)
https://doi.org/10.1007/978-981-99-6282-2

Jointly published with Beijing Institute of Technology Press
The print edition is not for sale in China (Mainland). Customers from China (Mainland) please order the print book from: Beijing Institute of Technology Press.

This Springer imprint is published by the registered company Springer Nature Singapore Pte Ltd.
The registered company address is: 152 Beach Road, #21-01/04 Gateway East, Singapore 189721, Singapore

Paper in this product is recyclable.

Contents

Part I
Fundamental Knowledge

Chapter 1
Introduction

1.1 Mission Heritage

1.1.1 Robotic Exploration Missions

As one of Earth's closest neighbors, Mars has attracted much attention for decades. Mars landing exploration missions have been conducted since the 1970s. Mars-2, launched by the Soviet Union in 1971, is the first lander to reach the surface of Mars, although it eventually crashed on the Martian surface [1]. Mars-3 and Mars-6 also reached the Martian surface in 1971 and 1973, respectively, but their signal disappeared soon after landing [1]. In 1976, the National Aeronautics and Space Administration (NASA) launched Viking-1 and Viking-2, both of which successfully landed on the surface of Mars [2]. The Mars entry, descent and landing (EDL) technologies developed by the Viking missions are the main heritage technologies for NASA's later Mars landing mission. On this basis, NASA's subsequent Mars landing missions, such as Mars Pathfinder (MPF) [3], Mars Exploration Rovers (MER) [4], Phoenix [5], Mars Science Laboratory (MSL) [6], InSight [7], and Mars 2020 [8], successfully landed on Mars. The European Space Agency (ESA) launched Mars Express in 2003, and its lander Beagle-2 successfully landed north of the Martian equator, but it fell silent after landing [9]. NASA also launched Mars Global Surveyor (MGS) [10], Mars Odyssey (MO) [11], and Mars Reconnaissance Orbiter (MRO) [12] in 1999, 2001, and 2005, which are representative orbital exploration missions. India launched its first Mars orbiter Mangalyaan in 2013 [13], which successfully reached Mars orbit in 2014. ESA also launched the ExoMars 2016 to implement the orbiting and landing mission [14]. Although the Trace Gas Orbiter successfully performed the aerobraking maneuver, the Landing Demonstrator Module (Schiaparelli) separated from the backshell earlier than expected, compromising the landing phase. InSight placed a stationary lander on the surface of Mars in 2018 [7], using an architecture similar to that of the Phoenix mission. The Mars 2020 entry, descent, and landing mission features three improvements over the previous 2012 Mars

S. Li et al., *Trajectory Optimization and Guidance Methods for Mars Entry*,
https://doi.org/10.1007/978-981-99-6282-2_1

Science Laboratory mission [8, 15], including correcting the touchdown velocity anomaly, modifying the parachute deployment trigger to reduce the landing ellipse, and adding Terrain Relative Navigation (TRN) to avoid hazards within the landing ellipse. In 2020, Tianwen-1, the first mission of China's planetary exploration project, achieved its goals of orbiting, landing and roving on Mars [16]. Future Mars missions include Mars Sample Return (MSR) [17] and human landing missions [18], which are expected to be launched in 2024 and 2040, respectively.

Mars entry, descent and landing (EDL) begins at the Martian atmosphere interface at approximately Mach 25 and ends with a safe touchdown, which includes the atmospheric entry phase, the parachute descent phase and the powered descent phase [19]. The entry, descent and landing phases are critical for a Mars landing exploration mission and directly determine the success of the entire mission. The Mars atmospheric entry phase, a particularly important sub-phase of the EDL, begins when the vehicle reaches the Mars atmospheric boundary (about 125 km altitude) and ends with the deployment of the supersonic parachute, which lasts about 4 minutes and suffers the worst aerodynamic heating environment of all three sub-phases of the EDL. The velocity of the entry vehicle is reduced from 4–7 km/s to about 400 m/s during the Mars atmospheric entry phase. Therefore, this phase is also called the hypersonic entry phase. 99% of the initial kinetic energy is dissipated during the atmospheric entry phase, and the peak overload and peak heat flux also occur during this phase, which presents a great challenge to the design of the aerodynamic deceleration, structure, and thermal protection system.

Based on differences in vehicle configurations, how lift is used, and whether an active guidance method is employed, the entry modes used by previous Mars landing missions can be broadly classified into two categories: unguided ballistic entry mode [3–5, 7], and guided ballistic-lifting entry mode [6, 15, 16]. A ballistic entry trajectory is a trajectory flown without lift. In a guided ballistic-lifting entry, the lift vector is oriented to shape the trajectory.

In the unguided ballistic entry, there is no aerodynamic control lift to adjust the trajectory. The accumulated navigation errors and uncertainties in the Martian atmosphere and aerodynamic parameters will result in a large landing dispersion. MER-A/B, Phoenix and MPF adopted the unguided ballistic entry mode and achieved the successful landing on Mars, which shows that the unguided ballistic entry mode has good robustness. Although an unguided ballistic entry configuration is adopted in the Viking and Phoenix missions, the lift is used to maintain the stability of the entry vehicles but not to adjust the entry trajectory. In a guided ballistic-lifting entry, the capsule's trajectory can be adjusted to reduce landing error by the reaction control system (RCS), which uses small thrusters to modify the vehicle's attitude and direct the lift vector to control the vehicle's trajectory. In addition to reducing the size of the landing ellipse, the lift can also be used to improve the elevation of landing sites. MSL is a typical representative of the guided ballistic lift entry mode. Compared with the first-generation unguided ballistic-lifting entry mode used in the Viking missions, the guided entry mode used in the MSL, Mars 200, and Tianwen-1 missions is capable of achieving a high-precision Mars landing with an error of ~2 km [6], ~1 km [20], and

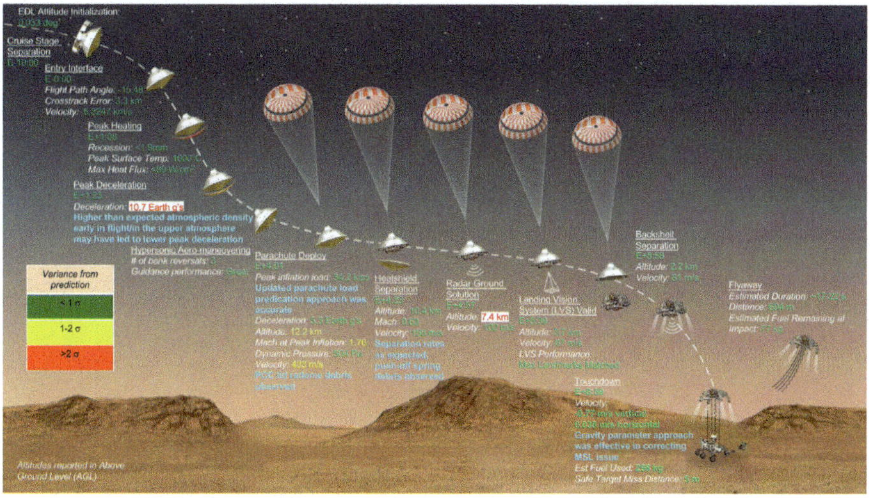

Fig. 1.1 Mars 2020 EDL sequence

3.1 km [21], respectively. The EDL sequence of the Mars 2020 is shown in Fig. 1.1 [15].

1. **MSL and Mars 2020 mission [15, 20]**

In order to land a larger rover on the surface of Mars and achieve a more precise high-altitude landing than any previous Mars landing mission, MSL/Curiosity is the first Mars mission to use guided entry with the goal of safely bringing the entry vehicle to a survivable parachute deployment state within 12.5 km of the pre-determined parachute deployment coordinates. The Entry Terminal Point Controller (ETPC) guidance algorithm is derived from the Apollo command module entry guidance [22, 23], and like Apollo, ETPC modulates bank angle to control range based on information about deviations in range, altitude rate, and drag acceleration from a reference trajectory. To use aerodynamic lift to control the entry trajectory, the entry capsule's center of mass is deliberately shifted from its axis of symmetry to allow the capsule to trim aerodynamically with a non-zero angle of attack. According to the navigation information from an IMU, the entry guidance algorithm plans the commanded aerodynamic force to follow the nominal trajectory, and the control algorithm adjusts the bank angle to follow the commanded aerodynamic force using the RCS thrusters. Then, the entry trajectory can be adjusted by changing the direction of the lift vector according to the control command. At the same time, three-axis stabilized attitude control is used to maintain entry attitude stability. Since only a thin Martian atmosphere can be used for deceleration during Mars entry, it is critical for the ETPC guidance algorithm to carefully balance the lift of the vehicle to minimize the range error while still maintaining a higher landing altitude and a desired safe parachute deployment altitude. The MSL guidance algorithm consists

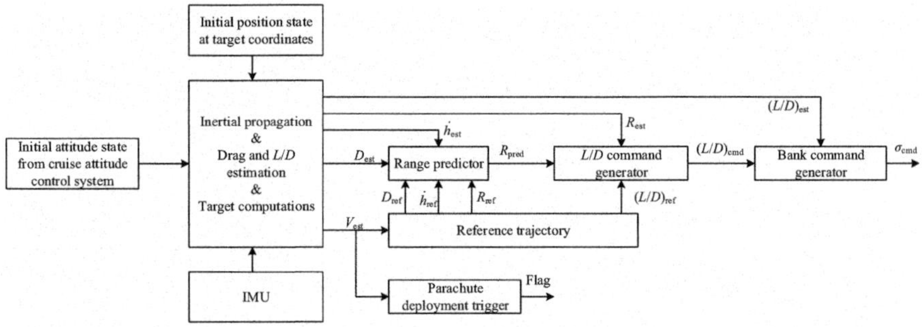

Fig. 1.2 Architecture of MSL entry guidance

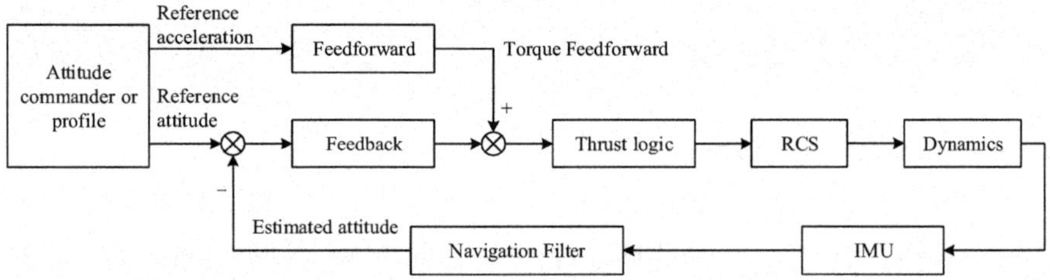

Fig. 1.3 Architecture of MSL Mars entry attitude control

of two parts: range control and attitude control. The structures of entry guidance and attitude control systems are shown in Fig. 1.2 [24] and Fig. 1.3 [25], respectively.

(1) **Range control**

The range control phase begins when the accelerometers measure 0.2 Earth-g's and ends when the estimated velocity is less than 1100 m/s. During the range control phase, a predictor–corrector algorithm commands the bank angle to adjust the lift force direction to control the range while keeping the cross-range error within a given corridor by performing, nominally, three bank reversals. To achieve this goal, the ETPC algorithm uses an onboard reference trajectory table and computes deviations from it according to state data from the inertial navigator. From this data, the predictor–corrector algorithm makes a prediction of the range-to-go until parachute deployment conditions are reached. The difference between this predicted range and the current range-to-go is the range error to be corrected. The commanded lift-up is then calculated based on the on-board reference trajectory plus a correction proportional to the range error. Finally, the commanded lift-up (or more specifically, *L/D*) is used to calculate the commanded bank angle.

Mars 2020 introduces a new local gravity correction to the on-board gravity model and adds this parameter to mitigate the MSL touchdown velocity anomaly, and it improves the Straighten Up and Flight Right (SUFR) maneuver [8, 26]. This maneuver completes the entry phase and eliminates the angle of attack by ejecting the entry balance masses and rolling the vehicle to point the TDS antennas toward the ground. For Mars 2020, the SUFR trigger changes from a velocity trigger to a velocity constraint downrange trigger. The trigger condition is given by:

$$V < V_{\max} \text{ and } R < R_{\text{threshold}}, \text{ or } V < V_{\min} \tag{1.1}$$

The velocity constraints are to protect against uncertainties. Triggering at too high a velocity could result in higher parachute loads or large attitude rates due to aero-oscillations. Triggering at too low a velocity could result in altitude loss, reducing the EDL timeline margin. The parachute deployment trigger has also been changed from a velocity trigger to a fixed time offset (14 s) from the SUFR trigger to allow for a possible maximum slew of 180°.

(2) **Heading alignment**

Heading alignment begins after range control and ends at parachute deployment. Once the estimated velocity drops below 1100 m/s, the guidance algorithm automatically stops range control and begins heading alignment. The entry guidance commands only the bank angle to reduce the cross-track position errors left over from the range control phase and those currently being introduced. The range error remains uncontrolled during this phase because the effectiveness of range control is reduced as the entry vehicle approaches the prescribed parachute deployment point. The dynamics of cross-range control in the heading alignment phase are analogous to the way an aircraft controls its heading. The magnitude of the commanded bank angle is limited within 30° to ensure that most of the supersonic lift counteracts gravity to prevent significant altitude loss at parachute deployment. Guided entry ends when the sequence of events for parachute deployment is commanded, which includes banking to 180° while jettisoning entry ballast to achieve a trim angle of attack near zero just prior to parachute deployment. The inertial propagator, which uses the attitude rate and acceleration measurements from the descent stage IMU, provides the estimated vehicle state to the entry guidance algorithm.

2. **Tianwen-1 [27–33]**

Similar to MSL and Mars 2020, Tianwen-1 also adopts the ballistic lifting entry mode shown in Fig. 1.4 [33]. At the entry interface, the inertial position and velocity of Tianwen-1 are [3163492.266471; −1,522,037.547769; 147,228.268912] km and [−215.964964069; 1333.853595255; −4605.398643736] m/s, including an initial altitude of 125 km and a velocity of 4.7 km/s. The vehicle passes through the Martian atmosphere with a trim angle of attack of −11.6° and maintains a bank angle of 52° until the acceleration exceeds 0.2 Earth g, at which time it enters lift control at an altitude of approximately 60 km and Mach number 24. In the lift control phase, a constrained adaptive trajectory planning and guidance method (as shown in Fig. 1.5 [21]) is used to compensate for the downrange error and meet the deployment conditions of the parachute.

After the lift control phase, the velocity of Tianwen-1 decreases to Mach 2.8, at which time the trim wing is deployed to zero the angle of attack, then the parachute

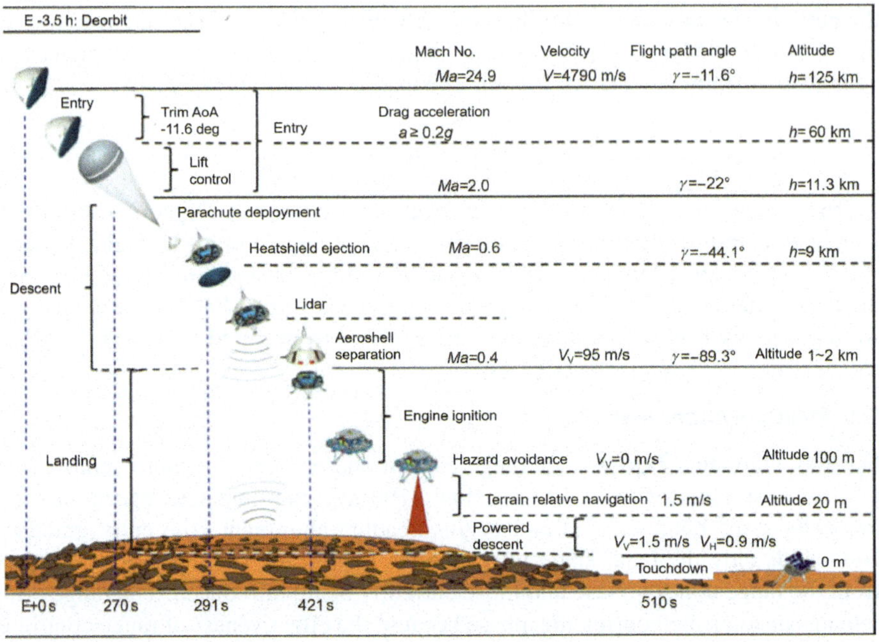

Fig. 1.4 Tianwen-1 EDL timeline

is deployed at Mach 1.8 at an altitude of 13 km. After the parachute is released for 0.5 s, the vehicle ejects the heat shield, and after 10 s the landing mechanism is deployed and the terrain relative navigation system is activated. When the vehicle decelerates to an altitude of 1.3 km and 0.25 Mach, the lander is released. After 1 s, the 7.5 kN thruster is ignited for descent maneuvers. When it reaches 20 m above the landing site, the vertical speed is 1.5 m/s and the horizontal speed is 0. The engine continues to work until at least two landing sensors are triggered. The final vertical speed is about 1.68 m/s and the horizontal speed is less than 0.16 m/s. The entire EDL process takes about 9 min. The latitude and longitude of the actual and target landing sites are (25.066°N, 109.925°E) and (25.1188°N, 109.9305°E) respectively, which results in a landing error of only 3.11 km.

3. **Summary of successful Mars EDL missions**

Table 1.1 shows the EDL parameters of successful Mars entry missions [7, 8, 19, 28].

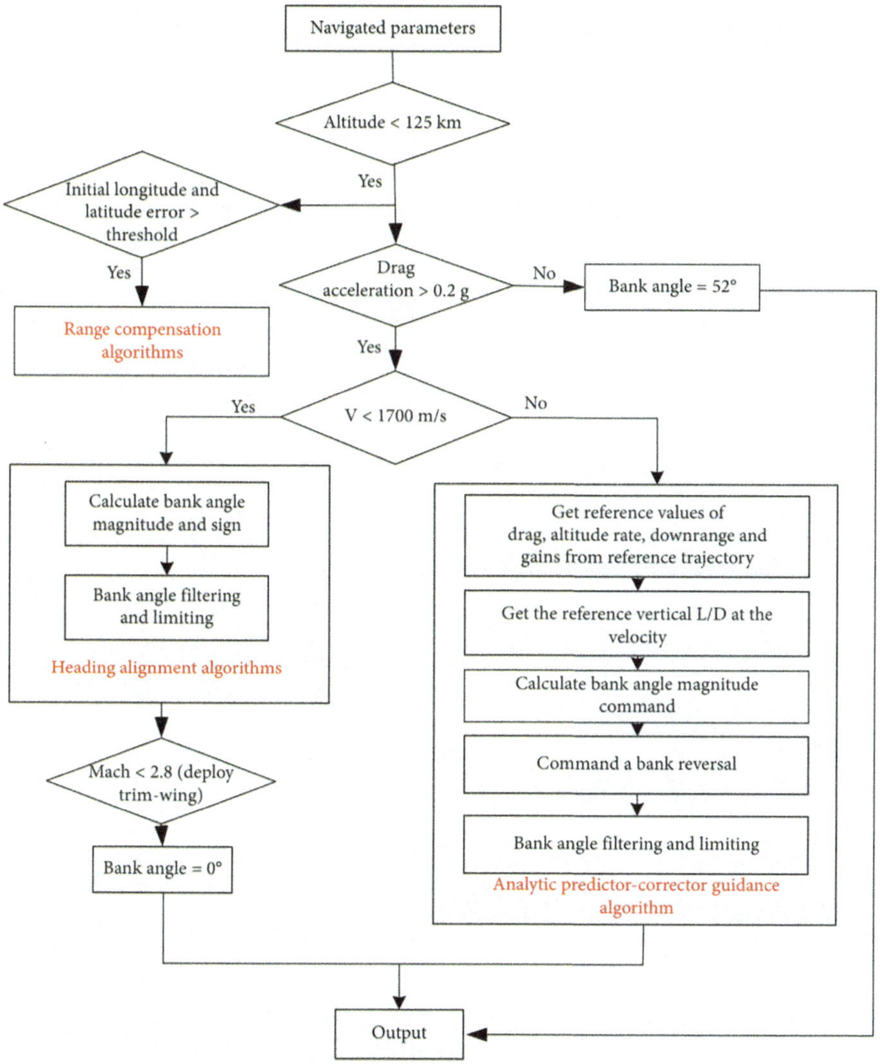

Fig. 1.5 Flow chart for Tianwen-1 entry guidance

1.1.2 Human Mars Landing Plans

Sample return and human landing missions are already on the agenda for future missions to Mars. The EDL technology for Mars Sample Return (MSR) is essentially the same as that for robotic missions. The main difficulty of the mission is to launch the Mars Ascent Vehicle from the surface of Mars into orbit [34]. However, the human Mars lander is expected to deliver a payload of more than 20 tons to Mars with a total launch mass of more than 40 tons [35]. In this case, the disk gap-band parachute

Table 1.1 Summary of successful Mars EDL missions

Parameters	Initial entry velocity (km/s)	Initial entry mass (kg)	Lift control	Guidance	*L/D*	Initial flight path angle (°)	Trim angle of attack (°)	Peak heating rate (W/cm^2)	Peak load (g_E)	Parachute deploy Mach No	Parachute deploy pressure (Pa)	Parachute deploy altitude (km)	Landing error ellipse (km)
Viking	4.7	922	CoM offset	No	0.18	−17	−11	26	7.4	1.1	350	5.79	280 × 100
MPF	7.2	584	No offset	No	0	−14.06	0	100	16.2	1.57	585	9.4	200 × 100
MER A/B	5.4, 5.5	827, 832	No offset	No	0	−11.47, −11.49	0	44	6.3	1.77	725, 750	7.4	80 × 12
Phoenix	5.67	600	No offset	No	0.06	−13	0	46	8.5	1.6	430	12.7	260 × 30
MSL	5.9	2800	CoM offset	ETP	0.22	−15.5	−15	120	12.5	2	750	6.5	20 × 20
InSight	5.5	608	No offset	No	0	−12	0	45.6	7.6	1.49	553.7	12	130 × 27
Mars 2020	5.3	3649	CoM offset	ETP	0.28	−15.48	−15	89	10.7	1.76	504	12.2	7.7 × 6.6
Tianwen-1	4.8	1238	Trim wing	ETP	0.15	−11.6	−10	67.86	10	1.8	500	13	50 × 20

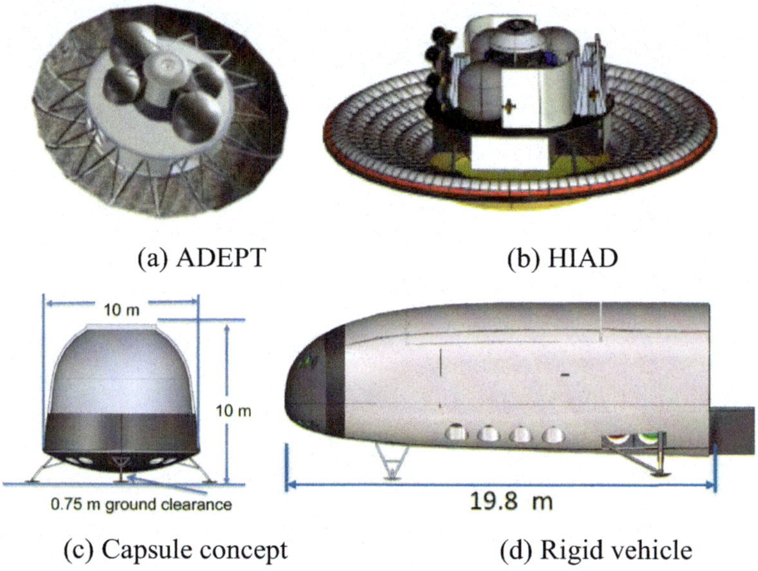

Fig. 1.6 Human-scale Mars lander

cannot be used for deceleration because the size and configuration requirements of the parachute for such a heavy vehicle are beyond the current state of the art.

Several potential entry systems are being designed and developed to support human surface missions in NASA's Mars Reference Architecture 5.0 (DRA5) [36] and Evolvable Mars Campaign (EMC) [37] projects, including the Hypersonic Inflatable Aerodynamic Decelerator (HIAD), Adaptable Deployable Entry and Placement Technology (ADEPT) [38], a mid-lift-to-drag rigid vehicle, and a low-lift-to-drag rigid capsule concept [39]. These options are paired with supersonic retro-propulsion for descent and landing, while different configurations are used for atmospheric entry. The four entry systems are shown in Fig. 1.6 [38–40].

1. **Rigid vehicle** [40, 41]

Two rigid vehicles are considered evolutionary technologies, meaning they are an extension of flight heritage vehicles from both human and robotic missions. The first, which draws some heritage from the Space Shuttle experience and is similar to the original DRA5 design but reduced in size to accommodate the smaller EMC payload, is a mid *L/D* rigid vehicle. The dimensions of the vehicle for this study are approximately 8.8×19.8 m (not including the aft body flaps) and it flies at an angle of attack of 55° with an *L/D* of 0.54. Figure 1.7b shows details of the sequence for the mid *L/D* vehicle [40]. At approximately Mach 2, the vehicle attitude is changed to a 90° angle of attack and the engines are initiated for descent and touchdown.

2. **Capsule concept** [40, 42]

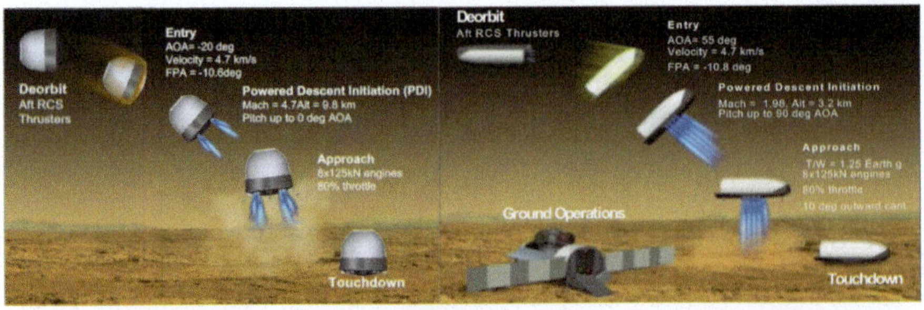

(a) The EDL sketchs of the capsule concept and the rigid vehicle

(b) The EDL sketchs of ADEPT and HIAD

Fig. 1.7 The EDL sequence for the human-scale Mars landers

The second evolutionary vehicle is the capsule concept, derived from robotic Mars missions, which looks like a scaled-up version of the Mars Science Laboratory (MSL) capsule. The vehicle flies at an *L/D* of 0.24, similar to the MSL. Both evolutionary designs include rigid shells covering the payload for aerocapture and entry and have ballistic coefficients greater than 300 kg/m^2.

After EDL is initiated with a deorbit burn using RCS thrusters, the vehicle reorients for entry and flies at an angle of attack of nearly negative 20°. At approximately Mach 4.7, the engines are engaged and the vehicle's angle of attack is reduced to 0°. The engines slow the vehicle for landing. A sketch of the EDL sequence is shown in Fig. 1.7a [40]. Once on the surface, the backshell separates into sections and is jettisoned. This allows access and deployment of other payloads such as surface power systems and rovers.

3. **ADEPT** [43]

The remaining two entry technologies that are considered revolutionary rather than evolutionary are deployable vehicles. Although deployable vehicles have not been extensively used in Mars or Earth missions, there have been many successful subscale test flights on Earth. Deployable technology addresses the challenge of low atmospheric density on Mars by increasing the area of drag beyond the limits of the launch vehicle envelope. Previous studies have also shown that deployable vehicles can save significant mass compared to rigid options. The first rigid deployable is the ADEPT configuration. Similar to an umbrella, ADEPT consists of a flexible 3-D woven TPS structural membrane over rib structures that are stowed during launch and deployed in space. The TPS is designed to survive both the aerocapture and entry thermal pulses. The original ADEPT design used an asymmetric 16 m diameter shape that provided the desired lift performance during aerocapture and entry, rather than relying on a moving mass or aerodynamic device to generate lift by changing the angle of attack. The asymmetric ADEPT had an *L/D* ratio of 0.27. An image of the EDL sequence of an asymmetric ADEPT is shown in Fig. 1.7b. However, difficulties with the asymmetric design led to a configuration change to a symmetrical 70° cone with an aerodynamic tab.

4. **HIAD** [43, 44]

The second deployable technology is the HIAD. The HIAD design utilizes an inflatable structure covered by a flexible TPS. The inflatable structure is a stacked torus design, meaning that it is constructed as a conical stack of pressurized rings that are interconnected and anchored to the central rigid nose by radial structural tapes. While HIAD flexible TPS samples have survived multiple heat pulse tests, the performance of the material deflated after aerocapture and re-inflated up to a year later has not been characterized, and it is not feasible to carry gas generators on board to keep the HIAD inflated during an extended stay in Mars orbit between aerocapture and entry. Therefore, it was decided to carry two separate HIADs, one for aerocapture and one for entry. The aerocapture HIAD will be jettisoned before the orbital circularization burn after the atmospheric pass. To maintain similar ballistic coefficients in current trajectory simulations, the aerocapture HIAD is slightly larger, 18.8 m, compared to the entry HIAD, which is 16.7 m in diameter. The nominal HIAD vehicle has an *L/D* of 0.2 and uses a symmetrical 70° cone shape. The diameter of both low *L/D* vehicles was chosen to maintain a ballistic coefficient similar to the MSL (less than 150 kg/m^2).

5. **EDL technology level**

Beginning in 2016, NASA directed the Entry, Descent and Landing Architecture Study (EDLAS) team to evaluate four candidate technologies for delivering a human-scale vehicle (carrying a 20-ton payload) to a precise location on the surface of Mars [38]. In this project, extensive packaging and three-degree-of-freedom (DOF) trajectory analyses were performed to determine the landing accuracy, propellant consumption, and guidance requirements of the four vehicles. Key technology gaps were identified in the areas of supersonic retropropulsion (powered descent), entry guidance, powered descent guidance, and navigation. As shown in Tables 1.2 and 1.3 [42], EDLAS found that the capsule concept did not meet the EMC packaging constraints and, due to the vehicle's high ballistic coefficient, required twice as much propellant to land the same 20-ton payload as the other three concepts. Therefore, the capsule concept was removed from further analysis. In addition, it was found that the two deployable low *L/D* concepts had similar pre-Phase A flight performance capabilities. As a result, it was decided to conduct the next level of analysis with

Table 1.2 The size of human-scale Mars landers

Name	*L/D*	Size, H-height, W-weight, D-diameter	Ballistic coefficient
Capsule	0.3	H-10 (m), W-10 (m)	500 (kg/m^2)
Mid *L/D*	0.55	L-22 (m), H-10 (m), W-10 (m)	380 (kg/m^2)
ADEPT	0.2	H-4.3 (m), D-18 (m)	155 (kg/m^2)
HIAD	0.2	H-4.3 (m), D-16.4 (m)	155 (kg/m^2)

Table 1.3 The mass of human-scale Mars landers

ID	Subsystem	Capsule	Mid *L/D*	ADEPT	HIAD
1	Structures (kg)	5422	14,836	5422	5422
2	Propulsion (kg)	5215	5190	4963	4916
3	Power (kg)	1568	1568	1568	1568
4	Avionics (kg)	333	333	333	333
5	Thermal (kg)	218	844	411	411
6	Aero decelerator (kg)	7025	5499	8658	7251
Dray mass (kg)		19,781	28,270	21,355	20,190
7	Cargo (kg)	20,000	20,000	20,000	20,000
8	Non-Propelled Fluids (kg)	1965	1523	1565	1479
Inert mass (kg)		41,746	49,793	42,920	41,669
9	Used propellant (kg)	26,531	16,399	17,515	15,564
Total stage gross mass (kg)		62,877	66,192	60,435	57,233

only one deployable concept. In 2018, EDLAS continued to increase the fidelity of the analysis to identify technology gaps using a medium *L/D* rigid and a low *L/D* flexible vehicle.

1.2 Advances in Mars Entry Trajectory Planning

Mars entry trajectory optimization is an important enabling technology for future Mars landing missions. On the one hand, a reasonable trajectory can make the operation of the entry process more flexible and reduce the requirements on the control system. On the other hand, the new online trajectory planning strategy can generate a new entry trajectory if necessary, improving the safety of a manned mission.

The Mars entry trajectory optimization involves solving an optimal control problem with complex states and boundary conditions. This task can be categorized into three primary types: heuristic, indirect, and direct methods. However, it is important to note that most of these methods do not account for the uncertainties in the Mars entry. Incorporating uncertainties into the trajectory optimization will significantly improve the safety and landing accuracy of Mars entry missions.

1.2.1 Indirect Method

The indirect method uses the maximum principle to solve the optimal control problem by introducing conjugate variables. The shooting or multiple shooting method is a typical representative of the indirect method. The advantage of the indirect method is the high accuracy of the solution, if obtained. At the same time, it has obvious disadvantages. The indirect method is sensitive to the initial guess, which leads to a small convergence radius. Since the initial values of the conjugate variables have no clear physical meaning, the initial guess is difficult to obtain. In order to find a proper initial guess, repeated iteration and integration of state and co-state equations are inevitable, which naturally makes it difficult to apply the indirect method to the online trajectory optimization of Mars atmospheric entry.

The indirect method has been widely used to solve the Mars entry problem because it satisfies the first-order optimality conditions. Jacob et al. [45] used Pontryagin's minimum principle (PMP) and the penalty function approach to transform the constrained TAM problem into a two-point boundary value problem (TPBVP), while PSO was used to estimate the initial values of the costate variables. Zheng et al. [46] used the exact penalty function method in conjunction with the homotopy method to solve the TAM problem, where several auxiliary problems are constructed to guess the value of the initial costate variable. Long et al. [47] proved that the optimal control profile for the constrained TAM problem has a bang-bang structure with at least two switches. Mall et al. [48] proposed a uniform trigonometry method (UTM) to approximate the bang-bang solution, although significant continuation sets were still required for homotopy or numerical continuation.

1.2.2 Direct Method

The direct method transforms the entry trajectory planning problem into a static parameter optimization problem with various path and boundary constraints, which can be easily solved as a nonlinear programming (NLP) or convex programming problem. Typical methods include the direct collocation method [49], the pseudospectral method [50], and the convex programming method [51]. Due to its ease of implementation, the direct method has been widely adopted in engineering practice, especially in the field of trajectory optimization.

One merit of the pseudospectral method is that the Karush–Kuhn–Tucker (KKT) conditions of the Gaussian pseudospectral method are equivalent to the first-order optimality conditions of Pontryagin's maximum principle [50, 52]. However, this method also relies heavily on the designer's experience to select the appropriate number of nodes because there are no definite rules for node selection. As for the convex optimization method, it has found increasing applications in aerospace trajectory optimization in recent years due to its strong theoretical properties on existence and uniqueness of the solution and appealing computational advantage ensured by polynomial complexity [53, 54]. Nevertheless, most real-world problems are non-convex, which leads to the sequential convex optimization method to approximate and solve these non-convex problems.

For the atmospheric entry problem, Liu et al. [55, 56] used the sine and cosine of the bank angle as new control inputs and introduced an additional second-order cone inequality constraint to replace the nonconvex equality constraint on the new controls, which can transform the constrained entry problem into a convex one and effectively suppress the high-frequency jitter of the bank angle. Then, the convex problem was solved by linearization, discretization, and successive convexification techniques using the well-established interior-point method. Owing to the advantages of fast convergence and low computational complexity, this strategy developed by Liu et al. has almost become the standard approach to the entry trajectory optimization problem via sequential convex programming (SCP). Inspired by this scheme, Zhao and Song [57] proposed a multi-phase convex programming algorithm for entry trajectory planning using the normalized lift coefficient and waypoint optimization. Wang and Grant [58] improved the work in [55] by using the bank angle rate as a new control input, which decouples the state and control in the entry dynamics to eliminate the high-frequency jitter in a simpler way. Furthermore, the backtracking line-search and dynamic trust-region techniques were introduced by Wang [59]. Sagliano and Mooij [60] proposed the pseudospectral convex programming approach, where the optimal control problem is discretized at flip-Radau collocation points and then solved in the framework of second-order cone programming (SOCP). Recently, Zhou et al. [61] investigated the atmospheric entry trajectory planning problem using SCP and an adaptive mesh refinement technique. Wan and Pei et al. [62, 63] proposed the customized Alternating Direction Method of Multipliers (ADMM) to solve the approximate polynomial optimal trajectory optimization problem.

1.2.3 Heuristic Method

To overcome the initial guess problem, intelligent global search algorithms such as genetic algorithms and particle swarm optimization are also introduced to solve the Mars entry trajectory optimization problem. The particle swarm optimization (PSO) and the genetic algorithm (GA) were utilized as the standard evolutionary approaches to plan the optimal Mars entry trajectory. Using the multi-objective PSO, Grant and Mendeck [64] evaluated the optimal Pareto fronts associated with the terminal altitude, error ellipse, and normal load of Mars entry. Lafleur and Cerimele [65, 66] subsequently investigated the optimal parametric configurations for the Mars entry problem using both bank and angle-of-attack modulations, accounting for factors such as mass, entry velocity, final Mach number, and altitude. Based on the GA, Sorgenfrei and Chester [67] developed the Parametric Entry, Descent, and Landing Synthesis (PEDALS) tool. This tool stochastically searches the design parameter space for the optimal entry trajectory that satisfies terminal conditions and path constraints. However, even after many iterations, the evolutionary technique may be trapped in a local optimum.

1.2.4 Uncertainty Optimization

Due to the large uncertainties of the state variables at the entry interface, the Martian atmospheric density, and the aerodynamic parameters, there is typically a large deviation between the actual landing site and the designated target [68]. To ensure a safe landing for the next generation of Mars exploration missions, which will require the ability to land spacecraft in hazardous areas with high scientific value, it is essential to assess the impact of the uncertainties on the state trajectories and determine the final landing error ellipse.

Uncertainty quantification is a mathematical method to quantitatively describe and evaluate the magnitude and propagation of uncertainties in a given system [69]. For Mars atmospheric entry dynamics, the uncertainties in the initial conditions and system parameters result in off-nominal trajectories. Traditional methods for trajectory optimization under uncertainty include the system sensitivity method, covariance-based methods, local linearization theory, and Monte Carlo simulation. Recently, researchers have proposed several approaches, including stochastic collocation technique, response surface method, kriging method, spectral method, and polynomial chaos, to quantify and propagate uncertainties in dynamical systems [70–72].

1.3 Progress in Mars Entry Guidance

The objective of Mars entry guidance is to steer an entry vehicle from atmospheric entry interface to a designated parachute deployment target at the end of the entry phase with certain accuracy [73–76]. However, there are many uncertainty factors and errors that will degrade the performance of the guidance algorithm in engineering practice. The most significant error sources leading to a larger Mars entry dispersion include the vehicle state estimation errors at the atmospheric entry point, the uncertainties in the atmospheric density and aerodynamic coefficients, and the winds and gusts. It was demonstrated that the most efficient way to increase the entry accuracy is to design a guidance law that is robust to uncertainties and errors.

Generally speaking, Mars atmospheric entry guidance methods can be divided into two categories: reference-trajectory tracking guidance [77–81] and predictor–corrector guidance [82–85]. Recently, the computational guidance concept and artificial intelligence based guidance law are also introduced and developed.

1.3.1 Reference Trajectory Tracking Guidance

In reference trajectory tracking guidance, an optimal reference trajectory is planned in advance according to specific performance indexes and constraints. Then, a controller is designed to track the reference trajectory. And most of the tracking controllers are constructed based on the advanced control theory [77–81], including the sliding mode control, model predictive control, backstepping control, fractional order theory, robust control, active disturbance rejection control, and so on.

The classical paradigm of constrained tracking guidance law is the Evolved Acceleration Guidance Logic for Entry (EAGLE) [78]. The EAGLE generates the drag energy and lateral acceleration profiles that satisfy all the constraints, and then the tracking guidance law based on the feedback linearization is designed to follow the planned drag acceleration and heading profiles. EAGLE has the advantage of compatibility, so it can be widely applied to different entry vehicles in different entry missions. Recently, Ridderhof [86] described the Mars entry tracking guidance as a stochastic optimal control problem and presented a linear feedback law. This algorithm chose to optimize the covariance of the entry trajectory rather than the state error, which improved the robustness and accuracy of the entry guidance. However, the real-world performance of the algorithm may be limited because the Gaussian distribution is used to describe the uncertainty.

1.3.2 Predictor–Corrector Guidance

In the predictor–corrector guidance law, the guidance system first predicts the future trajectory of the vehicle based on the current state and inputs. The predictor uses a mathematical model or numerical integration to predict the future trajectory and compares the predicted solution with the desired trajectory. Then, the on-board corrector generates the control command to compensate for these deviations or errors between the predicted and desired trajectories. In this step, the root-finding algorithm, such as the Newton–Raphson method, is used to determine the command. The motivation for predictor–corrector guidance is the requirement of the future Mars landing exploration mission to significantly improve landing accuracy and robustness to large errors at the entry interface [83, 85]. Compared to trajectory tracking methods, this method inherently overcomes the weakness of introducing errors from assumptions and the linearization process.

The ETPC algorithm [22] and its variants have been used to perform entry guidance for a number of Mars missions, including the Mars Science Laboratory [23], Mars 2020 [8], and Tianwen-1 [21]. In essence, ETPC guidance predicts the final downrange error at the reference final altitude and generates a feedback control correction for the guidance system to eliminate the downrange error, rather than simply following the reference trajectory. The ETPC, on the other hand, does not consider the path constraints in the guidance logic. The constrained predictor–corrector guidance law [87, 88] utilizes the modified quasi-equilibrium glide condition to compute a bound for the speed-dependent bank angle magnitude that satisfies all the path constraints during the entry of medium-to-high lift-to-drag ratio vehicles. In order to extend the range of applications of the constrained predictor–corrector algorithm, a unified entry guidance law for different vehicles with low and high lift-to-drag ratios has been proposed in [89]. The two key ideas are the numerical predictor–corrector logic and the predictive load relief strategy. The former generates the bank angle command by minimizing the downrange error, and the latter reduces the heating and load peaks through active altitude control. However, the constrained tracking and predictor–corrector guidance laws require significant onboard computing power because they involve online trajectory planning or numerical integration.

1.3.3 Computational Guidance

Computational guidance refers to a type of guidance algorithm that relies on computational methods, such as numerical optimization or mathematical modeling, to generate control commands to steer a vehicle [90, 91]. It involves the use of onboard computational techniques to solve complex mathematical problems and optimize the guidance process with the goal of minimizing a cost function or maximizing a performance metric. This cost function typically represents a trade-off between various factors such as accuracy, fuel consumption, time, or safety.

Computational guidance laws offer the advantages of flexibility and adaptability because they can handle complex systems and adapt to changing conditions. They can incorporate various factors, constraints, and objectives into the optimization process, allowing for improved accuracy, efficiency, and robustness in vehicle guidance [92].

As its merits of convergence and efficiency, the convex optimization is often used to design computational guidance laws. Wang and Grant proposed a closed-loop optimal feedback guidance law based on the QCQP to track the optimal entry trajectory generated by second-order cone programming [93]. Sagliano and Mooij [94] introduced two slack variables to convexify the nonconvex constraints and an objective function related to the inverse of drag acceleration. The optimal drag-energy tracking guidance law was then proposed via pseudospectral convex optimization. Subsequently, Bae et al. [95] designed an online trajectory optimization strategy for entry guidance by using virtual control, change of control variable, and second-order cone programming techniques. Similarly, Calabuig and Mooij [96] integrated the aforementioned techniques and developed an open-source Successive Convex Optimal Control Tool (SCOPT) for the orbital and descent abort of the re-entry vehicle SPHYNX. In addition, Tracy and Manchester [97] proposed a convex predictor–corrector guidance law, where the predictor updated the reference trajectory via Runge–Kutta integration and the corrector generated the optimal control by online sequential convex programming (SCP). Recently, Wan et al. [62] parameterized the Mars entry and landing guidance problem via polynomial approximation into a nonconvex polynomial programming problem and then proposed a customized alternating direction method of multipliers to find the local optimal solution of a QCQP problem that is equivalent to the resulting polynomial programming problem.

1.3.4 Artificial Intelligence-Based Guidance

The artificial intelligence (AI)-based guidance method mainly relies on the deep neural network and reinforcement learning technologies [98]. The deep neural networks have been used to model and predict the behavior of the input vehicle. They can learn from large data sets and capture complex relationships between input variables (e.g., vehicle state, atmospheric conditions) and output variables (e.g., control commands or trajectory corrections). Neural networks can be used for trajectory optimization, real-time control, and decision making [99–101]. Reinforcement learning algorithms allow vehicles to learn optimal control rules by interacting with the environment [102, 103]. They can adapt and improve their decision-making processes based on feedback from the environment.

AI-based guidance and control methods combine the advantages of well-developed trajectory optimization methods and deep learning techniques to form an integrated framework capable of generating optimal guidance and control commands in a relatively short time. Roughly speaking, an AI-based approach involves two steps. The first step is to create a large dataset of optimal flight trajectories for a given mission profile. Note that there is usually no open dataset for a specific space application. Therefore, this step is essential, and simulated results are used as an alternative. In the second step, (deep) neural networks are constructed and trained on the pre-generated dataset so that they can be used in later stages to directly represent the optimal relationship between state and control actions. To better demonstrate how a spacecraft-related mission can benefit from the implementation of AI-based guidance methods.

1.4 Discussions

From a guidance and control perspective, there are several reasons that limit the accuracy of Mars entry missions.

First, the mechanism of uncertainty propagation during Mars entry and its impact on guidance planning are not fully understood. Current Mars entry guidance methods mainly rely on advanced control techniques to identify, compensate for, or suppress the adverse effects of uncertainties on entry guidance accuracy. However, they rarely focus on and quantitatively analyze the propagation mechanism of uncertainties in the dynamic system. As a result, it is difficult to make significant breakthroughs in the performance of existing Mars entry guidance methods.

Second, the robustness of the nominal trajectory during the Mars entry under uncertain conditions needs to be improved. During Mars atmospheric entry, there are significant uncertainties in the initial state of the entry vehicle, the dynamic parameters, and the Martian atmospheric density. Traditional trajectory optimization methods often overlook the influence of uncertainties, leading to significant tracking errors in the nominal trajectory during uncertain Mars entry. Therefore, it is necessary to robustly optimize the Mars entry nominal trajectory by fully considering the effects of various uncertainties. This would ensure that the nominal trajectory reliably satisfies various constraints under uncertain conditions without significant performance degradation. Alternatively, online trajectory planning methods can be used to generate new nominal trajectories as needed and achieve high-precision entry guidance based on these trajectories.

Third, there is a lack of reported robust optimal guidance methods specifically designed for Mars entry under uncertainty. Most current robust entry guidance methods under uncertainty are based on traditional nominal trajectory guidance or predictor-corrector guidance, with the addition of uncertainty identification or compensation. However, these methods not only increase system complexity, but also cannot guarantee optimal guidance. In addition, the guidance accuracy is highly dependent on the accuracy of the uncertainty identification or compensation algorithms, which often depend on experience and ground experiments, limiting their application. Therefore, there is a need to develop robust optimal guidance methods specifically tailored for Mars atmospheric entry under uncertain conditions.

In response to the challenges faced in optimizing the trajectory and guidance of Mars entry, the author proposes corresponding methods to address the problem of optimizing the optimal trajectory and guidance for Mars entry from the following three aspects:

1. **Uncertainty quantification of Mars entry**

To address the problem of uncertainty in the Mars entry process, which severely limits the improvement of entry guidance performance, effective mathematical tools such as polynomial chaos, random sampling methods, and response surface methods can be used. This can be used to design computational methods and theoretical frameworks suitable for quantifying the uncertainty of Mars atmospheric entry, and to analyze the quantification laws and evolution mechanisms of Mars entry uncertainty.

2. **Robust optimization of nominal trajectory and online trajectory optimization for Mars entry**

To address the issue of robustness of the nominal trajectory for Mars entry to uncertainty, the robust flight envelope for Mars atmospheric entry can be calculated based on the uncertainty quantification combined with mission constraints, uncertainty propagation, reachable and controllable domains, etc. This provides a practical and feasible domain for subsequent nominal trajectory and guidance design. Furthermore, considering the impact of uncertainty parameters, tracking control capability, and landing accuracy requirements, robust optimization methods for Mars entry trajectory planning under uncertain conditions can be designed. The performance of the generated trajectory under uncertain conditions can be evaluated. Alternatively, optimization methods with online computational capabilities can be explored to plan new optimal trajectories based on real-time conditions during the entry process. Guidance methods can then be designed based on these trajectories to counteract the influence of various uncertainty factors.

3. **Robust guidance for Mars entry under uncertainties**

To deal with the uncertainties in the initial state and dynamic parameters faced during the Mars entry process, the computational guidance control (CGC) strategy can be introduced. It decomposes the Mars entry guidance problem into several simple optimal control sub-problems. A complete and easily solvable solution loop is designed according to certain rules. The real-time optimal control is solved based on the actual flight state and control effectiveness, and its robustness and accuracy in Mars entry guidance under uncertain conditions are evaluated. Finally, a relatively comprehensive theoretical framework for quantifying uncertainty in Mars entry and guidance planning is established.

Chapter 2
Preliminaries

2.1 Entry Vehicle and Environment Models

2.1.1 Mars Gravity Field

The Martian gravity field is represented as a spherical harmonic expansion with normalized coefficients (C_{nm}, S_{nm}), and is given by [104]

$$U = \frac{GM}{r}\left[1 + \sum_{n=2}^{\infty}\left(\frac{R_0}{r}\right)^n \sum_{m=0}^{n}(C_{nm}\cos m\theta + S_{nm}\sin m\theta)P_{nm}(\sin\phi)\right] \tag{2.1}$$

where GM is the gravitational constant times the mass of Mars, n is the degree, m is the order, P_{nm} are the fully normalized associated Legendre polynomials, $R_0 =$ 3396.2 km is the reference radius of Mars, ϕ is the latitude, and θ is the longitude (east positive).

For the Mars entry mission, model fidelity is guaranteed if the 2nd order spherical harmonic term is taken into account. Then, the gravitational accelerations in the radial and north–south directions are given by [105]:

$$\begin{cases} g_r = -\frac{\partial U}{\partial r} = \frac{GM}{r^2}\left[1 + \frac{3J_2}{2}\left(\frac{R_0}{r}\right)^2(1 - 3\sin^2\phi)\right] \\ g_\phi = -\frac{1}{r}\frac{\partial U}{\partial \phi} = \frac{GM}{r^2}\frac{3J_2}{2}\left(\frac{R_0}{r}\right)^2 \sin 2\phi \end{cases} \tag{2.2}$$

where $J_2 = 1.96 \times 10^{-3}$ is the coefficient corresponding to the second spherical harmonic of the gravitational field.

S. Li et al., *Trajectory Optimization and Guidance Methods for Mars Entry*,
https://doi.org/10.1007/978-981-99-6282-2_2

2.1.2 Mars Atmospheric Density

For real world Mars missions, the Mars Global Reference Atmospheric Model [106] developed by NASA and the Mars Climate Database [107] established by ESA are used to describe the Martian atmosphere. In numerical simulations, the atmospheric density is usually reduced to a function of altitude. Typical approximations to the Mars atmospheric density profiles are listed in the Table 2.1 [108–110], and these approximate solutions are shown in Fig. 2.1. In general, the second model is used in this book.

Table 2.1 Typical Martian atmospheric density model

Index	Model	
1	$\rho = \rho_0 e^{-h/h_s}$	$\rho_0 = 1.474 \times 10^{-2}$ kg/m^3, $h_s = 8805.7$ m
2	$\rho = \rho_0 e^{-h/h_s}$	$\rho_0 = 1.85 \times 10^{-2}$ kg/m^3, $h_s = 9354.5$ m
3	$\rho = \rho_0 e^{-(h-h_0)/h_s}$	$\rho_0 = 2 \times 10^{-4}$ kg/m^3, $h_s = 7500$ m, $h_0 = 40000$ m
4	$T = 1.4 \times 10^{-13} h^3 - 8.85 \times 10^{-9} h^2 - 1.245 \times 10^{-3} h + 205.3645$, $P = 559.351005946503 e^{-0.000105h}$, $\rho = P/188.95110711075/T$	

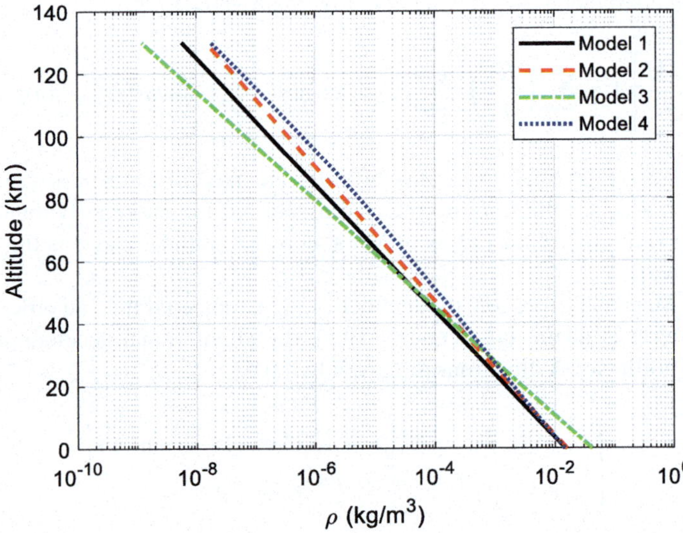

Fig. 2.1 Comparison of different models

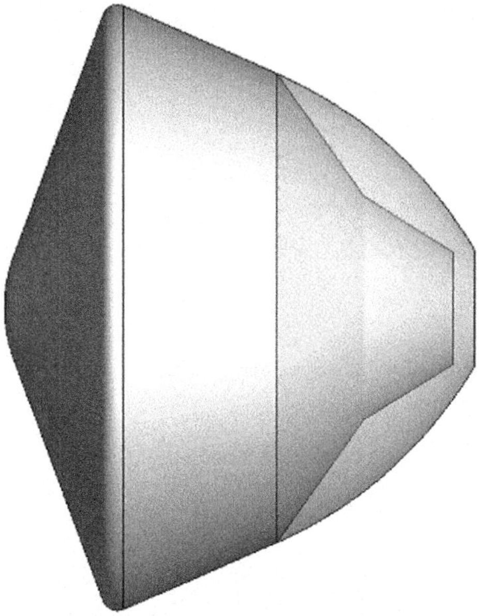

Fig. 2.2 MSL-type capsule for robotic missions

2.1.3 Mars Entry Vehicle

The well-known configuration for the Mars entry vehicle is the MSL-type capsule (as shown in Fig. 2.2) [111], which is a double-cone, with an aeroshell forebody shape of a 70° sphere-cone. By offsetting the center of gravity from the vehicle axis, the vehicle assumes an asymmetric orientation with respect to the incoming flow, which provides a small amount of lift. For this vehicle, the bank angle is used to control the lift vector, and the trim angle of attack provides the lift-to-drag ratio of about 0.25.

Another configuration is the mid-lift-to-drag ratio rigid vehicle, which is presented in Fig. 1.6d. The ellipsled vehicle is intended to be flown at $L/D \approx 0.58$ hypersonically and supersonically, and uses both aero-surfaces and a propulsive reaction control system (RCS) to affect longitudinal and lateral directional behavior.

2.1.4 Reference Frames

1. **Mars-centered inertial reference frame, index I**

As depicted in Fig. 2.3, the origin of the inertial reference frame (O_M-$X_I Y_I Z_I$) coincides with the CoM (O_M) of the Mars. Furthermore, the Martian equatorial plane concurs with the $O_M X_I Y_I$-plane, while the Z_I-axis is directed north. The direction of the X_I-axis is settled at the zero-longitude meridian at zero time and the right-handed system is completed by the Y_I-axis.

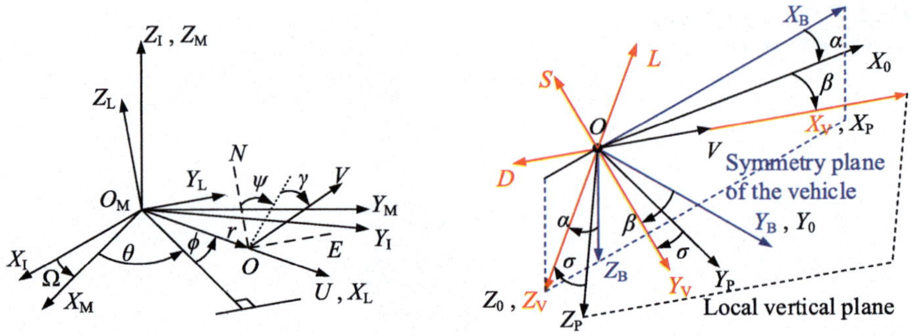

Fig. 2.3 Transformation of frames

2. **Mars-centered Mars-fixed reference frame, index M**

Like the inertial reference frame (O_M-$X_M Y_M Z_M$), the rotating frame is fixed to the Mars with the Z_M-axis pointing north. At zero longitude the X_M-axis intersects the equator and the right-handed system is withal completed by the Y_M-axis. In Fig. 2.3 the configuration of the rotating frame MCMF with respect to the inertial frame MCI is shown, with ω_0 being the rotational rate along the spin-axis of the Mars.

3. **Local east, north, up reference frame, index L**

The origin of the local frame (O_M-$X_L Y_L Z_L$) lies at the O_M with the X_L-axis pointing upwards towards the CoM (O) of the vehicle. The Z_L-axis and the Y_L-axis point in the local northern and eastern direction, respectively.

4. **Path reference frame, index P**

The origin of the path reference frame (O-$X_P Y_P Z_P$) is located at the CoM (O) of the vehicle, and the X_P-axis coincides with the relative velocity vector. The Z_P-axis points positively downwards in the local vertical plane and the Y_P-axis completing the right-handed reference frame.

5. **Wind reference frame, index V**

The origin of wind reference frame (O-$X_V Y_V Z_V$) is also at the CoM (O) of the vehicle, and the X_V-axis is collinear with the X_P-axis. The Z_V-axis points downwards towards the Mars, which is located in the longitudinal symmetry plane of the vehicle and perpendicular to the X_V-axis. The Y_V-axes finalizes the right-hand system.

6. **Body reference frame, index B**

Fixed to the vehicle, the X_B and Z_B axes of this frame (O-$X_B Y_B Z_B$) lie in the plane of symmetry being positive forward and downward, respectively. Again, the right-hand system is completed by the Y_B-axis.

Table 2.2 Transformation between reference frames

Frames	Transformation	Variables		
		Symbol	Unit	Description
I → M	$\boldsymbol{C}_{\mathrm{M/I}} = \boldsymbol{C}_3(\omega_0 t)$	ω_0	rad/s	Rotational rate of the Mars
		t	s	Time from epoch
M → L	$\boldsymbol{C}_{\mathrm{L/M}} = \boldsymbol{C}_2(-\phi)\boldsymbol{C}_3(\theta)$	θ	rad	Longitude
		ϕ	rad	Latitude
L → P	$\boldsymbol{C}_{\mathrm{P/L}} = \boldsymbol{C}_2(\gamma - \pi/2)\boldsymbol{C}_1(-\psi)$	γ	rad	Flight-path angle
		ψ	rad	Azimuth (or heading) angle
P → V	$\boldsymbol{C}_{\mathrm{V/P}} = \boldsymbol{C}_1(\sigma)$	σ	rad	Bank angle
V → B	$\boldsymbol{C}_{\mathrm{B/V}} = \boldsymbol{C}_2(\alpha)\boldsymbol{C}_3(-\beta)$	α	rad	Angle of attack
		β	rad	Angle of sideslip

7. **Transformation of frames**

The coordinate transformation is performed by a sequence of rotations of a reference frame about its own *X*, *Y*, or *Z* axes, which is given by:

$$
\begin{aligned}
&\boldsymbol{C}_1(a) = \begin{bmatrix} 1 & 0 & 0 \\ 0 & \cos a & \sin a \\ 0 & -\sin a & \cos a \end{bmatrix}, \quad \boldsymbol{C}_2(a) = \begin{bmatrix} \cos a & 0 & -\sin a \\ 0 & 1 & 0 \\ \sin a & 0 & \cos a \end{bmatrix}, \\
&\boldsymbol{C}_3(a) = \begin{bmatrix} \cos a & \sin a & 0 \\ -\sin a & \cos a & 0 \\ 0 & 0 & 1 \end{bmatrix}
\end{aligned}
\tag{2.3}
$$

Notice that the subscripts 1, 2 and 3 indicate the *X*, *Y*, and *Z* axes, respectively.

The total transformation from frame A to frame B is defined as a consolidation of the orthonormal unit rotation matrices $\boldsymbol{C}_1$, $\boldsymbol{C}_2$ and $\boldsymbol{C}_3$. The product of these matrices is also of an orthonormal nature and the inverse is the transpose, which is known as the direction cosine matrix $\boldsymbol{C}_{\mathrm{B/A}}$. The transformation between reference frames for Mars atmospheric entry is list in Table 2.2.

2.1.5 Translational Equations of Motion

1. **Time domain**

This section describes the motion of a vehicle over a rotating, non-spherical planet. We assume that the atmosphere rotates uniformly with the planet, so there is no wind. The state of the vehicle is defined with respect to the reference frames introduced in the previous section.

According to Newton's second law of motion in inertial frames I, we have

$$\frac{d\boldsymbol{r}}{dt} = \boldsymbol{v}, \quad \frac{d^2\boldsymbol{r}}{dt^2} = \frac{\boldsymbol{T} + \boldsymbol{A}}{m} + \boldsymbol{g} \tag{2.4}$$

where $\boldsymbol{r}$ and $\boldsymbol{v}$ are the inertial position and velocity vectors of the vehicle; $\boldsymbol{T}$, $\boldsymbol{A}$, and $\boldsymbol{g}$ are the thrust, aerodynamic force, and gravitational acceleration.

Note that the Mars entry vehicle is unpowered, then the time derivative of the position vector in a rotating system is

$$\begin{aligned} \frac{d\boldsymbol{r}}{dt} &= \frac{d[\boldsymbol{r}]_{\mathrm{M}}}{dt} + \boldsymbol{\omega}_0 \times [\boldsymbol{r}]_{\mathrm{M}} = \boldsymbol{V} + \boldsymbol{\omega}_0 \times [\boldsymbol{r}]_{\mathrm{M}} \\ \frac{d^2\boldsymbol{r}}{dt^2} &= \frac{d\boldsymbol{V}}{dt} + 2\boldsymbol{\omega}_0 \times \boldsymbol{V} + \boldsymbol{\omega}_0 \times (\boldsymbol{\omega}_0 \times [\boldsymbol{r}]_{\mathrm{M}}) = \frac{\boldsymbol{A}}{m} + \boldsymbol{g} \end{aligned} \tag{2.5}$$

where $\boldsymbol{\omega}_0 = \begin{bmatrix} 0 & 0 & \omega_0 \end{bmatrix}^{\mathrm{T}}$ is the angular velocity vector of the L frame relative to the I frame, operator $[\boldsymbol{a}]_{\mathrm{A}}$ denotes the coordinate of the vector $\boldsymbol{a}$ in the frame A, and $\boldsymbol{V}$ is the relative velocity vector of the vehicle.

According to the transformation of frames, the relative velocity can be rewritten as

$$\boldsymbol{V} = \frac{d[\boldsymbol{r}]_{\mathrm{M}}}{dt} = \frac{d(\boldsymbol{C}_{\mathrm{M/L}}[\boldsymbol{r}]_{\mathrm{L}})}{dt} = \frac{d(r\begin{bmatrix} \cos\phi\cos\theta & \cos\phi\sin\theta & \sin\phi \end{bmatrix}^{\mathrm{T}})}{dt} \tag{2.6}$$

Meanwhile, the coordinate of $\boldsymbol{V}$ in the L frame is

$$[\boldsymbol{V}]_{\mathrm{L}} = \boldsymbol{C}_{\mathrm{L/V}}[\boldsymbol{V}]_{\mathrm{V}} = \boldsymbol{C}_{\mathrm{L/V}}\begin{bmatrix} V & 0 & 0 \end{bmatrix} = \boldsymbol{C}_{\mathrm{L/M}}\boldsymbol{V} \tag{2.7}$$

Then, substituting Eq. (2.6) into Eq. (2.7), one has

$$\begin{bmatrix} V\sin\gamma & V\cos\gamma\sin\psi & V\cos\gamma\cos\psi \end{bmatrix}^{\mathrm{T}} = \begin{bmatrix} \frac{dr}{dt} & \frac{d\theta}{dt}r\cos\phi & \frac{d\phi}{dt}r \end{bmatrix}^{\mathrm{T}} \tag{2.8}$$

The second acceleration term in Eq. (2.5) can be expressed in the frame:

$$\frac{d[\boldsymbol{V}]_{\mathrm{P}}}{dt} = \frac{[\boldsymbol{T}]_{\mathrm{P}} + [\boldsymbol{A}]_{\mathrm{P}}}{m} + [\boldsymbol{g}]_{\mathrm{P}} - 2[\boldsymbol{\omega}_0]_{\mathrm{P}} \times [\boldsymbol{V}]_{\mathrm{P}} - [\boldsymbol{\omega}_0]_{\mathrm{P}} \times ([\boldsymbol{\omega}_0]_{\mathrm{P}} \times [\boldsymbol{r}]_{\mathrm{P}}) \tag{2.9}$$

And the Eq. (2.9) can be rewritten as the follows.

$$\frac{d[\boldsymbol{V}]_{\mathrm{P}}}{dt} = \frac{d(\boldsymbol{C}_{\mathrm{P/M}}\boldsymbol{V})}{dt} + \boldsymbol{\omega}_{\mathrm{P/M}} \times [\boldsymbol{V}]_{\mathrm{P}} \tag{2.10}$$

$$\begin{aligned} \boldsymbol{\omega}_{\mathrm{P/M}} = {} & \boldsymbol{C}_{\mathrm{P/L}}\boldsymbol{C}_{\mathrm{L/M}}\begin{bmatrix} 0 & 0 & \frac{d\theta}{dt} \end{bmatrix}^{\mathrm{T}} + \boldsymbol{C}_{\mathrm{P/L}}\boldsymbol{C}_2(-\phi)\begin{bmatrix} 0 & -\frac{d\phi}{dt} & 0 \end{bmatrix}^{\mathrm{T}} \\ & + \boldsymbol{C}_{\mathrm{P/L}}\begin{bmatrix} -\frac{d\psi}{dt} & 0 & 0 \end{bmatrix}^{\mathrm{T}} + \boldsymbol{C}_2(\gamma - \pi/2)\begin{bmatrix} 0 & \frac{d\gamma}{dt} & 0 \end{bmatrix}^{\mathrm{T}} \end{aligned} \tag{2.11}$$

where $\boldsymbol{\omega}_{\mathrm{P/M}}$ is the angular velocity vector of the P frame relative to the M frame.

The right-hand terms in the Eq. (2.9) can be expressed in the P frame as follows:

$$[\boldsymbol{r}]_{\mathrm{P}} = \boldsymbol{C}_{\mathrm{P/L}}[\boldsymbol{r}]_{\mathrm{L}} = \boldsymbol{C}_{\mathrm{P/L}}\begin{bmatrix} r & 0 & 0 \end{bmatrix}^{\mathrm{T}},\ [\boldsymbol{V}]_{\mathrm{P}} = \begin{bmatrix} V & 0 & 0 \end{bmatrix}^{\mathrm{T}} \tag{2.12}$$

$$[\boldsymbol{T}]_{\mathrm{P}} = \boldsymbol{C}_{\mathrm{P/V}}\boldsymbol{C}_{\mathrm{V/B}}[\boldsymbol{T}]_{\mathrm{B}} = \boldsymbol{C}_{\mathrm{P/V}}\boldsymbol{C}_{\mathrm{V/B}}\begin{bmatrix} T & 0 & 0 \end{bmatrix}^{\mathrm{T}} \tag{2.13}$$

$$[\boldsymbol{A}]_{\mathrm{P}} = \boldsymbol{C}_{\mathrm{P/V}}[\boldsymbol{A}]_{\mathrm{V}} = \boldsymbol{C}_{\mathrm{P/V}}\begin{bmatrix} -D & -S & -L \end{bmatrix}^{\mathrm{T}} \tag{2.14}$$

$$[\boldsymbol{g}]_{\mathrm{P}} = \boldsymbol{C}_{\mathrm{P/L}}[\boldsymbol{g}]_{\mathrm{L}} = \boldsymbol{C}_{\mathrm{P/L}}\begin{bmatrix} -g_r & 0 & -g_\phi \end{bmatrix}^{\mathrm{T}} \tag{2.15}$$

$$[\boldsymbol{\omega}_0]_{\mathrm{P}} = \boldsymbol{C}_{\mathrm{P/L}}\boldsymbol{C}_{\mathrm{L/M}}\begin{bmatrix} 0 & 0 & \omega_0 \end{bmatrix}^{\mathrm{T}} \tag{2.16}$$

Then we can rewrite Eqs. (2.10)–(2.16) as follows:

$$\begin{aligned}
\begin{bmatrix} \frac{dV}{dt} \\ V\frac{d\psi}{dt}\cos\gamma \\ -V\frac{d\gamma}{dt} \end{bmatrix} &= \frac{1}{m}\begin{bmatrix} -D \\ -S\cos\sigma + L\sin\sigma \\ -S\sin\sigma - L\cos\sigma \end{bmatrix} + \begin{bmatrix} -g_r\sin\gamma - g_\phi\cos\gamma\cos\psi \\ g_\phi\sin\psi \\ g_r\cos\gamma - g_\phi\sin\gamma\cos\psi \end{bmatrix} \\
&+ \begin{bmatrix} 0 \\ 2V(\omega_0\sin\phi\cos\gamma - \omega_0\cos\phi\sin\gamma\cos\psi) \\ -2V\omega_0\cos\phi\sin\psi \end{bmatrix} \\
&+ \begin{bmatrix} r\omega_0^2\cos\phi(\cos\phi\sin\gamma - \sin\phi\cos\gamma\cos\psi) \\ r\omega_0^2\sin\phi\cos\phi\sin\psi \\ -r\omega_0^2\cos\phi(\cos\phi\cos\gamma + \cos\psi\sin\phi\sin\gamma) \end{bmatrix} \\
&+ \begin{bmatrix} 0 \\ V^2\cos^2\gamma\sin\psi\tan\phi/r \\ -V^2\cos\gamma/r \end{bmatrix}
\end{aligned} \tag{2.17}$$

where *D*, *L*, and *S* are the drag, lift, and side forces, respectively.

Further, the translational equations of motion for the Mars entry is given by:

$$\begin{cases} \frac{dr}{dt} = V \sin\gamma \\ \frac{d\theta}{dt} = \frac{V\cos\gamma\sin\psi}{r\cos\phi} \\ \frac{d\phi}{dt} = \frac{V\cos\gamma\cos\psi}{r} \\ \frac{dV}{dt} = -D - g_r\sin\gamma - g_\phi\cos\gamma\cos\psi \\ \quad +\omega_0^2 r\cos\phi(\cos\phi\sin\gamma - \sin\phi\cos\gamma\cos\psi) \\ \frac{d\gamma}{dt} = \frac{L\cos\sigma}{V} + \left(\frac{V}{r} - \frac{g_r}{V}\right)\cos\gamma + \frac{g_\phi\sin\gamma\cos\psi}{V} \\ \quad +2\omega_0\cos\phi\sin\psi + \frac{\omega_0^2 r\cos\phi(\cos\phi\cos\gamma + \cos\psi\sin\phi\sin\gamma)}{V} \\ \frac{d\psi}{dt} = \frac{L\sin\sigma}{V\cos\gamma} + \frac{V\cos\gamma\sin\psi\tan\phi}{r} + \frac{g_\phi\sin\psi}{V\cos\gamma} \\ \quad +\frac{2\omega_0(\sin\phi - \cos\phi\tan\gamma\cos\psi)}{V} + \frac{\omega_0^2 r\sin\phi\cos\phi\sin\psi}{V\cos\gamma} \end{cases} \tag{2.18}$$

To avoid numerical singularity, the dimensionless dynamics are introduced. And considering that Mars entry phase lasts about 5 min, J_2-dependent terms are ignored. Thus the non-dimensional equations of motion are:

$$\begin{cases} \frac{dr}{d\tau} = V \sin\gamma \\ \frac{d\theta}{d\tau} = \frac{V\cos\gamma\sin\psi}{r\cos\phi} \\ \frac{d\phi}{d\tau} = V\cos\gamma\cos\psi / r \\ \frac{dV}{d\tau} = -D - \frac{\sin\gamma}{r^2} + \omega_0^2 r\cos\phi(\cos\phi\sin\gamma - \sin\phi\cos\gamma\cos\psi) \\ \frac{d\gamma}{d\tau} = \frac{L\cos\sigma}{V} + \left(\frac{V}{r} - \frac{1}{r^2 V}\right)\cos\gamma \\ \quad +2\omega_0\cos\phi\sin\psi + \frac{\omega_0^2 r\cos\phi(\cos\phi\cos\gamma + \cos\psi\sin\phi\sin\gamma)}{V} \\ \frac{d\psi}{d\tau} = \frac{L\sin\sigma}{V\cos\gamma} + \frac{V\cos\gamma\sin\psi\tan\phi}{r} \\ \quad +\frac{2\omega_0(\sin\phi - \cos\phi\tan\gamma\cos\psi)}{V} + \frac{\omega_0^2 r\sin\phi\cos\phi\sin\psi}{V\cos\gamma} \end{cases} \tag{2.19}$$

In dimensionless form, length is normalized by the equatorial radius of the Mars R_0, and time is normalized by $t_s = \sqrt{R_0 / g_0}$, where g_0 is the gravitational acceleration at R_0. The choices of the distance and time normalization lead to the velocity being normalized by $V_s = \sqrt{R_0 g_0}$. The differentiations in the previous equations are with respect to the dimensionless time $\tau = t / t_s$. The dimensionless drag and lift accelerations are:

$$L = \frac{R_0 \rho V^2 S_r C_L}{2m}, \quad D = \frac{R_0 \rho V^2 S_r C_D}{2m} \tag{2.20}$$

where S_r is the reference area of the entry vehicle, C_L and C_D are the lift and drag coefficients, ρ is the atmospheric density.

2. **Energy domain**

The dimensionless negative specific mechanical energy is defined as:

$$e = \frac{1}{r} - \frac{V^2}{2} \tag{2.21}$$

If the rotation term is ignored, it can be readily shown that

$$\frac{de}{d\tau} \approx DV \tag{2.22}$$

Accordingly, the velocity V is determined by

$$V = \sqrt{2\left(\frac{1}{r} - e\right)} \approx \sqrt{2(1 - e)} \tag{2.23}$$

Then, dynamic equations with respect to energy are

$$
\begin{cases}
\frac{dr}{de} = \frac{\sin\gamma}{D} \\
\frac{d\theta}{de} = \frac{\cos\gamma\sin\psi}{rD\cos\phi} \\
\frac{d\phi}{de} = \frac{\cos\gamma\cos\psi}{rD} \\
\frac{d\gamma}{de} = \frac{L\cos\sigma}{V^2D} + \left(\frac{1}{r} - \frac{1}{r^2V^2}\right)\frac{\cos\gamma}{D} \\
\quad + \frac{2\omega_0\cos\phi\sin\psi}{D} + \frac{\omega_0^2 r\cos\phi(\cos\phi\cos\gamma + \cos\psi\sin\phi\sin\gamma)}{V^2D} \\
\frac{d\psi}{de} = \frac{L\sin\sigma}{V^2D\cos\gamma} + \frac{\cos\gamma\sin\psi\tan\phi}{rD} \\
\quad + \frac{2\omega_0(\sin\phi - \cos\phi\tan\gamma\cos\psi)}{VD} + \frac{\omega_0^2 r\sin\phi\cos\phi\sin\psi}{V^2D\cos\gamma}
\end{cases}
\tag{2.24}
$$

2.1.6 *Path Constraints*

The ability of the entry vehicle to withstand heat is determined by the thermal protection system. The system's heat tolerance is defined in terms of the heat flux $\dot{Q}$. An approximation of the heat flux is given by the Sutton-Graves-Equation, based on the effective nose radius R_n of the heat shield and a parameter k_Q depending on the composition of the atmosphere. For the Mars mission, we have $k_Q = 1.9027 \times 10^{-4}\sqrt{\text{kg}}/\text{m}$. The typical dynamic pressure q, heat flux $\dot{Q}$, and load factor a constraints are given by:

$$
\begin{cases}
q = \rho(V_s V)^2/2 \le q_{\max} \\
\dot{Q} = k_Q(\rho/R_n)^{0.5}(V_s V)^{3.15} \le \dot{Q}_{\max} \\
a = \sqrt{L^2 + D^2} \le a_{\max}
\end{cases}
\tag{2.25}
$$

2.2 Convex Optimization

Convex optimization is a subfield of mathematical optimization that studies the problem of minimizing convex functions over convex sets [51]. Typically, the linear programming problem and the conic optimization problem are convex problems. In a convex optimization problem, the feasible region is also a convex region.

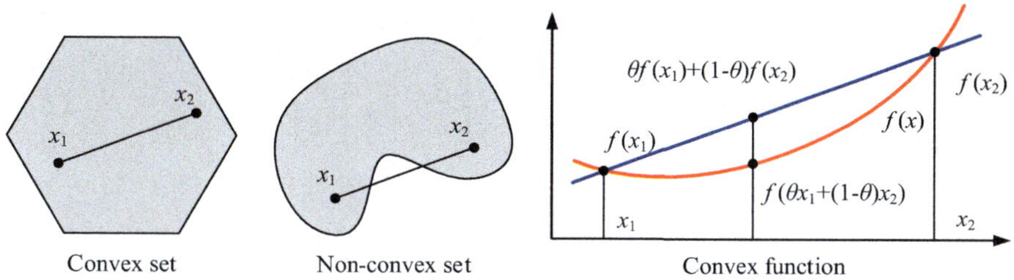

Fig. 2.4 Convex set, non-convex set, and convex function

1. **Convex set**

A set S is convex if the line segment between any two points in S lies in S, i.e., if for any $x_1, x_2 \in S$ and any θ with $0 \leq \theta \leq 1$, we have

$$\theta x_1 + (1 - \theta)x_2 \in S \tag{2.26}$$

2. **Convex function**

A function $f : \mathbb{R}^n \rightarrow \mathbb{R}$ is convex if **dom** f is a convex set and if for all $x_1, x_2 \in$ **dom** f, and θ with $0 \leq \theta \leq 1$, we have

$$f(\theta x_1 + (1 - \theta)x_2) \leq \theta f(x_1) + (1 - \theta)f(x_2) \tag{2.27}$$

Geometrically, this inequality means that the line segment between $(x_1, f(x_1))$ and $(x_2, f(x_2))$, which is the chord from x_1 to x_2, lies above the graph of f (as shown in Fig. 2.4).

A convex optimization problem is in standard form if it is written as

$$\begin{aligned} &\text{minimize } f_0(x) \\ &\text{subject to } f_i(x) \leq 0, \quad i = 1, \ldots, m \\ &Ax = b \end{aligned} \tag{2.28}$$

where $x \in \mathbb{R}^n$, $f_i : \mathbb{R}^n \rightarrow \mathbb{R}$, $i = 0, \ldots, m$ is convex, and equality constraint functions are affine transformations.

2.2.1 Second-Order Cone Programming, SOCP

The second-order cone is the norm cone for the Euclidean norm, i.e.,

$$\boldsymbol{C}_n = \left\{ \begin{bmatrix} \boldsymbol{x} \\ t \end{bmatrix} \in \mathbb{R}^n | \boldsymbol{x} \in \mathbb{R}^{n-1}, t \in \mathbb{R}, \|\boldsymbol{x}\| \leq t \right\} \tag{2.29}$$

And the second-order cone constraint is defined as:

$$\|\boldsymbol{Ax}+\boldsymbol{b}\| \leq \boldsymbol{c}^{\mathrm{T}}\boldsymbol{x}+d \Leftrightarrow \begin{bmatrix} \boldsymbol{Ax} \\ \boldsymbol{c}^{T} \end{bmatrix} + \begin{bmatrix} \boldsymbol{b} \\ d \end{bmatrix} \in \boldsymbol{C}_n \tag{2.30}$$

where $\boldsymbol{x} \in \mathbb{R}^n$, $\boldsymbol{A} \in \mathbb{R}^{(n-1)\times n}$, $\boldsymbol{b} \in \mathbb{R}^{n-1}$, $\boldsymbol{c} \in \mathbb{R}^n$, $d \in \mathbb{R}$.

The standard Second-Order Cone Programming (SOCP) problem has the form

$$\begin{aligned} &\underset{\boldsymbol{x}}{\text{minimize}}\ \boldsymbol{f}^{\mathrm{T}}\boldsymbol{x} \\ &\text{subject to} \|\boldsymbol{A}_i\boldsymbol{x}+\boldsymbol{b}_i\| \leq \boldsymbol{c}_i^{\mathrm{T}}\boldsymbol{x}+d_i,\ i=1,\ldots,m \\ &\boldsymbol{Fx}=\boldsymbol{g} \end{aligned} \tag{2.31}$$

where $\boldsymbol{x} \in \mathbb{R}^n$, $\boldsymbol{A}_i \in \mathbb{R}^{n_i\times n}$, $\boldsymbol{b}_i \in \mathbb{R}^{n_i}$, $\boldsymbol{c}_i \in \mathbb{R}^{n_i}$, $d_i \in \mathbb{R}$, $\boldsymbol{F} \in \mathbb{R}^{p\times n}$, $\boldsymbol{g} \in \mathbb{R}^p$

2.2.2 *Quadratically Constrained Quadratic Programming, QCQP*

In mathematical optimization, a quadratically constrained quadratic program (QCQP) is an optimization problem in which both the objective function and the constraints are quadratic functions. It has the form

$$\begin{aligned} &\underset{x}{\text{minimize}}\ \frac{1}{2}\boldsymbol{x}\boldsymbol{P}_0^{\mathrm{T}}\boldsymbol{x}+\boldsymbol{q}_0^{\mathrm{T}}\boldsymbol{x}+r_0 \\ &\text{subject to}\ \frac{1}{2}\boldsymbol{x}\boldsymbol{P}_i^{\mathrm{T}}\boldsymbol{x}+\boldsymbol{q}_i^{\mathrm{T}}\boldsymbol{x}+r_i \leq 0,\ i=1,\ldots,m \\ &\boldsymbol{Ax}=\boldsymbol{b} \end{aligned} \tag{2.32}$$

where $\boldsymbol{P}_i \in \mathbb{S}^n, i=0,\ldots,m$, $\boldsymbol{A} \in \mathbb{R}^{p\times n}$, $\mathbb{S}^n$ is the set of symmetric square matrices of order n.

If $\boldsymbol{P}_i \succcurlyeq 0$ for $i=0,\ldots,m$, QCQP is a convex programming problem that can be solved in polynomial time, but in general the QCQP problem is NP-Hard.

2.3 Uncertainty Quantification

Uncertainty quantification is a mathematical methodology to quantitatively describe and evaluate the magnitude and propagation of uncertainties in a given system [69]. It enables the assessment of risk and operational safety margin and the incorporation of this knowledge for the purpose of robust optimal design supported by statistical verification and validation. For Mars atmospheric entry dynamics, the uncertainties

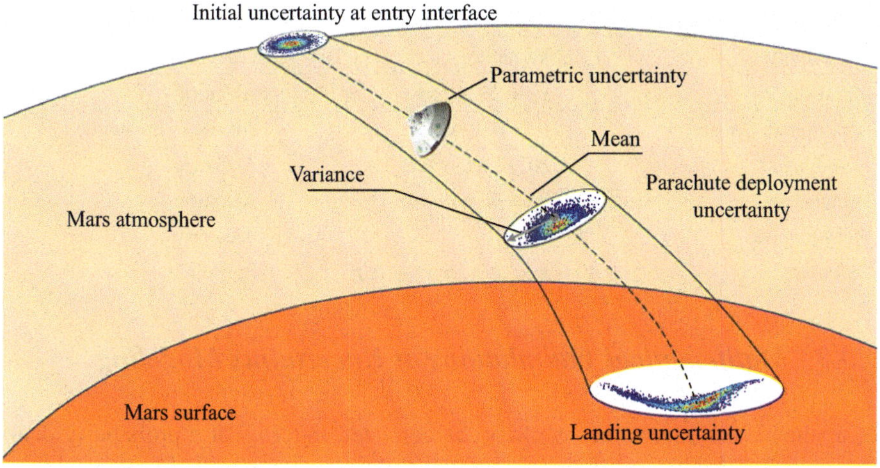

Fig. 2.5 Uncertainties of the Mars entry phase

in the initial conditions and system parameters result in off-nominal trajectories, as shown in Fig. 2.5 [71]. The evolution of these uncertainties can be described quantitatively using mathematical expectation and variance, etc.

2.3.1 *Definitions and Categories of Uncertainties*

In general, uncertainty can be classified into two categories [112]:

1. **Aleatoric uncertainty (or stochastic uncertainty)**

Aleatoric uncertainty refers to the notion of randomness caused by the stochastic variability inherent in the data. Owing to its inherent nature, this uncertainty is irreducible.

2. **Epistemic uncertainty (or systematic uncertainty)**

Epistemic uncertainty accounts for the uncertainty in a model caused by a lack of knowledge. This uncertainty can be reduced by collecting more information of the system.

Aleatoric and epistemic uncertainty can occur simultaneously in a Mars entry mission, and the sources of uncertainty are diverse. The main uncertainties are summarized in the Table 2.3 [70].

Table 2.3 Uncertainties in the Mars entry mission

Uncertainty	Variable	Category
Initial state	$\boldsymbol{x}_0 = \overline{\boldsymbol{x}}_0 + \Delta \boldsymbol{x}_0$	Aleatoric
Atmosphere density	$\rho = \overline{\rho} + \Delta\rho$	Aleatoric, epistemic
Aerodynamic parameters	$\Delta C_L, \Delta C_D, \Delta S_r$	Aleatoric
Vehicle mass	$m = \overline{m} + \Delta m$	Aleatoric
Measurement	$\delta \boldsymbol{x}_0$	Aleatoric
Dynamics	$\dot{\boldsymbol{x}} = \boldsymbol{f}(\boldsymbol{x}, \boldsymbol{u}, t) + \Delta \boldsymbol{f}$	Epistemic

2.3.2 *Mathematical Foundations of Uncertainty Modeling*

In mathematics, uncertainty is often characterized in terms of a probability distribution. From this perspective, epistemic uncertainty means not being sure what the relevant probability distribution is, and aleatoric uncertainty means not being sure what a random sample drawn from a probability distribution will be [113].

1. **Probability space and probability density function**

(1) **Sample space**

A sample space is a collection or a set of possible outcomes of a random experiment. The sample space is represented using the symbol Ω. The subset of possible outcomes of an experiment is called events.

(2) **Random variable**

A random variable X is a measurable function $X : \Omega \rightarrow \mathrm{E}$ from a sample space Ω as a set of possible outcomes to a measurable space E.

(3) **Stochastic processes**

Stochastic processes are probabilistic models for random quantities evolving in time or space. The evolution is governed by some dependence relationship between the random quantities at different times or locations.

(4) **Probability space**

A probability space is a measure space with total measure one. The standard notation is (Ω, F, Pr) where:

Ω is a set (sometimes called a sample space in elementary probability). Elements of Ω are denoted ω and are sometimes called outcomes.

F is a σ-algebra (or σ-field, we will use these terms synonymously) of subsets of Ω. Sets in F are called events.

Pr is a function from F to [0, 1] with Pr(Ω) = 1 and such that if $E_1, E_2, \ldots E_\infty \in F$ are disjoint,

$$\Pr\left(\bigcup_{j=1}^{\infty} E_j\right) = \sum_{j=1}^{\infty} \Pr(E_j) \tag{2.33}$$

And we say "probability of E" for $\Pr[E]$.

(5) **Probability density function, PDF**

PDF is used to define the random variable's probability coming within a distinct range of values, as opposed to taking on any one value. If a random variable $X \in \mathbb{R}^n$, then probability of $X \in \Omega$ is

$$\Pr(X \in \Omega) = \int_{\Omega} p(\boldsymbol{\xi}, t)\mathrm{d}\boldsymbol{\xi} = \int_{\Omega} p(\boldsymbol{\xi}, t)\mathrm{d}\xi_1 \mathrm{d}\xi_2 \cdots \mathrm{d}\xi_n \tag{2.34}$$

where $p(\boldsymbol{\xi}, t)$ is the PDF and it has

$$p(\mathbf{x}, t) \geq 0 \tag{2.35}$$

$$\int_{-\infty}^{+\infty} p(\boldsymbol{\xi}, t)\mathrm{d}\xi = 1 \tag{2.36}$$

$$\Pr(\mathbf{x} = \mathbf{const}) = 0,\ \mathbf{const} \in \mathbb{R}^n \tag{2.37}$$

2. **Numerical characteristics of random variables**

The distribution of a random variable or vector, which comprises the complete collection of related probabilities, contains all the information about its behavior. This detailed information can be summarized in a few important numerical characteristics that describe the average value, the most probable value of a random variable, its spread, its variability, and so on. These are numbers that provide some information about a random variable or about the relationship between random variables.

(1) **Mathematical expectation**

Consider a random variable X with a finite list $x_1, \ldots, x_n$ of possible outcomes, each of which (respectively) has probability $p_1, \ldots, p_n$ of occurring. The expectation of X is defined as

$$E(X) = x_1 p_1 + \cdots + x_k p_k \tag{2.38}$$

For a continuous case, the expectation is given by:

$$E(X) = \int_{-\infty}^{+\infty} x \operatorname{Pr}(x) \mathrm{d}x \tag{2.39}$$

where x is the possible value of X. Therefore, it is natural to interpret $E(X)$ as a weighted average of the all possible values, with weights given by their probabilities.

(2) **Variance**

The variance is a measure of variability. It is calculated by taking the average of squared deviations from the mean of a random variable. It is defined as

$$D(X) = \int_{-\infty}^{+\infty} [x - E(X)]^2 \operatorname{Pr}(x) \mathrm{d}x, \text{ or } D(X) = E(X^2) - [E(X)]^2 \tag{2.40}$$

Variance tells you the degree of spread in your data set. The more spread the data, the larger the variance is in relation to the mean. Moreover, the square root of the variance is the standard deviation

$$\sigma(X) = \sqrt{D(X)} \tag{2.41}$$

(3) **Covariance**

Covariance measures the direction of the relationship between two random variables. A positive covariance means that both variables tend to be high or low at the same time. A negative covariance means that when one variable is high, the other tends to be low.

For two jointly distributed real-valued random variables X and Y with finite second moments, the covariance is defined as the expected value (or mean) of the product of their deviations from their individual expected values:

$$\operatorname{Cov}(X, Y) = E([X - E(X)][Y - E(Y)]) \tag{2.42}$$

(4) **Moments**

Moments in probability theory are mathematical quantities used to describe the shape and characteristics of probability distributions. The moment of order k ($k > 0$ an integer) of a random variable X is defined as the mathematical expectation $E(X^k)$, if it exists. If F is the distribution function of the random variable X, then

$$E(X^k) = \int_{-\infty}^{+\infty} x^k dF(x) \tag{2.43}$$

The first-order moment (a statistical moment in mechanics) of a random variable X is the mathematical expectation $E(X)$. The value $E[(X-a)^k]$ is called the moment of order k relative to a, $E[(X-E(X))^k]$ is the central moment of order k. The second-order central moment $E[(X - E(X))^2]$ is called the variance $D(X)$(or the moment of inertia in mechanics).

3. **Probability distribution**

A probability distribution is the mathematical function that gives the probabilities of occurrence of different possible outcomes for an experiment. It is a mathematical description of a random phenomenon in terms of its sample space and the probabilities of events (subsets of the sample space).

(1) **Normal distribution (or Gaussian distribution)**

The general form of its probability density function is

$$p(x) = \frac{1}{\sqrt{2\pi}\,\sigma}\exp\left[-\frac{(x-\mu)^2}{2\sigma^2}\right], \quad x \in \mathbb{R} \tag{2.44}$$

where the parameter μ is the mean or expectation of the distribution (and also its median and mode), while the parameter σ is its standard deviation. The variance of the distribution is σ^2. Thus when a random variable X is normally distributed with mean μ and standard deviation σ, one may write $X \sim N(\mu, \sigma^2)$.

(2) **Uniform distribution**

The probability density function of the continuous uniform distribution is:

$$p(x) = \begin{cases} \frac{1}{b-a}, & a < x < b \\ 0, & \text{others} \end{cases} \tag{2.45}$$

The normal distribution is often referred to as $X \sim U(a, b)$.

4. **Stochastic Differential Equation**

A stochastic differential equation is a differential equation whose coefficients are random numbers or random functions of the independent variable (or variables). A typical equation is of the form

$$dX(t,\omega) = f(t, X(t,\omega))dt + g(t, X(t,\omega))dW(t,\omega) \tag{2.46}$$

where ω denotes that $X = X(t, \omega)$ is a random variable and possesses the initial condition $X(0, \omega) = X_0$ with probability 1, and $W(t, \omega)$ denotes a Wiener process (standard Brownian motion).

Table 2.4 Classical methods for the Mars entry uncertainty optimization

Index	Method	Subclass
1	Monte Carlo	
2	Stochastic expansion	Polynomial chaos expansions
		Stochastic collocation method
3	Gaussian sum	
4	Taylor series expansion	

2.3.3 Uncertainty Optimization Methods

For the Mars entry uncertainty optimization, the main purpose is to obtain the numerical characteristics and probability distribution of the state variable during the Mars entry [112]. Some classical methods for the Mars entry uncertainty optimization are listed in Table 2.4.

2.4 Robust Optimization

Robust optimization is an important subfield of optimization that deals with uncertainty in the data of optimization problems. Under this framework, the objective and constraint functions are only assumed to belong to certain sets in function space (the so-called "uncertainty sets"). The goal is to make a decision that is feasible no matter what the constraints turn out to be, and optimal for the worst-case objective function [70].

2.4.1 Deterministic Optimization, DO

A general optimization problem is defined as:

$$\begin{cases} \text{find} \quad \boldsymbol{u} \\ \min \quad J \\ \text{s.t.} \quad \dot{\mathbf{x}} = \mathbf{f}(\mathbf{x}, \boldsymbol{u}, t) \\ \qquad \mathbf{g}(\mathbf{x}, \boldsymbol{u}, t) \leq \mathbf{0} \end{cases} \tag{2.47}$$

where $\boldsymbol{x}$ and $\boldsymbol{u}$ are the state and control variables, respectively; J is the performance index, $\boldsymbol{f}$ is the model of a dynamical system, $\boldsymbol{g}$ denotes the constraint.

For a deterministic optimization problem, all functions are exactly known. However, in real life, this is most likely not the case. The objective and constraint functions are often not precisely known or at best known with some noise. To solve the uncertainty-based optimization problem, the reliability-based optimization (RBO) and robust optimization (RO) are introduced.

2.4.2 Robust Optimization, RO

The robust optimization is a methodology to optimize design which is insensitive to various variations. The mathematical formulation for RO with probability theory is

$$\begin{cases} \text{find} \quad \boldsymbol{u} \\ \min \quad \mu(J) + \nu \cdot \sigma(J) \\ \text{s.t.} \quad \dot{\mathbf{x}} = \mathbf{f}(\mathbf{x}, \mathbf{c}, \boldsymbol{u}, t) \\ \qquad \mathbf{g}(\mathbf{x}, \mathbf{c}, \boldsymbol{u}, t) \le \mathbf{0} \end{cases} \tag{2.48}$$

where $\mu(J)$ and $\sigma(J)$ are the mean and standard deviation of the original performance index J, respectively; the robustness level ν can be set to be $1, 2, \ldots, 6$; $\boldsymbol{c}$ is the random variable.

In Fig. 2.6, the x denotes the design variables and system parameters, which could be uncertain. The $J(x)$ denotes the optimization objective function. The classic optimum is given by x_1. Under the influence of uncertainty Δx, the value of the objective function varies in the region ΔJ_1. It can be obviously found that $J(x)$ may move into the infeasible region. Instead of finding the deterministic optimum, a suboptimal solution which is called the "robust optimum" will be sought. In Fig. 2.6, x_2 denotes the robust optimum which never goes into the infeasible region under a same magnitude of uncertainty.

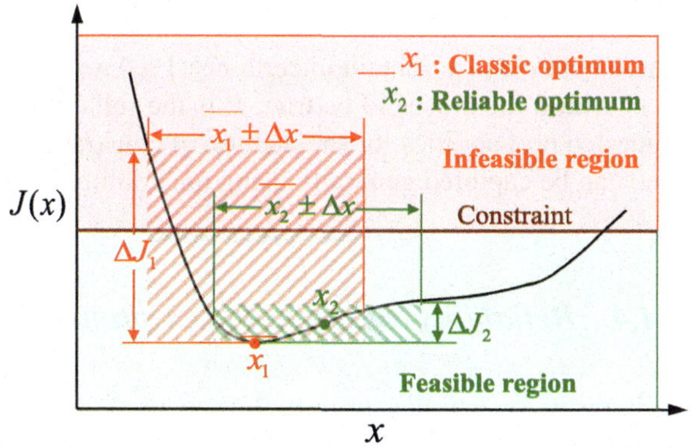

Fig. 2.6 Graphical illustration of robust optimization

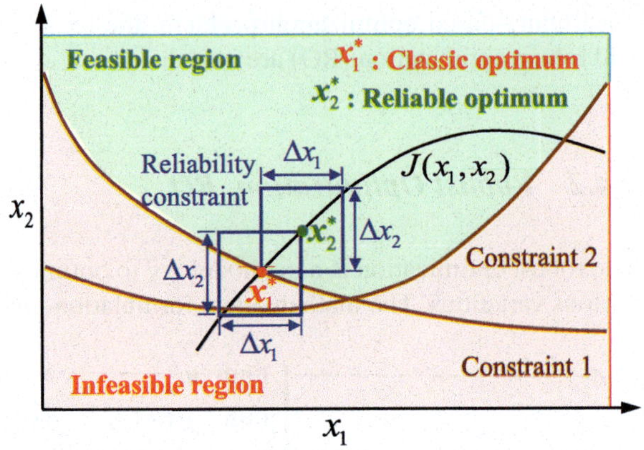

Fig. 2.7 Graphical illustration of reliability-based optimization

2.4.3 Reliability-Based Optimization, RBO

As shown in Fig. 2.7, the classic optimal solution $\boldsymbol{x}_1^*$ typically lies in the boundary of its feasible region. However, the optimal solution may be located in the infeasible region when uncertainties are considered. To obtain a feasible solution under uncertainties, the RBO chooses to find the solution $\boldsymbol{x}_2^*$ that is reliable with small chance of failure (i.e. violation of constraints) under a predefined acceptable level.

The mathematical formulation for RBO with probability theory is

$$\begin{cases} \text{find} \ \ \boldsymbol{u} \\ \min \ \ \mu(J) \\ \text{s.t.} \ \ \ \dot{\mathbf{x}} = \mathbf{f}(\mathbf{x}, \mathbf{c}, \boldsymbol{u}, t) \\ \qquad \Pr\{\mathbf{g}(\mathbf{x}, \mathbf{c}, \boldsymbol{u}, t) \leq \mathbf{0}\} \geq \mathbf{P} \end{cases} \tag{2.49}$$

where both $\boldsymbol{c}$ and $\boldsymbol{x}$ contain uncertainty, $\Pr(\cdot)$ represents the probability of the statement within the braces to be true; $\boldsymbol{P}$ is the reliability vector which is consist of the requested probabilities for satisfaction of constraints accordingly. Note that $\mu(\cdot)$ and $\Pr(\cdot)$ can be captured generally using uncertainty quantification techniques.

2.4.4 Reliability-Based Robust Optimization, RBRO

To improve system design in both robustness and reliability, RO and RBO can be combined and referred to as reliability-based robust design optimization (RBRO), which is formulated as

$$\begin{cases} \text{find} & u \\ \min & \mu(J) + \nu \cdot \sigma(J) \\ \text{s.t.} & \dot{\mathbf{x}} = \mathbf{f}(\mathbf{x}, \mathbf{c}, u, t) \\ & \Pr\{\mathbf{g}(\mathbf{x}, \mathbf{c}, u, t) \le \mathbf{0}\} \ge \mathbf{P} \end{cases} \tag{2.50}$$

This hybrid not only reduces the sensitivity of the objective function to uncertainty, but also ensures the reliability of the constraints.

2.4.5 Typical Methods

Typical methods of uncertainty-based optimization include sequential optimization and reliability assessment (SORA), robust collaborative optimization, and probabilistic analytic objective cascading, etc.

The SORA framework (as shown in Fig. 2.8) converts the robustness requirements into a set of probabilistic constraints, and then the optimal control problem is handled by the HSIGA. The SORA can find the robust optimum by solving the deterministic optimization problem homotopically and performing the robustness analysis iteratively. In this iterative process, the solution obtained in the previous iteration becomes the initial guess of the next iteration. In this framework, the influence of model uncertainty is taken into account during the optimization process. Thus, the robustness and feasibility of the optimized trajectory can be ensured.

Robust collaborative optimization is a multidisciplinary optimization (MDO), which provides the advantages of combining multiple optimization algorithms to solve multi-module, multi-objective, and multi-constraint optimization problems. By decomposing the complex optimization problem into several simple subsystem level and system level optimization problems, the existing simple and efficient optimization algorithms can be used to obtain the optimal solution of the robust collaborative optimization. The illustration of the robust collaborative optimization is given in Fig. 2.9.

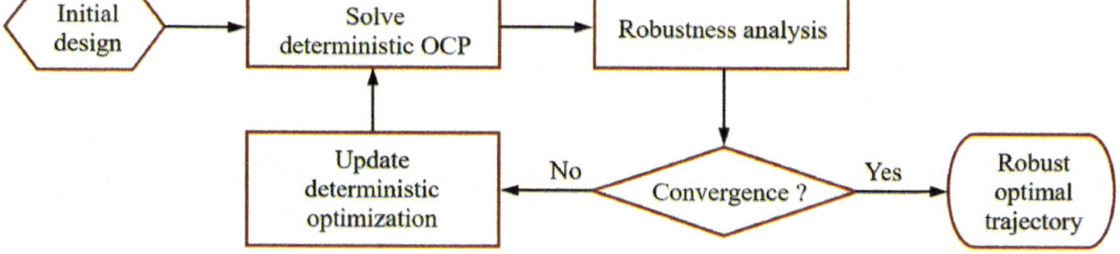

Fig. 2.8 The flowchart of sequential optimization and reliability assessment (SORA)

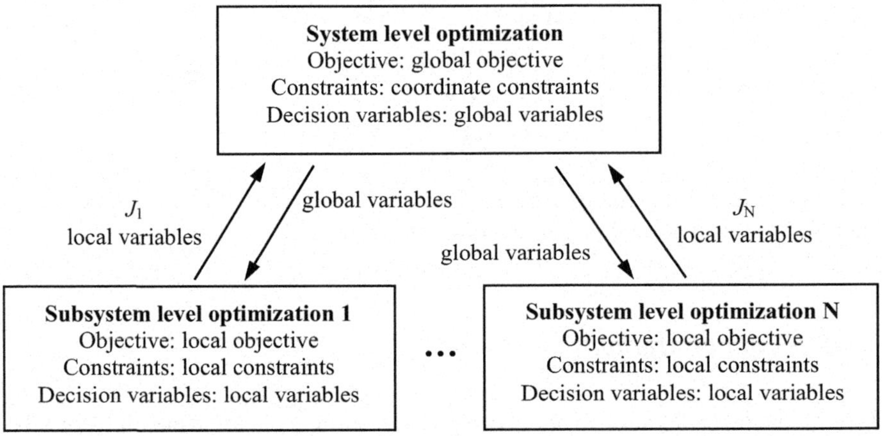

Fig. 2.9 The flowchart of robust collaborative optimization

2.5 Conclusions

This chapter mainly discusses the dynamics of Mars entry and theories related to convex optimization, uncertainty quantification, and robust optimization. Specifically, it includes: (1) the equations of motion are established to describe the Mars entry problem; (2) some basic concepts of convex optimization theory are introduced; (3) the definition of uncertainty modeling and the basic method of uncertainty quantification are explained. (4) The basic concepts and methods of robust optimization are presented.

Part II
Deterministic Optimization

Chapter 3
Improved Gauss Pseudospectral Method for Mars Entry Trajectory Planning

Aiming to improve the desired high efficiency under the initial state uncertainty, this chapter is to develop a hybrid trajectory optimization strategy for Mars entry by using PSO and Gauss Pseudospectral Method (GPM). As an intelligent optimization algorithm with a powerful global search capability, PSO is used to avoid the inefficient search through the neighborhood of the optimal solution. It is insensitive to initial guess and has good global optimization performance for some large-scale complicated nonlinear problems. Instead of using piecewise-continuous polynomials to interpolate between prescribed subintervals, Gaussian orthogonal polynomials can be used to approximate the control and state variables. These orthogonal polynomials are extensively used in spectral methods for solving Optimal Control Problem (OCP), and a new approach has been developed to transform OCP into Nonlinear Programing (NLP). Therefore, the GPM with Sequential Quadratic Programming (SQP) is able to solve the optimization problem quickly and accurately when a suitable initial guess is given.

This chapter is organized as follows: the Mars atmospheric entry trajectory optimization problem is formulated in the next section. Next, a hybrid optimization strategy is addressed. PSO is used for the global search of the appropriate initial guess, and the trajectory optimization problem is transformed into an NLP problem using GPM. Later, simulation system compositions are elaborated and a numerical simulation case is given to validate the hybrid optimization framework developed in this chapter. Finally, the last section contains the conclusion.

S. Li et al., *Trajectory Optimization and Guidance Methods for Mars Entry*,
https://doi.org/10.1007/978-981-99-6282-2_3

3.1 Formulation of Mars Entry Problem

3.1.1 Equations of Motion

The Mars atmospheric entry phase typically lasts from the time the vehicle enters the Martian atmosphere until the supersonic parachute deploys. The entry dynamics model is constructed under the following considerations. It is assumed that the entry vehicle is commanded to fly at a trim angle of attack and the guidance is achieved only by adjusting the bank angle of the vehicle using the reaction control system (RCS) or the internal moving mass actuator control system (IMMACS). A uniform gravitational field is used because the decrease in altitude during atmospheric entry is much less than the Martian radius and the phase lasts only a very short time. The dimensional dynamics for Mars atmospheric entry, defined with respect to the Mars inertial coordinate system, is given by:

$$\begin{cases} \dot{r} = v \sin\gamma \\ \dot{\theta} = \frac{v\cos\gamma\sin\psi}{r\cos\lambda} \\ \dot{\lambda} = \frac{v}{r}\cos\gamma\cos\psi \\ \dot{v} = -D - g_M\sin\gamma + \omega_M^2 r\cos\lambda(\sin\gamma\cos\lambda - \cos\gamma\sin\lambda\cos\psi) \\ \dot{\gamma} = \frac{L\cos\sigma}{v} + \left(\frac{v}{r} - \frac{g_M}{v}\right)\cos\gamma + 2\omega_M\cos\lambda\sin\psi \\ +\frac{\omega_M^2 r}{v}\cos\lambda(\cos\gamma\cos\lambda + \sin\gamma\sin\lambda\cos\psi) \\ \dot{\psi} = \frac{L\sin\sigma}{v\cos\gamma} + \frac{v}{r}\sin\psi\cos\gamma\tan\lambda \\ -2\omega_M(\tan\gamma\cos\psi\cos\lambda - \sin\lambda) + \frac{\omega_M^2 r}{v\cos\gamma}\cos\lambda\sin\lambda\sin\psi \end{cases} \tag{3.1}$$

where r is the distance from the center of Mars to the vehicle center of mass, v is vehicle's velocity, θ and λ are the longitude and latitude, respectively; γ is the flight path angle, ψ is the heading (azimuth) angle, and σ is the bank angle. μ is the gravitational constant of Mars, ω_M is the rotation rate of Mars. Gravitational acceleration g_M, lift and drag accelerations L and D are defined by

$$g_M = \mu/r^2,\quad L = \frac{1}{2}\rho v^2\frac{C_L S}{m},\quad D = \frac{1}{2}\rho v^2\frac{C_D S}{m},\quad \rho = \rho_0\exp\left(-\frac{r-r_0}{h_s}\right)$$

where S and m are the reference area and mass of entry vehicle, respectively. ρ_0 is atmospheric density on Martian surface, r_0 is the radius of Mars, and h_s is the scale altitude. The state variable and control variable are defined as $\boldsymbol{x} = [r, \theta, \lambda, v, \gamma, \psi]^{\mathrm{T}}$ and $u = \sigma$.

3.1.2 Constraints

The entry vehicle's normal overload, dynamic pressure, and aerodynamic heating rate should meet certain constraints during Mars atmospheric entry process

$$n_a = \sqrt{L^2 + D^2} \le n_{\max} \tag{3.2}$$

$$q = \frac{1}{2}\rho v^2 \le q_{\max} \tag{3.3}$$

$$\dot{Q} = c_q \sqrt{\frac{\rho}{r_n}} v^3 \le \dot{Q}_{\max} \tag{3.4}$$

$$Q = \int_{t_0}^{t_f} \dot{Q}\, dt \le Q_{\max} \tag{3.5}$$

where n_a, q, $\dot{Q}$, Q are the normal overload, dynamic pressure, and aerodynamic heating rate, total aerodynamic heat, respectively, $n_{\max}$, $q_{\max}$, $\dot{Q}_{\max}$, $Q_{\max}$ are the tolerable maximum values, c_q and r_n are the heat flux-related coefficient and the radius of the vehicle's nose.

The terminal constraints for parachute deployment are defined as follows:

$$r(t_f) - r_0 \ge h_{f\min} \tag{3.6}$$

$$v(t_f) \le v_{f\max} \tag{3.7}$$

where $h_{f\min}$ and $v_{f\max}$ denote the minimum terminal altitude and maximum terminal velocity for parachute deployment. Mars entry vehicle only has limited maneuverability, which usually leads to strict terminal constraints. In order to satisfy the requirements of control system, the control input $u = \sigma$ and the maneuver rate of bank angle $\dot{u} = \dot{\sigma}$ are constrained with prescribed bounds as follows

$$|u| \le u_{\max}, \quad |\dot{u}| \le \dot{u}_{\max} \tag{3.8}$$

3.1.3 Objective Function

Required capabilities for future Mars pin-point landing missions involve landing larger spacecraft with an accuracy of hundreds of meters. Once the entry phase comes to an end, the final landing accuracy can be hardly further improved. The velocity of entry vehicle with larger payload should be slowed down as soon as possible to meet the conditions of parachute deployment. Therefore, the main function of the entry guidance system is to accurately steer the vehicle from Mars atmospheric boundary to the predesigned site of deploying supersonic parachute. In addition, there are usually larger uncertainties in Mars atmospheric entry dynamics, which inevitably introduce some errors to entry guidance. Hence, the terminal position errors and velocity error of entry phase are included into performance function in this chapter.

$$
\begin{aligned}
J = {} & c_h \left| h(t_f) - h_f \right| + c_\lambda \left| \lambda(t_f) - \lambda_f \right| \\
& + c_\theta \left| \theta(t_f) - \theta_f \right| + c_v \left| v(t_f) - v_f \right|
\end{aligned}
\tag{3.9}
$$

where $h(t_f)$, $\lambda(t_f)$, $\theta(t_f)$, $v(t_f)$ are the terminal height, longitude, latitude, and velocity, respectively; h_f, λ_f, θ_f, v_f are the expected terminal height, longitude, latitude, and velocity, respectively; c_h, c_λ, c_θ, c_v are the corresponding weight coefficients, which can be determined by the desensitized optimal control (DOC) method.

3.1.4 *Optimal Mars Entry Problem*

As described above, the optimal Mars entry problem can be given as:

$$
\begin{cases}
\text{find} \quad u(t) \\
\min \quad J \\
\text{s.t.} \quad \dot{\boldsymbol{x}} = \boldsymbol{f}(\boldsymbol{x}, u, t) \\
\qquad\quad \boldsymbol{g}(\boldsymbol{x}, u, t) \le 0 \\
\qquad\quad \boldsymbol{x}(t_0) = \overline{\boldsymbol{x}}_0
\end{cases}
\tag{3.10}
$$

where $\boldsymbol{f}(\boldsymbol{x}, u, t)$ and $\boldsymbol{g}(\boldsymbol{x}, u, t)$ are the differential equations and inequality constraints, respectively. $\overline{\boldsymbol{x}}_0$ is the nominal initial state.

3.2 Particle Swarm Optimization

As one of the swarm intelligence optimization methods, PSO was first proposed by Kennedy and Eberhart [115], who tried to simulate the unpredictable movements of flocks of birds searching for food. It is a gradient free and probabilistic search algorithm. In this algorithm, the initial states of the particles in the first iteration are generated randomly throughout the design space. After that, the particles are driven to the optimal solution through a series of iterative computations. These calculations are performed by taking advantage of current swarm information, which includes the best value for fitness function found by each particle heretofore (i.e. particle's knowledge) and the corresponding best position in the swarm so far (i.e. swarm's knowledge). In each iteration, the new position and velocity vector of the i-th particle in the swarm are respectively determined by

$$\boldsymbol{x}_i^{k+1} = \boldsymbol{x}_i^k + \boldsymbol{v}_i^{k+1} \tag{3.11}$$

$$\boldsymbol{v}_i^{k+1} = \omega \boldsymbol{v}_i^k + c_1 r_1 (\boldsymbol{p}_i - \boldsymbol{x}_i^k) + c_2 r_2 (\boldsymbol{p}_g^k - \boldsymbol{x}_i^k) \tag{3.12}$$

where r_1 and r_2 are independent random numbers between 0 and 1, $\boldsymbol{p}_i$ is the best position found by the i-th particle so far, $\boldsymbol{p}_g^k$ is the best position in the swarm at the k-th iteration, c_1 and c_2 are two trust parameters, c_1 is the self-cognitive coefficient, and c_2 is the social coefficient. ω is the inertia factor that facilitates more global behavior by larger value and more local behavior by smaller value. To enhance the global search capability of the hybrid optimization strategy described above, PSO algorithm needs to trust swarm more than single particle, so the parameters values of $c_1 = 1.5$ and $c_2 = 2.5$ are utilized here. In order to converge to global optimal solution faster, PSO algorithm needs to adjust the value of ω dynamically. Hence, the value of ω is slightly decreased as shown in equation below, when there is no more performance improvement within three consecutive iterations. The initial value $\omega = 1.4$ is used here

$$\omega_{\text{new}} = f_w \omega_{\text{old}} \tag{3.13}$$

where the value of f_w is eventually chosen as 0.95 through a trial and error on the basis of the empirical value mentioned in [64, 115, 116]. And the performance of this algorithm doesn't degrade when the value of f_w is set to be greater than 0.9. Therefore, the value of f_w is set to gradually reduce from 0.975 for rapid convergence and good performance. The lower limit of ω is chosen as 0.35 to ensure that this algorithm has sufficient global search capability. To avoid the case that the design variables overstep the design space, the position vector is reset to the nearest point on the boundary once the new position vector $\boldsymbol{x}_i^{k+1}$ is beyond the desired bounds. The convergence of this algorithm can be confirmed when the performance index corresponding to $\boldsymbol{p}_g^k$ is not further improved in the following fifteen iterations [116].

The initial swarm of the PSO is generated according to following two equations, and the particles generated by this method disperse randomly throughout the design space. Both r_3 and r_4 are random numbers between 0 and 1. $\boldsymbol{x}_{\min}$ and $\boldsymbol{x}_{\max}$ are the lower limit and upper limit of the design variables, respectively.

$$\mathbf{x}_i^0 = \mathbf{x}_{\min} + r_3(\mathbf{x}_{\max} - \mathbf{x}_{\min}) \tag{3.14}$$

$$\mathbf{v}_i^0 = \mathbf{x}_{\min} + r_4(\mathbf{x}_{\max} - \mathbf{x}_{\min}) \tag{3.15}$$

3.3 Gauss Pseudospectral Method

As an excellent pseudospectral optimization algorithm, GPM is slightly better than Legendre pseudospectral method and Radau pseudospectral method in solution accuracy and convergence speed. It also belongs to direct optimization methods category. In this algorithm, the time span, state variable, and control variable of the trajectory optimal problem are discretized at a series of Legendre–Gauss (LG) points which satisfy certain constrains [117]. These LG points are treated as collocation points, and then they are further processed according to the prescribed steps. Finally, the trajectory optimal problem can be transcribed into a nonlinear programming problem, which can be conveniently solved by some appropriate algorithms, such as SQP [118].

First, the time variable $t \in [t_0, t_f]$ is mapped to another span $\tau \in [-1, 1]$, and discretized to a group of LG points.

$$\tau = \frac{2t}{t_f - t_0} - \frac{t_f + t_0}{t_f - t_0} \tag{3.16}$$

The state variable is approximated via $N + 1$ Lagrange interpolating polynomials $L_i(\tau), i = 0, \ldots, N$ as follows

$$\boldsymbol{x}(\tau) \approx \boldsymbol{X}(\tau) = \sum_{i=0}^{N} \boldsymbol{X}(\tau_i)\boldsymbol{L}_i(\tau), \quad \boldsymbol{L}_i(\tau) = \prod_{j=0, j\neq i}^{N} (\tau - \tau_j)/(\tau_i - \tau_j) \tag{3.17}$$

where τ_k is the Legendre–Gauss points and $\tau_0 = -1$. The LG points are the roots of the Legendre polynomials P_k:

$$P_k(\tau) = \frac{1}{2^k k!} \frac{\mathrm{d}^k(\tau^2 - 1)^k}{\mathrm{d}\tau^k} \tag{3.18}$$

The Fig. 3.1 shows the normalized position distribution of LG points with the different given number.

Similarly, the control variable is approximated via N Lagrange interpolating polynomials

$$\boldsymbol{u}(\tau) \approx \boldsymbol{U}(\tau) = \sum_{i=0}^{N} \boldsymbol{U}(\tau_i)\boldsymbol{L}_i^*(\tau), \; \boldsymbol{L}_i^*(\tau) = \prod_{j=1, j\neq i}^{N} (\tau - \tau_j)/(\tau_i - \tau_j) \tag{3.19}$$

The dynamical constrains can be obtained by differentiating the state expression in Eq. (3.17).

$$\dot{\boldsymbol{x}}(\tau) \approx \dot{\boldsymbol{X}}(\tau) = \sum_{i=0}^{N} \boldsymbol{X}(\tau_i)\dot{\boldsymbol{L}}_i(\tau) \tag{3.20}$$

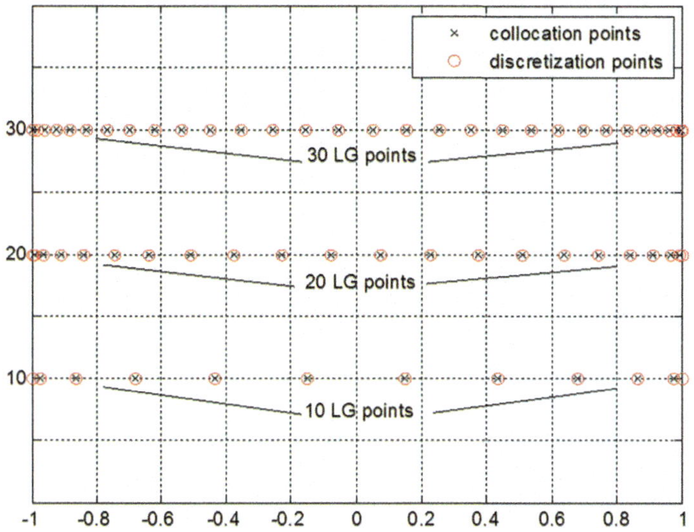

Fig. 3.1 Normalized position distribution of LG points

In GPM algorithm, the system should satisfy the dynamical constrains at each LG point according to Eq. (3.20). The differential of each Lagrange polynomial at the LG points can be expressed in a differential approximation matrix $\boldsymbol{D} \in \mathbb{R}^{N\times N+1}$ and its elements are calculated via the following formula.

$$\boldsymbol{D}_{ki} = \dot{\boldsymbol{L}}_i(\tau_k) = \begin{cases} \frac{(1+\tau_k)\dot{P}_k(\tau_k)+P_k(\tau_k)}{(\tau_k-\tau_i)[(1+\tau_i)\dot{P}_k(\tau_i)+P_k(\tau_i)]}, & i \neq k \\ \frac{(1+\tau_i)\ddot{P}_k(\tau_i)+2\dot{P}_k(\tau_i)}{2[(1+\tau_i)\dot{P}_k(\tau_i)+P_k(\tau_i)]}, & i = k \end{cases} \tag{3.21}$$

where $i = 0,\ 1,\ \ldots,\ N$ and $k = 1, \ldots, N$.

By taking use of the differential approximation matrix given above, the dynamical constrains are transformed into algebraic constraints below.

$$\sum_{i=0}^{N} \boldsymbol{D}_{ki}\boldsymbol{X}(\tau_i) - \frac{t_f - t_0}{2} f\big(\boldsymbol{X}(\tau_k), \boldsymbol{X}(\tau_k), \tau_k; t_0, t_f\big) = 0 \tag{3.22}$$

The terminal state is obtained according to the Gauss integral expression as

$$\boldsymbol{X}_f = \boldsymbol{X}_0 + \frac{t_f - t_0}{2} \sum_{k=1}^{N} w_k f\big(\boldsymbol{X}(\tau_k), \boldsymbol{U}(\tau_k), \tau_k; t_0, t_f\big) \tag{3.23}$$

where $\boldsymbol{X}_0 = \boldsymbol{X}(-1)$ and w_k is the Gauss weight and can be computed by the following expression

$$w_k = \int_{-1}^{1} \boldsymbol{L}_i(\tau)\mathrm{d}\tau = \frac{2}{(1-\tau_i^2)\big(\dot{P}_k(\tau_i)\big)^2} \tag{3.24}$$

Similarly, the cost function can also be approximated as

$$J = \phi(\boldsymbol{X}_0, t_0, \boldsymbol{X}_f, t_f) + \frac{t_f - t_0}{2}\sum_{k=1}^{N} w_k f\big(\boldsymbol{X}(\tau_k), \boldsymbol{X}(\tau_k), \tau_k; t_0, t_f\big) \tag{3.25}$$

The boundary constrains are defined as follows

$$\boldsymbol{\Phi}\big(\mathbf{X}_0, t_0, \mathbf{X}_f, t_f\big) = \mathbf{0} \tag{3.26}$$

The path constrains are described as follows

$$\boldsymbol{C}\big(\boldsymbol{X}(\tau_k), \boldsymbol{U}(\tau_k), \tau_k; t_0, t_f\big) \le \mathbf{0} \tag{3.27}$$

At this point, complete description of nonlinear programming problem is finished. Equation (3.25) defines the cost function. Equations (3.23) and (3.26) are equality constrains. Equation (3.27) is the inequality constrains. As usually, this parameter optimization problem can be easily solved via SQP method.

3.4 Hybrid Optimization Strategy

The aim of this chapter is to develop a general framework for rapid design and optimization of the Mars atmospheric entry trajectory for future Mars pinpoint landing missions. The flowchart of the general framework proposed in this chapter is depicted in Fig. 3.2. The integrated optimization process involves the closely related three steps. First, PSO algorithm is adopted to perform global coarse optimization, which provides a suitable initial guess for subsequent precise optimizing operation. Secondly, GPM is used to transform the trajectory optimization problem into a nonlinear programming problem, which can be easily solved by well-developed nonlinear optimization algorithms. Thirdly, SQP is utilized to solve the nonlinear programming problem, and obtain the optimal Mars atmospheric entry trajectory.

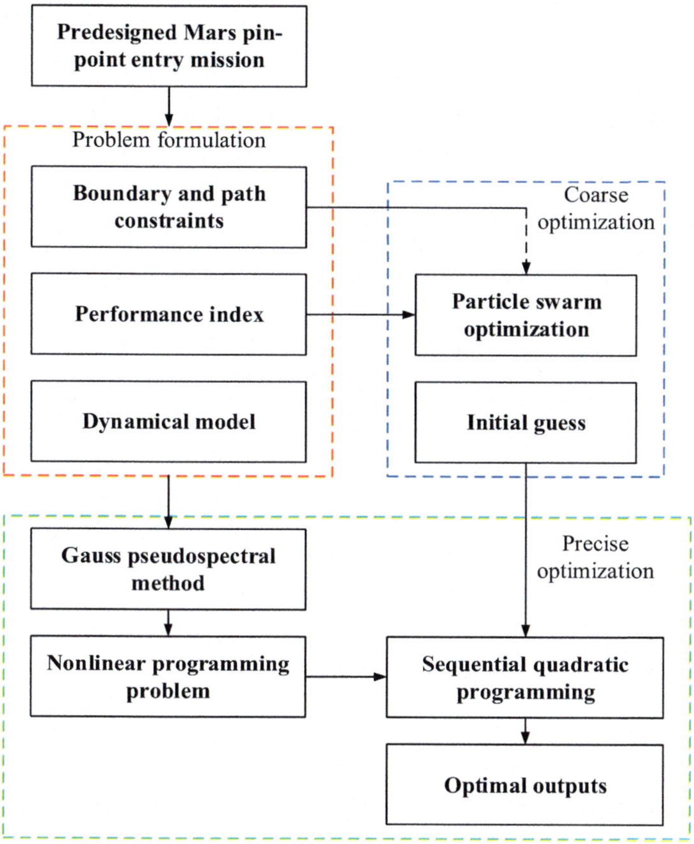

Fig. 3.2 Optimization framework of Mars atmospheric entry trajectory

3.5 Numerical Demonstration

In order to confirm the validity of hybrid optimization strategy for Mars atmospheric entry trajectory developed in this chapter, MATLAB/Simulink environment is adopted to implement computer simulation and analysis. The Mars entry trajectory optimization problem defined above is solved using the open-source software Gauss pseudospectral optimization software (GPOPS) [119] with the NLP solver SNOPT. GPOPS is modified to connect the PSO algorithm described in this chapter. SNOPT is an effective software package for solving large-scale linear and nonlinear optimization problems, which is based on SQP method. The entry vehicle parameters used in Mars atmospheric entry dynamics equations are inherited from the MSL mission.

The Mars entry dynamics uncertainties are taken into account. In following simulation cases, the Mars atmospheric density, vehicle's ballistic coefficient, and lift-to-drag ratio are calculated by $\rho' = \rho \cdot (1 + \Delta\rho)$, $\beta' = \beta \cdot (1 + \Delta\beta)$,

$(L/D)' = (L/D) \cdot (1 + \Delta(L/D))$, respectively; $\Delta\rho, \Delta\beta, \Delta(L/D)$ denote the percentage of uncertain deviation in Mars atmospheric density, vehicle's ballistic coefficient, and lift-to-drag ratio, respectively.

The configuration and physical parameters of MSL entry vehicle are utilized as an example to perform simulation and analysis in this chapter. The detailed parameters of entry vehicle and Mars environment are set as Table 3.1. The vehicle's initial states at the entry interface are provided by the deep space network (DSN), Table 3.2 presents the initial states and the desired target states used in our simulation. The conditions and optimal results are shown in Table 3.3. The trajectory optimization results are given in Fig. 3.3.

As shown in Fig. 3.3, the required terminal conditions of entry phase are achieved perfectly, and all constraints are satisfied. Because only one reversal is conducted in the entire process, bank angle is easily controlled. At the same time, the entry control operation is energy efficient when compared with those reported in [64, 196]. Because there are uncertainties in Mars atmospheric density, vehicle's ballistic coefficient,

Table 3.1 Parameters of entry vehicle and Mars environment

Parameters	Symbol	Value	Unit
Entry vehicle mass	m	2920	kg
Reference area	S_{ref}	15.9	m^2
Drag coefficient	C_D	1.45	–
Lift-to-drag ratio	L/D	0.24	–
Radius of vehicle's nose	r_n	0.66	m
Heating rate coefficient	c_q	1.9027×10^{-4}	$\sqrt{kg}/m$
Mars gravitational constant	μ	4.282829×10^{13}	m^3/s^2
Mars reference radius	R_M	3386.6	km
Rotation rate of the Mars	ω	7.0882×10^{-5}	rad/s
Martian surface atmospheric density	ρ_0	0.0158	kg/m^3
Scale height	h_s	9354.5	m
Martian surface gravitational acceleration	g_M	3.7132	m/s^2

Table 3.2 Initial state and desired terminal state

Parameters	Symbol	Value	Unit
Initial height	h_0	125	km
Initial velocity	v_0	6	km/s
Initial longitude	θ_0	0	°
Initial latitude	λ_0	0	°
Desired terminal height	h_f	7.5	km
Desired terminal velocity	v_f	400	km/s
Desired terminal longitude	θ_f	11	°
Desired terminal latitude	λ_f	0	°

Table 3.3 Comparison of constraints and simulation results

Parameters	Allowable values	Optimal results
Δh_f (m)	±500	213
Δv_f (m/s)	±50	31.8
$\Delta \theta_f$ (°)	±0.01	-2.86×10^{-6}
$\Delta \lambda_f$ (°)	±0.01	0
n_{max} (g)	16	10.78
q_{max} (psf)	18	12.26
$\dot{Q}_{max}$ (Btu/ft^2 s)	90	58.89

and lift-to-drag ratio, which directly affect flight velocity, flight path angle, and azimuth angle, and indirectly affect the distance from the center of Mars to the vehicle center of mass, longitude, and latitude according to the mathematical relationship in Mars atmospheric entry dynamics model described in Eq. (3.1). Therefore, some fluctuations may be found in all those profiles mentioned above. However, only latitude profile and azimuth angle profile have some slight fluctuations as shown in Fig. 3.3, which apparently benefit from the application of desensitized optimal control method.

In our simulation, the optimal result is obtained after 15 iterations, which only takes 1.75 s. PSO runs 10 iterations for coarse optimization, and GPM runs 5 iterations for accurate optimization. Figure 3.4 clearly depicts the convergence process performance function with the number of iterations. It is easily observed that convergence rate of performance index decreases significantly near the 10-th iteration, which illustrates that the particles of PSO have approached the global optimum point in the feasible area. GPM runs from the 11-th iteration for local accurate optimization and obtains the global optimal solution. Then, we can safely arrive at the conclusion that the relay hybrid optimization strategy developed in this chapter has a faster convergence rate, and run time has been notably reduced, when compared with those reported in [64, 116]. The improvement of optimization performance is attributed to the hybrid optimization strategy succeeds in avoiding the defect of the traditional PSO that the expected solution converges extremely slow for a long iteration period when it is closer to the global optimum point in the feasible area, and exerting the superiority of GPM that accurate optimization solution can be obtained rapidly when proper initial guess is inputted.

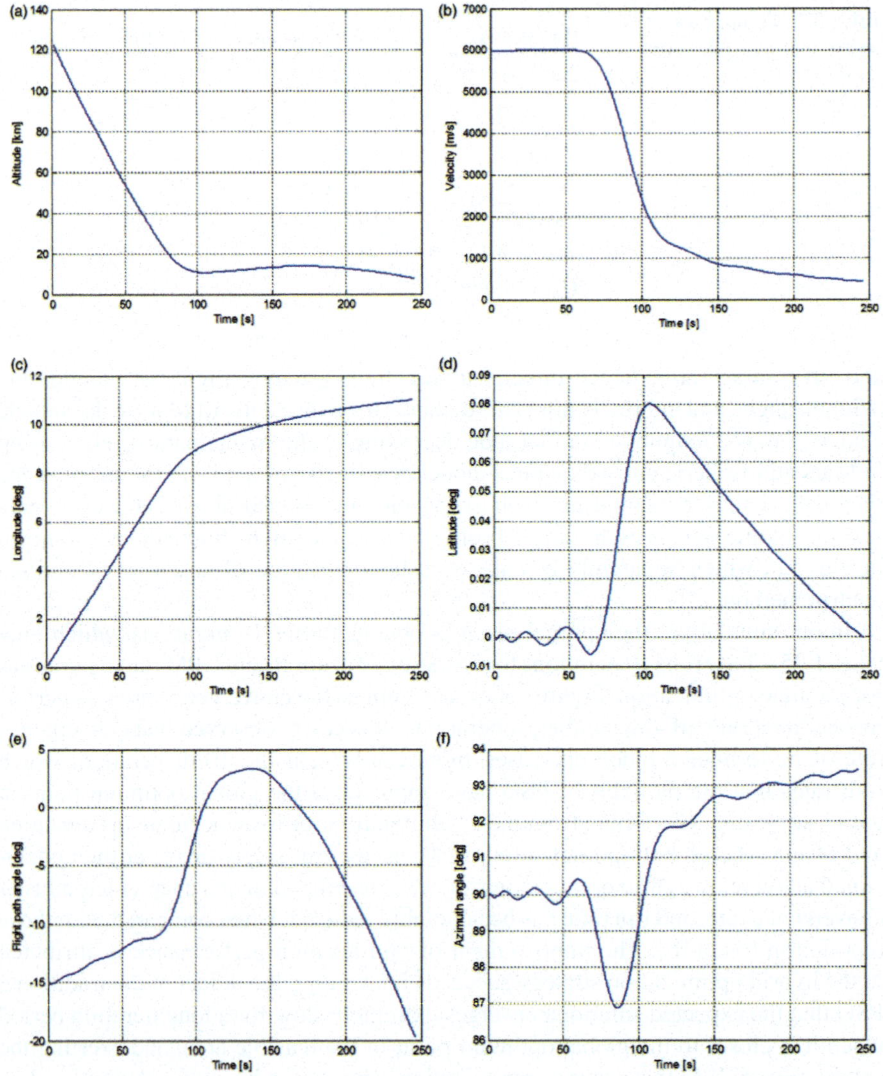

Fig. 3.3 Optimization results of Mars atmospheric entry trajectory: **a** altitude profile; **b** velocity profile; **c** longitude profile; **d** latitude profile; **e** flight path angle profile; **f** Azimuth angle profile; **g** controlled bank angle profile; **h** Overload profile; **i** dynamic pressure profile; **j** heating rate profile

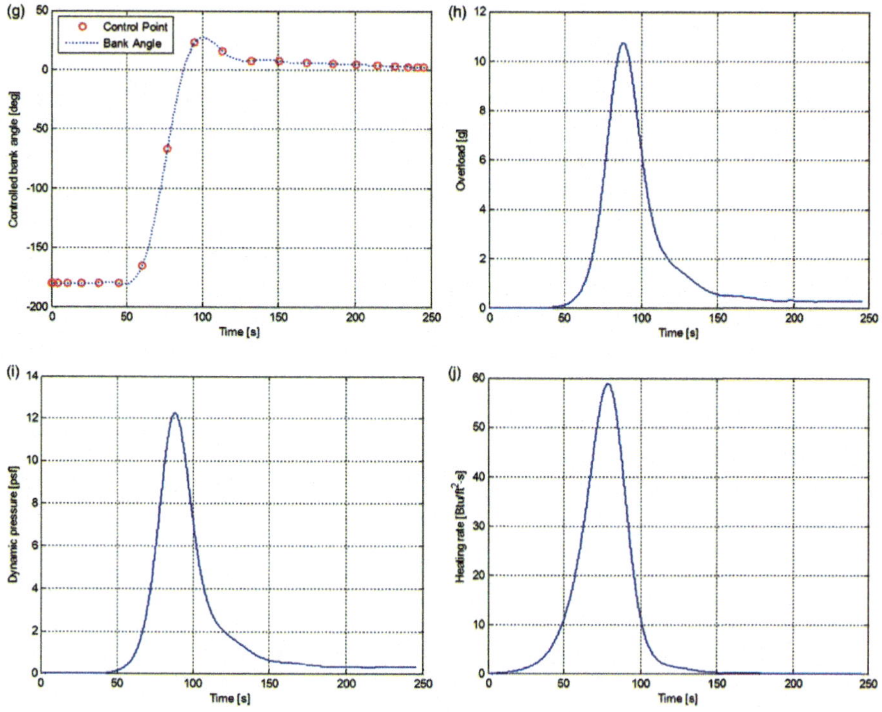

Fig. 3.3 (continued)

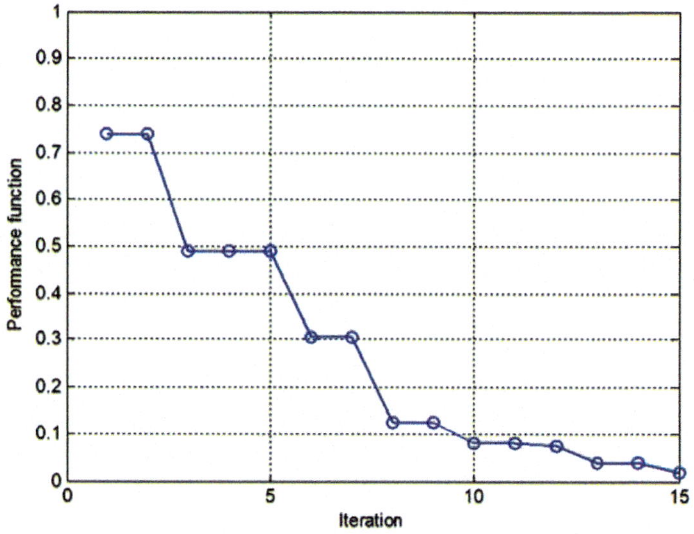

Fig. 3.4 Convergence process of performance function

3.6 Discussions

A new trajectory hybrid optimization strategy is proposed for Mars entry using PSO and GPM in this chapter. Equations of motion are normalized for optimization to reduce the difficulty of iterative convergence. In the coarse optimization, the PSO algorithm with global searching capability is adopted to provide the initial guess for subsequent precise optimization. The precise trajectory optimization problem is transformed to nonlinear programming problem using GPM, and then solved by SQP according to the initial guess provided by the coarse optimization. Simulation results clearly show that all the terminal constraints are met with high accuracy and all the process constraints are completely satisfied. It is evident that the desired entry performances are excellently achieved and the hybrid optimization strategy developed in this chapter is reliable.

Chapter 4
Improved Sequential Convex Optimization for Mars Entry Trajectory Planning

Most existing literatures study the optimal control problem with a fixed final time rather than the free final time. To address this issue, some papers [53, 120, 121] employed a bi-level optimization strategy to find the optimal final time. Another commonly used method is to transform the time interval $[0, t_f]$ into a normalized interval $[-1, 1]$ in the pseudospectral transcription paradigm [121, 123] or into the interval [0, 1] in the classical optimal control framework [57, 124]. A new method to solve the optimal control problem (OCP) with free final time is to choose a new independent variable for dynamics formulation [125]. And the altitude was used as an independent variable for trajectory optimization of aerodynamically controlled missiles and powered descent and landing. Besides, the transfer angle was adopted to rewrite the Earth-to-Mars low-thrust planar transfer dynamics [126].

In [55], the first-order linearization approximation technique was used to convexify the integral performance index, which in turn introduces linearization error into the performance index. Moreover, artificial infeasibility may occur during the successive optimization due to an improper linearization. To mitigate this issue, Mao and Szmuk [127] introduced the virtual control and trust-region techniques to compensate for the linearized dynamics. Wang [128, 129] further improved the trust-region algorithm with adaptive techniques, where the trust-region radius was refined by evaluating the exact decrease of the performance index.

In this chapter, a range discretization-based sequential convex programming (r-SCP) approach is proposed. The main contributions can be summarized as follows:

(1) The Mars entry trajectory planning problem is reformulated by using the range of entry trajectory as an independent variable rather than the energy or time. This can avoid numerical inaccuracy of dynamics propagation that occurs if the problem is evenly discretized with respect to the independent variable energy, especially at the high altitude.
(2) The minimum-time performance index is convexified by taking the flight time as one of the state variables; and the constraints on bank angle along with its rate are addressed by introducing a new control input and inequality relaxation.

S. Li et al., *Trajectory Optimization and Guidance Methods for Mars Entry*,
https://doi.org/10.1007/978-981-99-6282-2_4

These two methods suggest an effective strategy to deal with non-convex state variables or control input constraints, that is, to convexify constraints by change of variables, inequality relaxation, or state-space augmentation.

(3) With these new formulations, range discretization and successive convexification approaches are applied for Mars entry trajectory planning in the existing convex optimization framework with virtual control and adaptive trust-region techniques. Furthermore, performance comparison of the proposed algorithm with some classical optimization methods is conducted for the Mars entry problem.

4.1 Reformulation of Entry Dynamics by Using Downrange Angle

It is common to use the non-dimensional flight time $\tau = t/t_s$ or negative specific mechanical energy (abbreviated as energy in the following) $e = 1/r - V^2/2$ as an independent variable in entry trajectory optimization problems, where t is the flight time and $t_s = \sqrt{R_0/g_0}$; R_0 and g_0 denote Mars reference radius and gravitational acceleration at R_0, respectively. r is the dimensionless radial distance from Martian center to the vehicle normalized by R_0. The relative velocity V is normalized by velocity scaling coefficient $V_s = \sqrt{R_0 g_0}$. Since the flight time is not a typical concern during entry mission as it is not related to boundary and path constraints, the energy is preferred to time as an independent variable. The energy monotonically increases when the effect of self-rotation is ignored:

$$\frac{de}{d\tau} = DV > 0 \tag{4.1}$$

where D is the non-dimensional drag and will be defined later. Nevertheless, the derivative $de/d\tau$ will no longer be in this simple form if the self-rotation-dependent terms are considered:

$$\frac{de}{d\tau} = DV - \Omega^2 rV\cos\phi(\cos\phi\sin\gamma - \sin\phi\cos\gamma\cos\psi) \tag{4.2}$$

where Ω is the Mars self-rotation rate non-dimensionalized by t_s. ϕ and γ are the latitude and fight-path angle, respectively. ψ is the heading angle measured clockwise in the local horizontal plane from the north. As can be seen from Eq. (4.2), the modeling accuracy of dynamics with respect to the energy variable is not guaranteed if the self-rotation-dependent term is ignored. To demonstrate this, propagation results of the two dynamics with respect to energy and time are listed in Table 4.1, in which h is the flight altitude and s denotes the range traveled by the vehicle after entry defined by $ds/d\tau = V\cos\gamma$.

Table 4.1 Numerical integration results of entry dynamics with time and energy as independent variables

Parameters	Initial value	Final value	
		Time	Energy
h (m)	125,000	16,232.04	16,203.80
θ (deg)	−90	−71.9758	−71.9794
ϕ (deg)	−45	−41.1426	−41.1441
V (m/s)	5500	752.6682	724.6368
γ (deg)	−13.5	−11.4325	−11.4560
ψ (deg)	85	48.2555	48.2447
σ (deg)	−50	−50	−50
s (m)	0	830,182.57	829,997.33

As shown in Table 4.1, the final altitude integrated by using energy-based dynamics is about 28.24 m lower than the one using the time-based dynamics. The terminal difference in terms of parachute deployment longitude and latitude is about $[-3.6281, -1.444] \times 10^{-3}$ deg or $[-215.12, -85.63]$ m, which exceeds the tolerable 100 m error for future human-scale Mars mission. Consequently, the energy-based model for Mars entry is not accurate enough. Moreover, the flight altitude h decreases dramatically at the beginning with a large magnitude of flight path angle because the thin atmosphere at high altitude cannot provide sufficient deceleration. By contrast, the energy changes slowly. Thus, the altitude-energy profile is featured by a steep slope during the initial phase of entry. In order to ensure numerical integration accuracy, the step size should be set small enough to capture such a rapid dynamic behavior if the energy is evenly discretized. Accordingly, the number of discretization nodes and the scale of optimization increase significantly. However, a large-scale optimization problem is tricky. Moreover, modeling with respect to energy also requires a fixed final altitude and speed to determine the final energy. However, the altitude and speed are usually constrained within a parachute box rather than constants in Mars entry.

It is worth noting that the range s increases monotonically since its derivative $ds/d\tau = V\cos\gamma$ is positive during the whole entry mission. In addition, the derivative of the range is simple enough to be an alternative for replacing the approximate derivative of energy. Thus, the Mars entry dynamics can be reformulated without losing accuracy by using the range as an independent variable. Moreover, the downrange angle $\overline{s} = \int_0^{\tau_f} V\cos\gamma / r d\tau$ shown in Fig. 4.1, where τ_f is the non-dimensional final time, can also be chosen as the independent variable because it has the same monotonicity property and its initial and final boundaries are specified in an entry mission with the given target latitude and longitude.

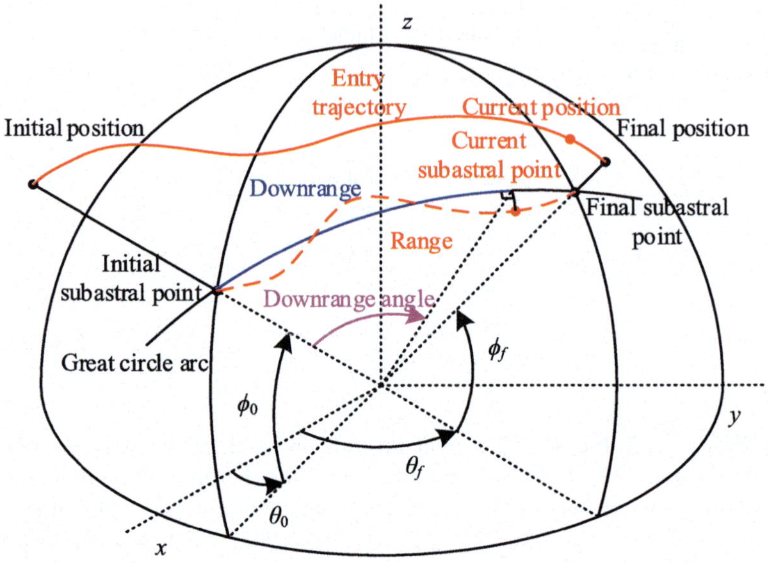

Fig. 4.1 The illustration of range, downrange, and downrange angle

4.2 Sequential Convex Programming Algorithm

4.2.1 Formulation of Problem

The equations of motion can be rewritten with respect to s as follows:

$$\frac{dr}{ds} = \tan\gamma \tag{4.3}$$

$$\frac{d\theta}{ds} = \frac{\sin\psi}{r\cos\phi} \tag{4.4}$$

$$\frac{d\phi}{ds} = \frac{\cos\psi}{r} \tag{4.5}$$

$$\frac{dV}{ds} = -\frac{D}{V\cos\gamma} - \frac{\tan\gamma}{r^2 V} + \frac{\Omega^2 r\cos\phi(\cos\phi\tan\gamma - \sin\phi\cos\psi)}{V} \tag{4.6}$$

$$\begin{aligned}\frac{d\gamma}{ds} =& \frac{L\cos\sigma}{V^2\cos\gamma} - \frac{1}{r^2V^2} + \frac{1}{r} + \frac{2\Omega\cos\phi\sin\psi}{V\cos\gamma} \\ &+ \frac{r\Omega^2\cos\phi(\cos\phi + \cos\psi\sin\phi\tan\gamma)}{V^2}\end{aligned} \tag{4.7}$$

$$\frac{d\psi}{ds} = \frac{L\sin\sigma}{V^2\cos^2\gamma} + \frac{\sin\psi\tan\phi}{r} + \frac{2\Omega(\sin\phi - \cos\phi\tan\gamma\cos\psi)}{V\cos\gamma} + \frac{\Omega^2 r\sin\phi\cos\phi\sin\psi}{V^2\cos^2\gamma} \tag{4.8}$$

where θ is the longitude and σ is the bank angle measured positive to the right from the view inside the vehicle. The non-dimensional drag and lift accelerations D and L are given by:

$$D = R_0\rho V^2 S_r C_D/(2m), L = R_0\rho V^2 S_r C_L/(2m) \tag{4.9}$$

where ρ is the atmospheric density, S_r denotes the reference area, and m is the mass of the unpowered vehicle. C_L and C_D are drag and lift coefficients of the vehicle, respectively. The Mars atmospheric density model is approximated by an exponential function of altitude as:

$$\rho = \rho_0 e^{-R_0(r-1)/h_s} \tag{4.10}$$

where ρ_0 is atmospheric density on the planetary surface, and h_s is the scale altitude. The Mars entry vehicle keeps a nominal trim angle of attack and a desired C_L/C_D via offsetting its center of gravity. Therefore, the C_L and C_D nearly remain unchanged in the hypersonic flight, and the bank angle σ is the only control variable.

Three typical path constraints during Mars entry are heat rate $\dot{Q}$ with respect to the dimensional time t, dynamic pressure q, and normal load a:

$$\dot{Q} = k_Q\sqrt{\rho/R_n}(VV_s)^{3.15}, q = \rho(VV_s)^2/2 \le q_{\max}, a = \sqrt{L^2 + D^2} \le a_{\max} \tag{4.11}$$

where k_Q is the heat rate related coefficient, R_n is vehicle's nose radius, $\dot{Q}_{\max}$, $q_{\max}$ and $a_{\max}$ are their tolerable maximum values.

The initial conditions and desired final conditions are assumed to be specified:

$$[r(0), \theta(0), \phi(0), V(0), \gamma(0), \psi(0)]^T = [r_0, \theta_0, \phi_0, V_0, \gamma_0, \psi_0]^T \tag{4.12}$$

$$\boldsymbol{E}[r(s_f), \theta(s_f), \phi(s_f), V(s_f), \gamma(s_f), \psi(s_f)]^T = \boldsymbol{E}[r_f, \theta_f, \phi_f, V_f, \gamma_f, \psi_f]^T \tag{4.13}$$

where $(\circ)_0$ and $(\circ)_f$ denote the initial and final values of non-dimensional variables, respectively, $\boldsymbol{E}$ is a diagonal matrix with diagonal entries being 0 and 1 to determine the final boundary constraints according to different performance indices.

Constraints on the bank angle are given by:

$$|\sigma| \le \sigma_{\max}, |\dot{\sigma}| \le \dot{\sigma}_{\max} \tag{4.14}$$

where $\dot{\sigma}$ is the bank angle rate with respect to the dimensional time t. The operators $(\circ)_{\min}$ and $(\circ)_{\max}$ denote the corresponding lower and upper bound of a variable, respectively.

For the atmospheric entry problem, the following performance indices are generally considered:

$$J_1 = -r(s_f),\ J_2 = V(s_f),\ J_3 = \int_0^{s_f} \frac{1}{V\cos\gamma} ds \tag{4.15}$$

which represent maximizing the final altitude, minimizing the speed, and minimizing the flight time at the instant of parachute deployment, respectively. Note that the final range s_f is an unconstrained variable, which leads to the difficulty in solving an optimal control problem in the framework of mathematical programming. To circumvent this issue, the final range is regarded as a dilation coefficient to normalize the interval from $[0, s_f]$ into $[0, 1]$ before linearization. Then, this problem can be viewed as a fixed-final-range optimization problem because the final normalized dimensionless range is always equal to 1. Then, one can obtain the derivative according to the chain rule:

$$\boldsymbol{x}' \triangleq \frac{d\boldsymbol{x}(S)}{dS} = \frac{d\boldsymbol{x}(s)}{ds}\frac{ds}{dS} = \frac{d\boldsymbol{x}(s)}{ds} s_f = s_f \boldsymbol{F}(\boldsymbol{x}(S)),\ s/s_f \triangleq S \in [0, 1] \tag{4.16}$$

where $\boldsymbol{x} = [r, \theta, \phi, V, \gamma, \psi]^T \in \mathbb{R}^6$, and $\boldsymbol{F}(\boldsymbol{x}(S))$ consists of the right-hand side terms in Eqs. (4.3–4.8). Then, the final boundary constraints in Eq. (4.13) and performance indices in Eq. (4.15) are rewritten as:

$$\boldsymbol{E}_i[r(1), \theta(1), \phi(1), V(1), \gamma(1), \psi(1)]^T = \boldsymbol{E}_i[r_f, \theta_f, \phi_f, V_f, \gamma_f, \psi_f]^T, i = 1, 2, 3 \tag{4.17}$$

$$J_1 = -r(1),\ J_2 = V(1),\ J_3 = \int_0^1 s_f \frac{1}{V\cos\gamma} dS \tag{4.18}$$

where $\boldsymbol{E}_1 = diag([0, 1, 1, 0, 0, 0])$, $\boldsymbol{E}_2 = \boldsymbol{E}_3 = diag([1, 1, 1, 0, 0, 0])$ make the final boundary constraints match each of the three performance indices, respectively.

Thus far, the constrained nonlinear optimal trajectory planning problem for Mars entry can be formulated as:

$$\begin{gathered} \text{P4} - 0: \min_{\boldsymbol{x}, u, s_f} \ \text{Eq. (4.18)} \\ \text{subject to Eqs. (4.11), (4.12), (4.14), (4.16), (4.17)} \end{gathered}$$

4.2.2 Convexification and Discretization of Problem

1. **Linearization**

Note that the high-frequency jitters will occur due to the coupling between states and control input after linearization of the nonlinear entry dynamics with respect to σ. Therefore, it is necessary to modify the dynamics before linearization. To address the issue of coupling, the method of introducing a new control variable u and augmenting the bank angle σ into the state vector is employed. The redefined control input u with respect to the new state variable σ is given by:

$$u = s_f^{-1} \frac{d\sigma}{dS} \tag{4.19}$$

Accordingly, the constraint defined in Eq. (4.14) can be relaxed into the following form:

$$\begin{aligned} |\sigma| \leq \sigma_{\max}, |\dot{\sigma}| &= \left|\frac{d\sigma}{dt}\right| = \left|\frac{d\sigma}{dS}\frac{dS}{ds}\frac{ds}{d\tau}\frac{d\tau}{dt}\right| \\ &= \left|us_f \frac{1}{s_f} V\cos\gamma \frac{1}{t_s}\right| = \left|\frac{uV\cos\gamma}{t_s}\right| \leq |u|\frac{V_{\max}}{t_s} \leq \dot{\sigma}_{\max} \end{aligned} \tag{4.20}$$

Note that a relaxation is used to transform this nonlinear inequality into a linear one since $V\cos\gamma \leq V_{\max}$ holds, and the predetermined constant $V_{\max} = 6000/V_s$ is the maximum speed that a vehicle can fly in a Mars entry mission. This relaxation guarantees the maximum control input meets the constraint on the bank angle rate.

Furthermore, a convex performance index is necessary for the successive convexification detailed in the next subsection. Note that the time-optimal performance index shown in Eq. (4.18) is a non-convex one. Generally, the first-order Taylor series expansion is used to deal with the performance index with a nonlinear integrant. However, this relaxation method introduces the approximation error. In this chapter, a new and simpler method is proposed. To this end, the dimensionless time τ is incorporated into the state vector, and its derivative is specified as:

$$\frac{d\tau}{dS} = s_f \frac{1}{V\cos\gamma} \tag{4.21}$$

Then, the state vector augmented by τ becomes:

$$\boldsymbol{x} = [r, \theta, \phi, V, \gamma, \psi, \sigma, \tau]^T \in \mathbb{R}^8 \tag{4.22}$$

and the new equivalent minimum-time performance index becomes $J_3 = \tau(1)$. As such, the decoupled dynamics with a convex performance index is obtained, and the reformulated dynamics can be rewritten as follows:

$$\boldsymbol{x}' = s_f(\boldsymbol{f}(\boldsymbol{x}) + \boldsymbol{B}u + \boldsymbol{f}_\Omega(\boldsymbol{x})) \tag{4.23}$$

where $\boldsymbol{B} = [0, 0, 0, 0, 0, 0, 1, 0]^T$; $\boldsymbol{f}(\boldsymbol{x}) \in \mathbb{R}^8$ and $\boldsymbol{f}_\Omega(\boldsymbol{x}) \in \mathbb{R}^8$, respectively, denote the terms without and with the self-rotation on the right-hand side of Eqs. (4.3 ~ 4.8, 4.19) and (4.21).

In order to solve the entry trajectory planning problem in a convex optimization framework, the non-convex dynamics and constraints can be relaxed into linear ones by the first-order Taylor series approximation at the point $(\boldsymbol{x}^k, u^k, s_f^k)$ of a reference trajectory, which is obtained at the k-th iteration in the successive convexification procedure.

$$\begin{aligned}\boldsymbol{x}' &= s_f\left[\boldsymbol{f}(\boldsymbol{x}) + \boldsymbol{B}u + \boldsymbol{f}_\Omega(\boldsymbol{x})\right] \\ &\approx s_f^k \left.\frac{\partial \boldsymbol{f}(\boldsymbol{x})}{\partial \boldsymbol{x}}\right|_{\boldsymbol{x}^k, u^k} (\boldsymbol{x} - \boldsymbol{x}^k) + s_f^k \left.\frac{\partial \boldsymbol{f}_\Omega(\boldsymbol{x})}{\partial \boldsymbol{x}}\right|_{\boldsymbol{x}^k, u^k} (\boldsymbol{x} - \boldsymbol{x}^k) + s_f^k \boldsymbol{B}(u - u^k) \\ &\quad + \left[\boldsymbol{f}(\boldsymbol{x}^k) + \boldsymbol{B}u^k + \boldsymbol{f}_\Omega(\boldsymbol{x}^k)\right](s_f - s_f^k) + s_f^k\left[\boldsymbol{f}(\boldsymbol{x}^k) + \boldsymbol{B}u^k + \boldsymbol{f}_\Omega(\boldsymbol{x}^k)\right] \\ &= \boldsymbol{A}^k \boldsymbol{x} + \boldsymbol{B}^k u + \boldsymbol{c}^k s_f + \boldsymbol{d}^k\end{aligned} \tag{4.24}$$

where $\boldsymbol{A}^k = s_f^k \left.\frac{\partial \boldsymbol{f}}{\partial \boldsymbol{x}}\right|_{\boldsymbol{x}^k, u^k}$, $\boldsymbol{B}^k = s_f^k \boldsymbol{B}$, $\boldsymbol{c}^k = \boldsymbol{f}(\boldsymbol{x}^k) + \boldsymbol{B}u^k + \boldsymbol{f}_\Omega(\boldsymbol{x}^k)$, and $\boldsymbol{d}^k = -s_f^k \left.\frac{\partial \boldsymbol{f}}{\partial \boldsymbol{x}}\right|_{\boldsymbol{x}^k, u^k} \boldsymbol{x}^k - s_f^k \boldsymbol{B}u^k$. It should be noted that the first-order term $s_f^k \left.\frac{\partial \boldsymbol{f}_\Omega(\boldsymbol{x})}{\partial \boldsymbol{x}}\right|_{\boldsymbol{x}^k, u^k} (\boldsymbol{x} - \boldsymbol{x}^k)$ in Eq. (4.24) is omitted since its Euclidean norm is much smaller than the other terms in $s_f^k \boldsymbol{f}_\Omega(\boldsymbol{x}^k)$.

After the above steps, the path constraint is the only source of non-convexity. Analogous to the linearization of dynamics, the path constraint can be relaxed as the following form:

$$f_i \approx f_i(r^k, V^k) + f_i'(r^k, V^k) \cdot [r - r^k, V - V^k]^T \leq f_{i,\max} \text{ for } i = 1, 2, 3 \tag{4.25}$$

where $f_1 \triangleq \dot{Q}$, $f_2 \triangleq q$, $f_3 \triangleq a$, $f_{1,\max} \triangleq \dot{Q}_{\max}$, $f_{2,\max} \triangleq q_{\max}$, $f_{3,\max} \triangleq a_{\max}$ and

$$\begin{aligned}f_1'(r^k, V^k) &= \left.\left[\frac{\partial f_1}{\partial r}, \frac{\partial f_1}{\partial V}\right]\right|_{r^k, V^k} \\ &= \left.\left[-0.5 R_0 k_Q (V V_s)^{3.15} \sqrt{\rho / R_n} \Big/ h_s, 3.15 k_Q V_s^{3.15} V^{2.15} \sqrt{\rho / R_n}\right]\right|_{r^k, V^k}\end{aligned} \tag{4.26}$$

$$f_2'(r^k, V^k) = \left[\frac{\partial f_2}{\partial r}, \frac{\partial f_2}{\partial V}\right]\bigg|_{r^k, V^k} = \left[-0.5R_0\rho(VV_s)^2/h_s, V_s^2\rho V\right]\big|_{r^k, V^k} \tag{4.27}$$

$$\begin{aligned} f_3'(r^k, V^k) &= \left[\frac{\partial f_3}{\partial r}, \frac{\partial f_3}{\partial V}\right]\bigg|_{r^k, V^k} \\ &= \left[-0.5R_0^2 S_r(\sqrt{C_L^2 + C_D^2})\rho V^2 \Big/ (mh_s), R_0 S_r(\sqrt{C_L^2 + C_D^2})\rho V \Big/ m\right]\bigg|_{r^k, V^k} \end{aligned} \tag{4.28}$$

Now all the equality and inequality constraints as well as the performance index are convex, then the entry trajectory planning problem can be solved in the framework of convex programming.

To avoid artificial infeasibility and unbounded relaxation, an additional control input $\boldsymbol{v}$ called virtual control and the adaptive trust-region technique are employed. Specifically, the virtual control is introduced into linearized dynamics to compensate the linearization error, thereby preventing the artificial infeasibility. Additionally, the trust-region constraint is enforced on state variables to avoid unbounded relaxation, where the adaptive trust-region radius is used to adaptively determine the step size in each iteration and reduce the total number of optimization steps. The modified dynamics and augmented constraints are given by:

$$\boldsymbol{x}' = s_f(\boldsymbol{f}(\boldsymbol{x}) + \boldsymbol{B}u + \boldsymbol{f}_\Omega(\boldsymbol{x})) \approx \boldsymbol{A}^k\boldsymbol{x} + \boldsymbol{B}^k u + \boldsymbol{c}^k s_f + \boldsymbol{d}^k + \boldsymbol{v} \tag{4.29}$$

$$\left\|\boldsymbol{x} - \boldsymbol{x}^k\right\|_2 \le \delta_x, \left|s_f - s_f^k\right| \le \delta_s \tag{4.30}$$

where the virtual control input $\boldsymbol{v} \in \mathbb{R}^8$ together with relaxation boundaries $\delta_x, \delta_s \in \mathbb{R}_+$ are new variables to be optimized, and the inequality involving the l_2-norm in Eq. (4.30) is a convex cone constraint that can be solved via SOCP. In turn, the performance index is accordingly modified by adding the penalty terms corresponding to the virtual control, trust-region and final range:

$$J = w_i J_i + w_v\|\boldsymbol{v}\|_1 + w_x\delta_x + w_s\delta_s, i = 1, 2, 3 \tag{4.31}$$

where $w_i, w_v, w_x, w_s \in \mathbb{R}_+$ are the weight parameters prescribed by users. Similar augmented performance indices can also be found in existing literatures. Among the three additional performance indices, the l_1-norm is adopted on $\boldsymbol{v}$ since it enables the virtual control to work only on discretization nodes with large errors, and not to compensate on others. As a result, less virtual control inputs are necessary when the algorithm converges. Note that l_2-norm can be used as well, but it makes these virtual controls converge to zero uniformly, suggesting that more variables and calculations are involved in l_2-norm optimization. The other terms about the trust-region and final range guarantee the bounded relaxation on state variables and the convergence of final range, respectively. The robustness of SCP is greatly improved by the virtual control and adaptive trust-region techniques. It is worth noting that these advantages are achieved at the expense of some optimality due to the augmented performance index. More discussions on this feature will be detailed based on the simulation results in Sect. 4.2.4.

2. **Discretization**

After linearization, it is still a continuous-time optimal control problem. To obtain the numerical solution, this problem is discretized at $N+1$ evenly distributed points over the interval [0, 1], and constraints are imposed at each point. The discretization points set is defined as $\{S_0, S_1, \cdots, S_N\}$ with a step $\Delta S = 1/N$ and $S_{n+1} = S_n + \Delta S$, $n = 0, 1, \cdots, N-1$. Correspondingly, the discretized state and control vectors at the k-th iteration are denoted as $\boldsymbol{x}_n^k \triangleq \boldsymbol{x}^k(S_n)$, and $u_n^k \triangleq u^k(S_n)$. Then the dynamics can be propagated numerically via the trapezoidal rule over the reference trajectory $\{\boldsymbol{x}_0^k, \boldsymbol{x}_1^k, \cdots, \boldsymbol{x}_N^k, u_0^k, u_1^k, \cdots, u_N^k, s_f^k\}$:

$$\boldsymbol{x}_{n+1} = \boldsymbol{x}_n + \frac{\Delta S}{2}(\boldsymbol{A}_n^k \boldsymbol{x}_n + \boldsymbol{B}_n^k u_n + \boldsymbol{c}_n^k s_f + \boldsymbol{d}_n^k + \boldsymbol{v}_n + \boldsymbol{A}_{n+1}^k \boldsymbol{x}_{n+1} + \boldsymbol{B}_{n+1}^k u_{n+1} + \boldsymbol{c}_{n+1}^k s_f + \boldsymbol{d}_{n+1}^k + \boldsymbol{v}_{n+1}) \tag{4.32}$$

where $\boldsymbol{A}_n^k = s_f^k \frac{\partial \boldsymbol{f}}{\partial \boldsymbol{x}}\Big|_{\boldsymbol{x}_n^k, u_n^k}$, $\boldsymbol{B}_n^k = \boldsymbol{B}_{n+1}^k = s_f^k[0, 0, 0, 0, 0, 0, 1, 0]^T$, $\boldsymbol{c}_n^k = \boldsymbol{f}(\boldsymbol{x}_n^k) + \boldsymbol{B}u_n^k + \boldsymbol{f}_\Omega(\boldsymbol{x}_n^k)$, and $\boldsymbol{d}_n^k = -s_f^k \frac{\partial \boldsymbol{f}}{\partial \boldsymbol{x}}\Big|_{\boldsymbol{x}_n^k, u_n^k} \boldsymbol{x}_n^k - s_f^k \boldsymbol{B} u_n^k$. Then Eq. (4.32) can be further rewritten as:

$$\boldsymbol{H}_n \boldsymbol{x}_n + \boldsymbol{H}_{n+1} \boldsymbol{x}_{n+1} + \frac{\Delta S}{2}\boldsymbol{B}^k(u_n + u_{n+1}) + \frac{\Delta S}{2}(\boldsymbol{v}_n + \boldsymbol{v}_{n+1}) + \boldsymbol{G}_{n+1} s_f = -\frac{\Delta S}{2}\left(\boldsymbol{d}_n^k + \boldsymbol{d}_{n+1}^k\right) \tag{4.33}$$

where $\boldsymbol{H}_n = \left(\boldsymbol{I} + \frac{\Delta S}{2}\boldsymbol{A}_n^k\right)$, $\boldsymbol{H}_{n+1} = -\left(\boldsymbol{I} - \frac{\Delta S}{2}\boldsymbol{A}_{n+1}^k\right)$, $\boldsymbol{G}_{n+1} = \frac{\Delta S}{2}\left(\boldsymbol{c}_n^k + \boldsymbol{c}_{n+1}^k\right)$.

For notational convenience, we concatenate all decision variables $\boldsymbol{x}_n$, u_n, s_f, and $\boldsymbol{v}_n$ into a lumped vector $\boldsymbol{y} = \left[\boldsymbol{x}_0; \boldsymbol{x}_1; \cdots; \boldsymbol{x}_N; u_0; u_1; \cdots; u_N; \boldsymbol{v}_0; \boldsymbol{v}_1; \cdots; \boldsymbol{v}_N; s_f\right] \in \mathbb{R}^{17(N+1)+1}$, and the dynamics is rewritten as a linear equality constraint on $\boldsymbol{y}$:

$$\left[\boldsymbol{M}_1\ \boldsymbol{M}_2\ \boldsymbol{M}_3\ \boldsymbol{M}_4\right]\boldsymbol{y} = \boldsymbol{P} \tag{4.34}$$

$$\boldsymbol{M}_1 = \begin{bmatrix} \boldsymbol{I} & 0 & 0 & \cdots & 0 & 0 \\ \boldsymbol{H}_0 & \boldsymbol{H}_1 & 0 & \cdots & 0 & 0 \\ \vdots & \vdots & \vdots & \ddots & \vdots & \vdots \\ 0 & 0 & 0 & \cdots & \boldsymbol{H}_{N-1} & \boldsymbol{H}_N \end{bmatrix}, \quad \boldsymbol{M}_2 = \boldsymbol{B}^k \boldsymbol{M}_3$$

$$
\boldsymbol{M}_3 = \frac{\Delta S}{2}\begin{bmatrix} 0 & 0 & 0 & \cdots & 0 & 0 \\ \boldsymbol{I} & \boldsymbol{I} & 0 & \cdots & 0 & 0 \\ \vdots & \vdots & \vdots & \ddots & \vdots & \vdots \\ 0 & 0 & 0 & \cdots & \boldsymbol{I} & \boldsymbol{I} \end{bmatrix}, \quad \boldsymbol{M}_4 = \begin{bmatrix} 0 \\ \boldsymbol{G}_1 \\ \vdots \\ \boldsymbol{G}_N \end{bmatrix} \tag{4.35}
$$

$$
\boldsymbol{P} = -\frac{\Delta S}{2}\begin{bmatrix} -\frac{2}{\Delta S}\boldsymbol{x}_0 \\ \boldsymbol{d}_0^k + \boldsymbol{d}_1^k \\ \vdots \\ \boldsymbol{d}_{N-1}^k + \boldsymbol{d}_N^k \end{bmatrix} \tag{4.36}
$$

All constraints and modified performance index can be reformulated as:

$$
f_{i,j} \le f_{i,\max}, \ |\sigma_j| \le \sigma_{\max}, \ |u_j| \le \dot{\sigma}_{\max} t_s / V_{\max} \tag{4.37}
$$

$$
\boldsymbol{x}(0) = \boldsymbol{x}_0, \ \boldsymbol{E}_i \boldsymbol{x}_N = \boldsymbol{E}_i \boldsymbol{x}(1) \tag{4.38}
$$

$$
\left\|\boldsymbol{x}_j - \boldsymbol{x}_j^k\right\|_2 \le \delta_{x,j}, \ \left|s_f - s_f^k\right| \le \delta_s \tag{4.39}
$$

$$
J = w_i J_i + w_v \|\overline{\boldsymbol{v}}\|_1 + w_x \left\|\overline{\boldsymbol{\delta}}_x\right\|_2 + w_s \delta_s \tag{4.40}
$$

where $i = 1, 2, 3$, $j = 0, 1, \cdots, N$, $\overline{\boldsymbol{\delta}}_x = [\delta_{x,0}; \delta_{x,1}; \cdots; \delta_{x,N}]$, $\overline{\boldsymbol{v}} = [\boldsymbol{v}_0; \boldsymbol{v}_1; \cdots; \boldsymbol{v}_N]$ and the trust-region constraint is a second-order cone. Note that decision variables are $\boldsymbol{y}$, $\overline{\boldsymbol{\delta}}_x$ and δ_s for the convexified optimal control problem.

4.2.3 Summary of Algorithm

With convexified dynamics, constraints and performance index, the r-SCP algorithm is summarized below:

(1) Let $k = 0$ and initialize the state vector with $[r_0, \theta_0, \phi_0, V_0, \gamma_0, \psi_0, \sigma_0, \tau_0]$. Propagate the entry dynamics with a constant control input to obtain the initial guess trajectory $\{\boldsymbol{x}_0^0, \boldsymbol{x}_1^0, \cdots, \boldsymbol{x}_N^0, u_0^0, u_1^0, \cdots, u_N^0, s_f^0\}$.

(2) For $k \ge 1$, solve the following optimal control problem to find the optimal decision variable $\boldsymbol{y}^k$ (or optimal trajectory $\{\boldsymbol{x}_0^k, \boldsymbol{x}_1^k, \cdots, \boldsymbol{x}_N^k, u_0^k, u_1^k, \cdots, u_N^k, s_f^k\}$ and virtual control $[\boldsymbol{v}_0^k; \boldsymbol{v}_1^k; \cdots; \boldsymbol{v}_N^k]$) with the previous solution $\{\boldsymbol{x}_0^{k-1}, \boldsymbol{x}_1^{k-1}, \cdots, \boldsymbol{x}_N^{k-1}, u_0^{k-1}, u_1^{k-1}, \cdots, u_N^{k-1}, s_f^{k-1}\}$ as the guess.

$$
\begin{gathered}
\text{P4} - 1: \min_{\boldsymbol{y}, \overline{\boldsymbol{\delta}}_x, \delta_s} \ \text{Eq.}\ (4 \cdot 40) \\
\text{subject to Eqs.}\ (4 \cdot 34),\ (4 \cdot 37) \sim (4 \cdot 39)
\end{gathered}
$$

(3) If $\left\| \overline{\boldsymbol{\delta}}_x \right\|_2 \le \delta_{tot}$ and $\|\overline{\boldsymbol{v}}\|_1 \le v_{tot}$, go to Step 4), otherwise set $k = k + 1$ and go to Step 2). Note that δ_{tot} and $v_{tot} \in \mathbb{R}_+$ are the given convergence tolerance specified by users.
(4) The solution of the Mars entry trajectory planning problem is found to be $\{\boldsymbol{x}_0^k, \boldsymbol{x}_1^k, \cdots, \boldsymbol{x}_N^k, u_0^k, u_1^k, \cdots, u_N^k, s_f^k\}$.

4.2.4 Numerical Demonstration

To validate the effectiveness of the proposed approach, performances of three trajectory planning algorithms are compared in MATLAB environment. Both r-SCP introduced here and SCP are formulated and solved by the modeling toolbox YALMIP [130] and solver MOSEK [131]. These solutions are compared with the solution generated by GPOPS using the default nonlinear programming solver IPOPT.

Simulation scenarios are set up as follows. The initial simulation conditions are given in Table 4.1, and the boundary conditions are given as: $0 \le h \le 120$ km, $-180° \le \theta \le 180°$, $-180° \le \phi \le 180°$, $0 \le V \le 6{,}000$ m/s, $-80° \le \gamma \le 80°$, $-180° \le \psi \le 180°$, $-80° \le \sigma \le 80°$, $700 \le s \le 1{,}100$ km, $-10°$ /s $\le \dot{\sigma} \le 10°$/s. The reference radius, gravitational acceleration, rotation rate, and atmosphere model parameters are specified as $R_0 = 3397.2$ km, $g_0 = 3.7114$ m/s^2, $\Omega = 7.0882 \times 10^{-5}$ rad/s, $\rho_0 = 0.0158$ kg/m^3 and $h_s = 9{,}354.5$ m, respectively. The specifications of MSL-class vehicle is adopted, where $C_D = 1.45$, $C_L = 0.36$, $m = 2{,}804$ kg, $S_r = 15.9$ m^2, $k_Q = 1.9027 \times 10^{-4}$ kg$^{0.5}$s$^{0.15}$/m$^{1.15}$, $R_n = 6.476$ m, $\dot{Q}_{\max} = 70$ W/cm^2, $q_{\max} = 8.5$ kPa, $a_{\max} = 18\, g_0$.

In the simulation, $w_{1,2,3}$, w_v, w_x, w_s are set as 20, 20, 20, 5×10^5, 0.1, 0.1, respectively, and the convergence tolerances δ_{tot} and v_{tot} are set as 10^{-4} and 10^{-5}, which ensures that the solved trajectory satisfies the condition that the virtual control approaches zero approximately. Moreover, the initial guess trajectory for r-SCP is generated by propagating the entry dynamics Eqs. (4.3~4.8) with a zero initial bank angle, a constant control input $u \triangleq d\sigma \big/ ds = -3$, an initial guess of range 900 km, and 201 uniform discretization points. Also, the initial simulation conditions for GPOPS are given in Table 4.1 except σ_0, which is a free variable within its definition domain. Then the global collocation method is performed by regarding $\dot{\sigma}$ as the control input and setting tolerance of mesh refinement as 10^{-3}. In addition, since only the fixed-final-time entry is investigated in previous work, the flight time of SCP is prescribed as 355 s in cases of minimizing terminal velocity and maximizing terminal altitude. The initial guess for SCP is obtained by propagating dynamics at a constant control input $u \triangleq d\sigma \big/ d\tau = -2.6$ and $\sigma_0 = 0$. In this way, a trajectory close to the initial guess trajectory of r-SCP is obtained for a fair comparison. This trajectory is chosen because of its good convergence for different planning problems with different performance indices while using SCP. The trust-region and stopping criterion of SCP are selected as

$$\boldsymbol{\delta} = [1000/R_0, 10\pi/180, 10\pi/180, 100/V_s, 10\pi/180, 10\pi/180, 10\pi/180]$$

$$\boldsymbol{\varepsilon} = [100/R_0, 0.05\pi/180, 0.05\pi/180, 10/V_s, 0.05\pi/180, 0.05\pi/180, 1\pi/180] \tag{4.41}$$

where the two parameters $\boldsymbol{\delta}$ and $\boldsymbol{\varepsilon}$ are respectively the trust region radius for linearization of nonlinear dynamics and the prescribed convergence tolerance of solution. Likewise, 201 uniform discretization points are also used for SCP.

It is worth noting that the SCP algorithm converges to the solution with unsatisfying accuracy in subsequent simulations because the fixed trust-region cannot guarantee the relaxation error of dynamics constraints, although this strategy is effective in the Earth entry. This issue stems from sensitivity to the initial reference trajectory. In addition, the infeasible situation can be encountered during the first few iterations of the SCP in the absence of virtual control, because some final states with the initial guess may deviate largely from the expected values. The limited bank angle and its rate can also aggravate this infeasibility. To prevent the occurrence of inaccuracy and infeasibility, the virtual control technique is used in SCP to avoid the deviation of linearized dynamics, where the weight of the original performance index w_i and the penalty weight w_v on the virtual control in the performance index Eq. (4.31), are set as 1 and 10^6, respectively.

1. **Mars entry with minimum terminal velocity**

The convergence histories of altitude profiles and performance index (speed) values for minimum terminal velocity entry using r-SCP and SCP are shown in Fig. 4.2. The two left subplots in Fig. 4.2 demonstrate some intermediate profiles during the convergence process and the optimal solution marked in red. Also, the numerical solution shown in the blue dash line is obtained by integrating the original nonlinear dynamics Eqs. (4.3–4.8) using the Runge–Kutta method with the optimal control generated by r-SCP or SCP. It can be seen that the numerical integration results are consistent with the optimal solutions, and the final flight altitude deviation is about only 50 m for r-SCP. From the right two subplots in Fig. 4.2, the r-SCP converges to the optimal solution in only 7 iterations, while SCP takes 34 steps. The trajectory profiles generated by the r-SCP, SCP and GPOPS are presented in Fig. 4.3 and Fig. 4.4. As can be seen, flight states among the three methods have the same trend, which indicates that these numerical algorithms all converge to the neighborhood of the optimal solution. However, the bank angle and its rate profiles of the three approaches shown in Fig. 4.4a are different. The bank angle and control input profiles generated by GPOPS exhibit more oscillations than those of SCP and r-SCP algorithms. In other words, the bank angle and its rate profiles of r-SCP are smoother, thus less control energy is required. The possible reason is that a smaller allowable control magnitude of u than its upper bound is imposed by replacing $V\cos\gamma$ with a larger $V_{\max}$ in Eq. (4.20) assuming a fixed maximum bank angular rate. Besides, at the beginning, bank angles of r-SCP and SCP oscillate around zero whereas the bank angle of GPOPS keeps unchanged at zero. Near the end, the bank angles of GPOPS and SCP tend to zero while the bank angle with r-SCP decreases. These differences are mainly

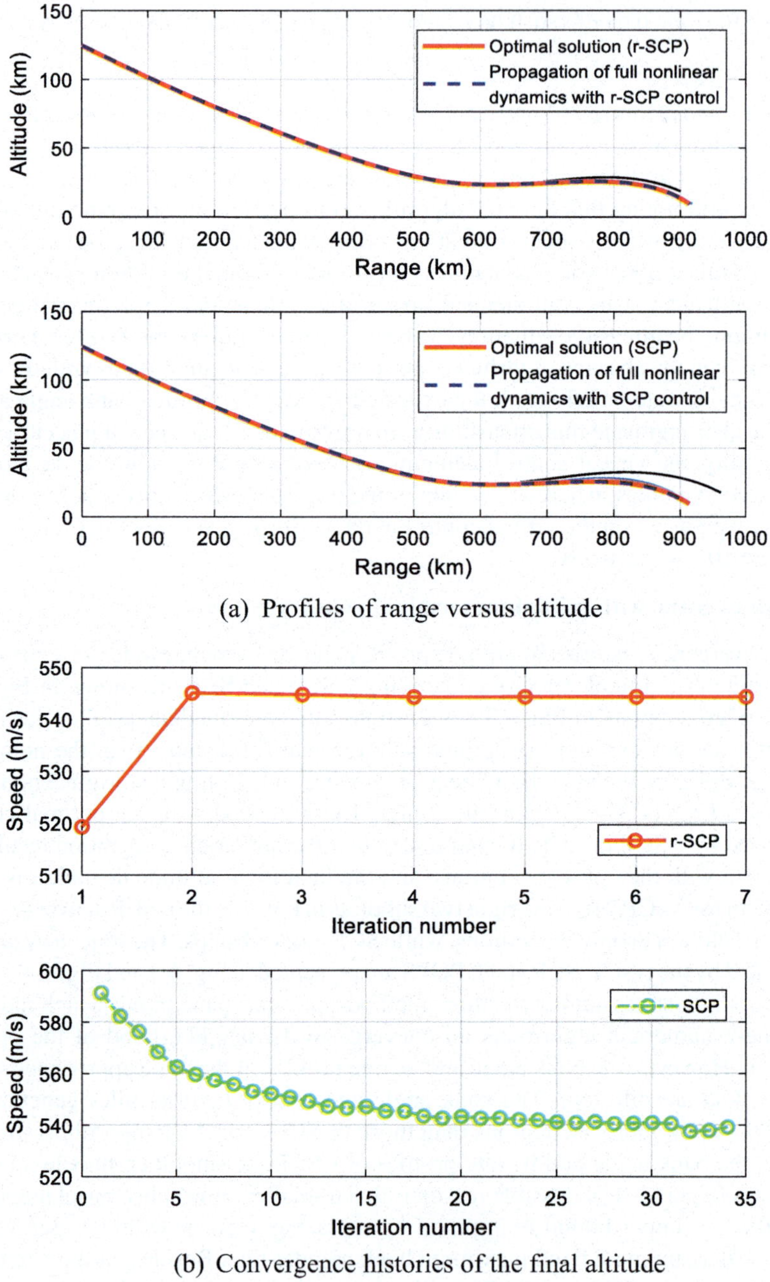

Fig. 4.2 Convergence histories of altitude, and optimal performance index values of minimum terminal velocity entry using r-SCP and SCP

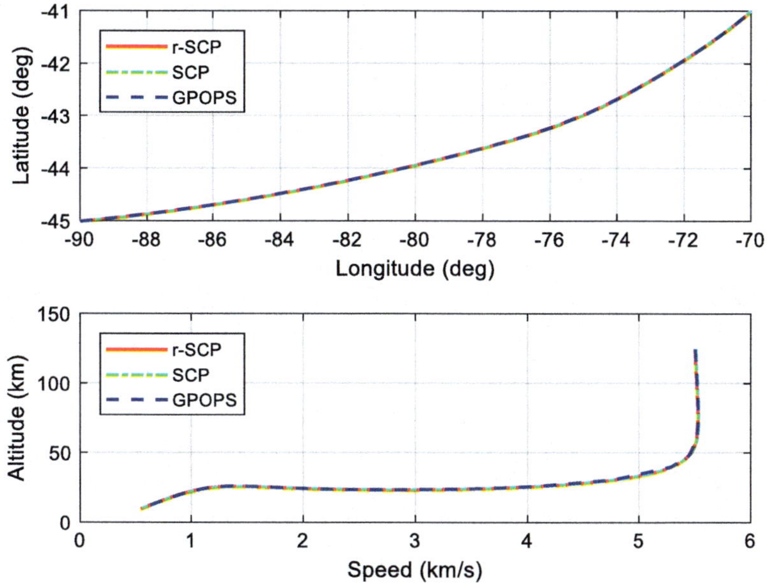

(a) Profiles of longitude versus latitude and speed versus altitude

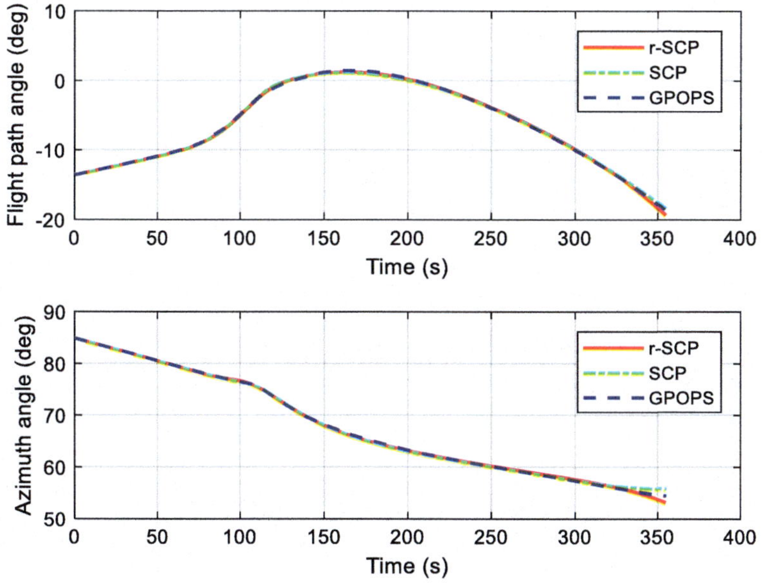

(b) Profiles of flight path angle and azimuth angle

Fig. 4.3 Longitude-latitude, speed-altitude profiles, and flight path angle and azimuth angle histories of minimum terminal velocity entry problem

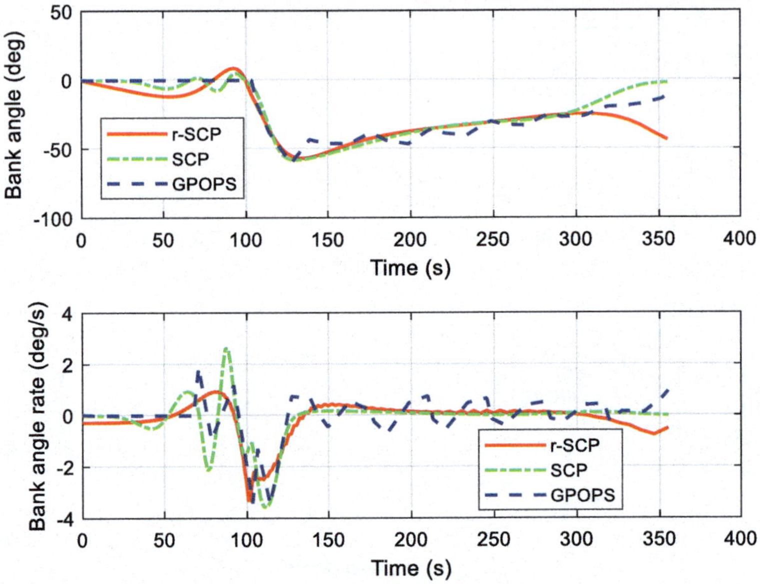

(a) Profiles of bank angle and its rate

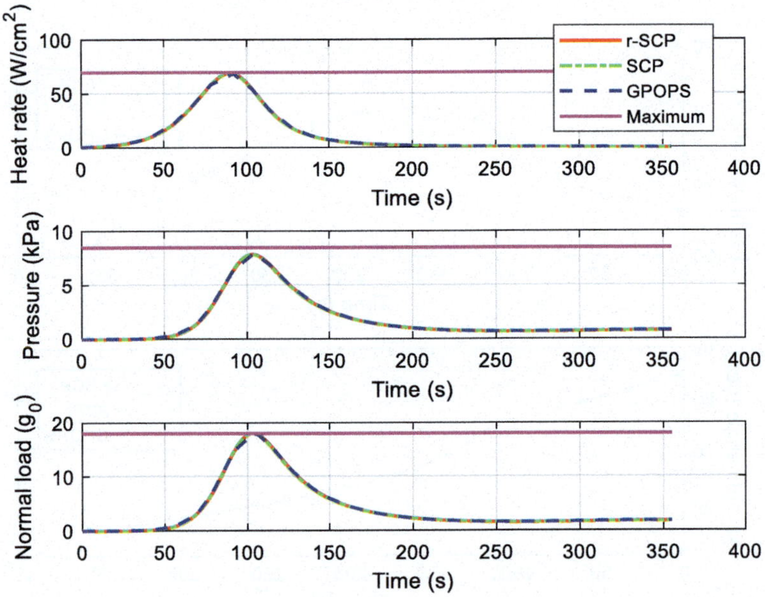

(b) Profiles of path constraints

Fig. 4.4 Bank angle and its rate time histories, and path constraints histories of minimum terminal velocity entry problem

caused by the differences in linearization, discretization and relaxation procedures. It is worth noting that the high-frequency jitter of the bank angle is avoided by choosing the bank angle rate as the system input in all the three trajectory planning methods. Additionally, as shown in Fig. 4.4b, all the path constraints are satisfied.

To demonstrate the performance of the algorithms quantitatively, the final states, path constraints, and CPU time consumption are summarized in Table 4.2. In terms of optimality, the final terminal velocities of r-SCP, SCP and GPOPS are 543.82 m/s, 539.39 m/s and 544.49 m/s, respectively. Thus, the terminal velocities of r-SCP and GPOPS are very close, and that of SCP is the minimum. Note that the optimality of r-SCP slightly degrades compared with that of SCP. As listed in Table 4.3, the original performance index accounts for about 94% of the new performance index for r-SCP, while the virtual control takes about 6%. For SCP, though, the original performance index accounts for nearly 100% with the virtual control only 0.68 ‱. This explains that the penalty on the virtual control for r-SCP does contribute to the change of optimality.

Despite optimality, r-SCP takes only 4.556 s to reach the optimal solution, whereas SCP and GPOPS take 16.58 s and 11.45 s, respectively. The trust-region radius is used by r-SCP to adaptively determine the step size in each iteration and reduce the total number of optimization steps. Additionally, it should be noted that the robustness of the SCP is greatly improved by introducing the virtual control, otherwise this algorithm may not converge at the prescribed accuracy ε shown in Eq. (4.41) even after considerable iterations. The differences of the states in the first ten iterations are shown in Table 4.4. Note that only the longitude, latitude and speed meet the convergence criterion. By contrast, the r-SCP and SCP with the virtual control technique converge in only a few iterations.

Table 4.2 Comparison of solutions for minimum terminal velocity entry

Parameters	r-SCP	SCP	GPOPS	Parameters	r-SCP	SCP	GPOPS
h_f(km)	9.999	10.00	10	t_f(s)	355	355	355
θ_f(deg)	−70	−70	−70	σ_f(deg)	−43.49	−1.962	−10.71
ϕ_f(deg)	−41	−41	−41	$\lvert\dot{\sigma}\rvert_{max}$(deg/s)	2.966	3.580	3.437
V_f(m/s)	543.82	539.39	544.49	$\dot{Q}_{max}$(W/cm^2)	69.34	69.35	68.44
γ_f(deg)	−19.22	−18.38	−18.57	q_{max}(kPa)	7.886	7.882	7.885
ψ_f(deg)	53.36	55.90	54.53	$a_{max}(g_0)$	17.99	17.99	18.00
s_f(km)	913.18	913.21	913.26	CPU time (s)	4.556	16.58	11.45

Table 4.3 Components of converged new performance index

Methods	$w_i J_i$	$w_v \lVert\overline{\boldsymbol{v}}\rVert_1$	$w_x \left\lVert\overline{\boldsymbol{\delta}}_x\right\rVert_2$	$w_s \delta_s$	J
r-SCP	3.0830	0.1975	6.1787×10^{-5}	0	3.2806
SCP	1.0036	6.8959×10^{-5}	/	/	1.0036

Table 4.4 Differences of the states between consecutive iterations for SCP without a virtual control

Iteration	$\|\Delta r\|$(m)	$\|\Delta\theta\|$(°)	$\|\Delta\phi\|$(°)	$\|\Delta V\|$(m/s)	$\|\Delta\gamma\|$(°)	$\|\Delta\psi\|$(°)	$\|\Delta\sigma\|$(°)
1	6862.43	0.9787	0.1615	251.7393	4.8358	5.2426	21.8927
2	1501.39	0.0925	0.0594	91.9612	1.4209	0.7964	21.9099
3	308.74	0.0132	0.0131	11.7171	0.4698	0.3495	18.8904
4	471.76	0.0320	0.0441	14.5576	0.5901	1.14020	21.6865
5	319.97	0.0118	0.0120	10.6519	0.2693	0.636	17.7101
6	155.95	0.0078	0.0062	8.1664	0.1776	0.424	14.2666
7	124.56	0.0058	0.0038	4.2608	0.1662	0.2693	14.9714
8	140.13	0.0079	0.0046	6.3911	0.1146	0.3037	21.4687
9	115.61	0.0063	0.0032	3.9057	0.0974	0.275	16.0256
10	115.76	0.0067	0.0019	6.0361	0.212	0.6818	22.3167
ε	100	0.05	0.05	10	0.05	0.05	1

2. **Mars entry with maximum terminal altitude and minimum final time**

Analogously, the simulation results of maximum terminal altitude entry and minimum final time entry can be obtained. Similar to the results in previous section, the optimal trajectories generated by the three methods are nearly the same, except for slight differences in the bank angle and its rate that show smoother responses with r-SCP. The terminal results of these two cases are summarized in Table 4.5 and Table 4.6, respectively. Note that only r-SCP and GPOPS are compared in Table 4.6 because SCP is designed for a fixed-final-time mission.

In Table 4.5, GPOPS uses the longest time (5.620 s) to find the optimal solution and generates the entry trajectory with the minimum altitude of 12.03 km, while the terminal altitudes are 12.19 km with r-SCP and 12.56 km with SCP, respectively. Although the terminal altitude of r-SCP is better than GPOPS, it is not as good as the altitude obtained by SCP. This is the price paid by the additional penalty terms $w_v\|\overline{v}\|_1 + w_x\left\|\overline{\delta}_x\right\|_2 + w_s\delta_s$ in the performance index as discussed in the previous subsection.

Table 4.5 Comparison of solutions for maximum terminal altitude entry

Parameters	r-SCP	SCP	GPOPS	Parameters	r-SCP	SCP	GPOPS
h_f(km)	12.19	12.56	12.03	t_f(s)	355	355	355
θ_f(deg)	−70	−70	−70	σ_f(deg)	−45.10	−15.53	−11.42
ϕ_f(deg)	−41	−41	−41	$\|\dot{\sigma}\|_{max}$(deg/s)	6.729	4.379	10.00
V_f(m/s)	626.36	619.54	605.14	$\dot{Q}_{max}$(W/cm^2)	69.26	69.31	67.83
γ_f(deg)	−18.44	−17.05	−17.26	q_{max}(kPa)	7.886	7.883	7.885
ψ_f(deg)	55.30	58.41	57.46	$a_{max}(g_0)$	17.99	17.99	18.00
s_f(km)	912.44	912.54	912.73	CPU time (s)	5.321	4.886	5.620

Table 4.6 Comparison of solutions for minimum final time entry

Parameters	r-SCP	GPOPS	Parameters	r-SCP	GPOPS
h_f (km)	9.999	10	t_f(s)	332.75	331.08
θ_f (deg)	−70	−70	σ_f(deg)	−47.66	−69.08
ϕ_f (deg)	−41	−41	$\lvert\dot{\sigma}\rvert_{\max}$(deg/s)	1.561	7.722
V_f (m/s)	626.35	651.60	$\dot{Q}_{\max}$(W/cm^2)	69.34	67.76
γ_f (deg)	−18.65	−20.37	$q_{\max}$(kPa)	7.886	7.845
ψ_f (deg)	48.88	48.65	$a_{\max}(g_0)$	17.99	17.91
s_f (km)	913.66	913.49	CPU time (s)	5.779	16.89

For the minimum time entry, r-SCP only performs 9 successive iterations to generate the converged solution, taking 11 s shorter than GPOPS. The final time achieved by GPOPS and r-SCP are 331.08 s and 332.75 s, respectively, as listed in Table 4.6. These results demonstrate that the optimality of r-SCP in terms of minimum final time is very close to GPOPS, while the gain in computation efficiency with r-SCP is significant.

4.3 LGL Pseudospectral Sequential Convex Programming

Although a dynamic model is established in Sect. 4.2 with the range as the independent variable, and an improved sequential convex optimization method is proposed to solve the Mars atmosphere entry trajectory optimization problem, it is still necessary to optimize the final range, and the use of uniform discrete nodes leads to a large number of variables to be optimized and a large amount of calculation. In order to avoid solving the problem with a free terminal time/energy/range, a dynamic model is established with the downrange angle as the independent variable, and the Legendre–Gauss-Lobatto (LGL) pseudospectral sequential convex optimization method is used to solve the Mars entry trajectory optimization problem.

The Legendre Pseudospectral Convex Programming (LPCP) method combining the pseudospectral method and convex optimization method discretizes state variables and control variables at LGL coordination points, and approximates state variables and control variables by Lagrange interpolation polynomials, thus converting integral operations in state differential equations and objective functions into algebraic operations. Combined with the first-order linearization method, the optimal control problem is transformed into a convex optimization problem, and the approximate solution of the original optimal control problem is obtained by solving the convex optimization problem.

4.3.1 Formulation of Problem

Before defining the Mars entry trajectory problem with maximum final altitude, the bank angle σ and dimensionless time τ are taken as new state variables, and the bank angle rate is defined as the control variable to avoid the high-frequency jitters of the bank angle. Then, the Mars entry dynamic model with respect to the downrange angle is rewritten as:

$$\frac{d\boldsymbol{x}}{d\overline{s}} = \boldsymbol{F}(\boldsymbol{x}) + \boldsymbol{B}(\boldsymbol{x})u \tag{4.42}$$

where $\boldsymbol{x} = [r, \theta, \phi, V, \gamma, \psi, \sigma, \tau]^{\mathrm{T}}$, $\boldsymbol{F}(\boldsymbol{x}) = [F_1, F_2, \cdots, F_8]^{\mathrm{T}}$, $\boldsymbol{B}(\boldsymbol{x}) = [b_1, b_2, \cdots, b_8]^{\mathrm{T}}$ and

$$\begin{aligned} F_1 = r\tan\gamma,\ F_2 = \tfrac{\sin\psi}{\cos\phi},\ F_3 = \cos\psi,\ F_4 = -\tfrac{rD}{V\cos\gamma} - \tfrac{\tan\gamma}{rV} \\ F_5 = \tfrac{rL\cos\sigma}{V^2\cos\gamma} - \tfrac{1}{rV^2} + 1,\ F_6 = \tfrac{rL\sin\sigma}{V^2\cos^2\gamma} + \sin\psi\tan\phi \\ F_7 = 0,\ F_8 = \tfrac{r}{V\cos\gamma},\ b_{1,2,3,4,5,6,8} = 0,\ b_7 = \tfrac{r}{V\cos\gamma} \end{aligned}$$

The optimal control problem for the Mars entry terminal altitude maximization problem is:

$$\text{P4} - 2: \text{Find}\, u,\ \text{minimize}\ J = -r(\overline{s}_f) \tag{4.43}$$

$$\text{subject to}\ (4\cdot 42), (4\cdot 11) \sim (4\cdot 14) \tag{4.44}$$

4.3.2 Convexification of Problem

The non-convexity of the Mars entry terminal height maximization problem is derived from the dynamics and the path constraints. Similarly, the two constraints can be linearized using the first-order Taylor expansion approximation. First, the dynamics is linearized at the reference trajectory $\{\boldsymbol{x}^*, u^*\}$, and one has:

$$\frac{d\boldsymbol{x}}{d\overline{s}} = \boldsymbol{F}(\boldsymbol{x}) + \boldsymbol{B}(\boldsymbol{x})u \approx \boldsymbol{A}(\boldsymbol{x}^*)\boldsymbol{x} + \boldsymbol{B}(\boldsymbol{x}^*)u + \boldsymbol{C}(\boldsymbol{x}^*) \tag{4.45}$$

$$\boldsymbol{A}(\boldsymbol{x}^*) = \left.\frac{\partial \boldsymbol{F}}{\partial \boldsymbol{x}}\right|_{\boldsymbol{x}=\boldsymbol{x}^*},\ \boldsymbol{C}(\boldsymbol{x}^*) = \boldsymbol{F}(\boldsymbol{x}^*) - \boldsymbol{A}(\boldsymbol{x}^*)\boldsymbol{x}^* \tag{4.46}$$

where the elements of matrix are

$$a_{11}=\tan\gamma,\ a_{15}=\frac{r}{\cos^2\gamma},\ a_{23}=\frac{\sin\psi\tan\phi}{\cos\phi},\ a_{26}=\frac{\cos\psi}{\cos\phi},\ a_{36}=-\sin\psi$$
$$a_{41}=-\frac{D+rD_r}{V\cos\gamma}+\frac{\tan\gamma}{r^2V},\ a_{44}=-\frac{rD_V}{V\cos\gamma}+\frac{rD}{V^2\cos\gamma}+\frac{\tan\gamma}{rV^2},\ a_{45}=-\frac{rD\tan\gamma}{V\cos\gamma}-\frac{1}{rV\cos^2\gamma}$$
$$a_{51}=\frac{L\cos\sigma+rL_r\cos\sigma}{V^2\cos\gamma}+\frac{1}{r^2V^2},\ a_{54}=0+\frac{2}{rV^3}=\frac{2}{rV^3},\ a_{55}=\frac{rL\cos\sigma}{V^2\cos\gamma}\tan\gamma,\ a_{57}=-\frac{rL\sin\sigma}{V^2\cos\gamma}$$
$$a_{61}=\frac{L\sin\sigma+rL_r\sin\sigma}{V^2\cos^2\gamma},\ a_{63}=\frac{\sin\psi}{\cos^2\phi},\ a_{65}=2\tan\gamma\frac{rL\sin\sigma}{V^2\cos^2\gamma},\ a_{66}=\cos\psi\tan\phi,\ a_{67}=\frac{rL\cos\sigma}{V^2\cos^2\gamma}$$
$$a_{81}=\frac{1}{V\cos\gamma},\ a_{84}=-\frac{r}{V^2\cos\gamma},\ a_{85}=\frac{r\tan\gamma}{V\cos\gamma}$$

and the other elements are 0. In addition, the partial derivatives of the dimensionless lift and drag accelerations with respect to dimensionless height and velocity, as well as the first-order linearized expressions of the path constraints, are consistent with results in Sect. 4.2.

4.3.3 Discretization of Problem

In order to apply the sequential convex optimization method to solve the Mars entry trajectory problem with maximum final altitude, this optimal control problem is then discretized at the LGL collocation point. Using $N+1$ LGL coordination points $-1=\xi_0<\xi_1<\cdots<\xi_N=1$ as interpolation nodes, the N-order Lagrange interpolation polynomials $\boldsymbol{X}(\xi)$ and $U(\xi)$ are constructed to approximate $\boldsymbol{x}(\xi)$ and $u(\xi)$, respectively, that is, the values of the state and the control variables at the coordination point satisfy $\boldsymbol{X}(\xi_i)=\boldsymbol{x}(\xi_i)$, $U(\xi_i)=u(\xi_i)$, $\boldsymbol{y}$, and there are:

$$\boldsymbol{x}(\xi)\approx\boldsymbol{X}(\xi)=\sum_{i=0}^{N}L_i(\xi)\boldsymbol{x}(\xi_i),\ u(\xi)\approx U(\xi)=\sum_{i=0}^{N}L_i(\xi)u(\xi_i) \tag{4.47}$$

where Lagrange interpolating basis function $L_i(\xi)$ satisfies

$$L_i(\xi)=\prod_{j=0,j\neq i}^{N}\frac{\xi-\xi_j}{\xi_i-\xi_j},\ L_i(\xi_k)=\begin{cases}1\,k=i\\0\,k\neq i\end{cases} \tag{4.48}$$

Then the elements of the LGL pseudospectral first-order differential matrix are:

$$D_{ki}\triangleq\dot{L}_i(\xi_k)=\sum_{l=0,l\neq i}^{N}\frac{\sum_{j=0,j\neq i,j\neq l}^{N}\xi_k-\xi_j}{\sum_{j=0,j\neq i}^{N}\xi_i-\xi_j} \tag{4.49}$$

Therefore, after transforming the define domain of problem P4-2 into the interval $[-1,1]$, the equations of motion can be transformed into a series of equality constraints at LGL collocation points:

$$\sum_{i=0}^{N} D_{ki}\boldsymbol{X}_i = \frac{s_f}{2}\left[\boldsymbol{A}(\boldsymbol{X}_k^*)\boldsymbol{X}_k + \boldsymbol{B}(\boldsymbol{X}_k^*)U_k + \boldsymbol{C}(\boldsymbol{X}_k^*)\right] \tag{4.50}$$

where $\boldsymbol{X}_k \triangleq \boldsymbol{X}(\xi_k) = \boldsymbol{x}(\xi_k)$, $U_k \triangleq U(\xi_k) = u(\xi_k)$, $\boldsymbol{X}_k^* \triangleq \boldsymbol{X}^*(\xi_k) = \boldsymbol{x}^*(\xi_k)$.

At the same time, in order to avoid "artificial infeasibility" caused by linearization errors, virtual control $\boldsymbol{v}$ is introduced to compensate the linearization error:

$$\sum_{i=0}^{N} D_{ki}\boldsymbol{X}_i = \frac{s_f}{2}\left[\boldsymbol{A}(\boldsymbol{X}_k^*)\boldsymbol{X}_k + \boldsymbol{B}(\boldsymbol{X}_k^*)U_k + \boldsymbol{C}(\boldsymbol{X}_k^*) + \boldsymbol{v}_k\right] \tag{4.51}$$

And to ensure that the virtual control is as small as possible when the algorithm converges, it is necessary to impose a penalty term in the objective function:

$$J = -r(\xi_N) + w\|\overline{\boldsymbol{v}}\|_1 \tag{4.52}$$

where w is the given virtual control penalty weight. The 1-norm of the penalty function is chosen so that the virtual controller has as many 0 elements as possible, and the 2-norm can also be chosen as the penalty function, that is, to ensure that all elements are as close to 0 as possible. Note that the objective function in Eq. (4.52) does not contain an integral term, so no additional processing is required. If there is an integral term, it can be approximated using the Gauss–Lobatto quadrature formula.

Similarly, the convex optimization method of LGL pseudo-spectral sequence needs to impose trust region constraints on the variables to ensure that the algorithm is optimized in a small range, so as to avoid the linearization problem from the original nonlinear problem in the process of solving. Due to the high precision of the LGL discrete scheme, the fixed trust region constraints can be applied in this section. Let $\overline{\boldsymbol{X}} = [\boldsymbol{X}_0, \boldsymbol{X}_1, \cdots, \boldsymbol{X}_N]$ and $\overline{U} = [U_0, U_1, \cdots, U_N]$, then there are:

$$\left\|\boldsymbol{X}_k - \boldsymbol{X}_k^*\right\|_2 \le \Delta, k = 0, 1, \cdots, N \tag{4.53}$$

where Δ is the prescribed radius of trust-region.

Finally, the algorithm converges when there is little difference between the solutions of the last two iterations. The virtual control must also be small enough to ensure that the linearized dynamic model is close enough to the original nonlinear model. Then the convergence criterion can be defined as follows:

$$\max_{k=0,1,\cdots,N}\left\|\boldsymbol{X}_k - \boldsymbol{X}_k^*\right\|_2 \le \delta,\ \|\overline{\boldsymbol{v}}\|_1 \le \varepsilon \tag{4.54}$$

where δ and ε are the given convergence tolerances. The remaining path constraints and the constraints related to state and control variables do not involve differential or integral terms, so these constraints remain unchanged after discretization, and it is only necessary to change the continuous form into the corresponding discrete

Table 4.7 Algorithm flow of convex optimization of LGL pseudo-spectral sequences

Step	Procedure
1	Let $n = 0$, propagate the nonlinear dynamics (4.42) with $\overline{U}^0$ to generate the initial guess solution $\overline{X}^0$
2	For $n \geq 1$, initialize P3-3 with the solution $\{\overline{X}^{n-1}, \overline{U}^{n-1}\}$, and solve P3-3 to obtain the solution $\{\overline{X}, \overline{U}\}$ and $\overline{v}$
3	If Eq. (4.54) holds, then go to Step 4; otherwise let $n = n + 1$, $\overline{X}^n = \overline{X}$, $\overline{U}^n = \overline{U}$, and go to Step 2
4	Output the optimal solution $\{\overline{X}, \overline{U}\}$, and the algorithm terminates

form. For specific expressions, see Eq. (4.37). At the same time, the LGL pseudospectral sequential convex optimization method directly selects $u = d\sigma / d\tau$ as the control variable, so the bank angle rate constraint is still a convex constraint, and the relaxation of the inequality constraint is similar to the process in Sect. 4.2.

So far, the Legendre pseudospectral convex optimization problem in discrete form (P4-3) can be defined as follows:

$$\text{P4} - 3 : \text{Find}\,\overline{X}, \overline{U}, \overline{v}\,\text{minimize}\,(4 \cdot 52) \tag{4.55}$$

$$\text{subject to}\,(4 \cdot 51),\ (4 \cdot 53),\ (4 \cdot 37) \tag{4.56}$$

4.3.4 *Summary of Algorithm*

To sum up, the LGL pseudospectral sequential convex optimization method can be summarized in Table 4.7.

4.3.5 *Numerical Demonstration*

To demonstrate the performance of LGL pseudospectral sequential convex programming (LSCP) in solving the Mars entry trajectory problem with maximum final altitude, the results of the LSCP, the hp-adaptive pseudospectral method (GPOPS), and the SCP are compared in this section. It should be noted that the virtual control is also used in the SCP to ensure the convergence. And in simulations, there are $R_0 = 3386.6$ km, $g_0 = 3.734246$ m/s^2, $\rho_0 = 0.0158$ kg/m^3, $h_s = 9354.5$ m, $m = 2920$ kg, $S_r = 15.9$ m^2, $k_Q = 7.47682 \times 10^{-5}$, $C_L = 0.348$, $C_D = 1.45$, $\dot{Q}_{\max} = 80$ W/cm^2, $q_{\max} = 10$ kPa, $a_{\max} = 18\ g_0$. The other parameters are given in Table 4.8:

Table 4.8 Parameters for simulation

Parameters	Initial value	Final value	Boundaries
h (km)	125	Free	[0, 150]
θ (°)	−65	−45	[−180, 180]
φ (°)	−50	−44	[−90, 90]
V (m/s)	5500	550	[0, 6000]
γ (°)	−13.2	Free	[−80, 80]
ψ (°)	74	Free	[−180, 180]
σ (°)	Free	Free	[−80, 80]
$\dot{\sigma}$ (°/s)	Free	Free	[−10, 10]

In addition, LPCP and SCP have 56 and 501 discretization points, respectively. The initial guess trajectory for both methods is the trajectory obtained by numerical integration, where the constant bank angle of $-35°$ is the control input of the integration of the original nonlinear system. Additionally, penalty weight $w = 1 \times 10^5$, radius of the trust region $\Delta = 0.5$, convergence threshold $\delta = 0.05$, $\varepsilon = 1 \times 10^{-3}$. In both LPCP and SCP, the YALMIP toolbox and MOSEK solver are used to describe the solve the convex optimization problem, respectively. The hp-adaptive pseudospectral method is solved by using GPOPS toolbox, with the solver IPOPT, and the mesh convergence accuracy of 1×10^{-3}. The simulation results are shown in Fig. 4.5.

From the above simulation results, it can be seen that the three methods can generate the trajectories that satisfy the path and boundary constraints. Among them, the optimal trajectories generated by LPCP and SCP are very close, and the profiles of state variables are almost the same, only a slight difference is shown in the profiles of the bank angle and its rate. In addition, SCP and LPCP meet the convergence condition after 8 and 10 iterations, respectively. Also, the corresponding $||\overline{v}||_1$ is 3.028×10^{-4} and 2.817×10^{-6}, both of which are less than the convergence threshold $\varepsilon = 1 \times 10^{-3}$.

At the same time, although the optimal trajectory generated by GPOPS is consistent with that of LPCP and SCP, the profiles of bank angle and its rate are somewhat different from those of the other two methods, and the oscillation of the bank angle rate profile is also larger. Algorithmically, GPOPS transforms the optimal control problem into a nonlinear programming problem, while LPCP and SCP approximate the original optimal control problem through sequential convex optimization. The discretization strategies in the two methods are also different, so there are some differences in the approximate solutions of the two methods.

In order to further illustrate the advantages of the LPCP algorithm, the results of the three methods are compared in Table 4.9. As can be seen from Table 4.9, the CPU time of LPCP is the least, which is 6.195 s, while the CPU time of SCP and GPOPS is 10.92 s and 48.66 s, respectively. Compared with LPCP and SCP, the only difference between the two methods is the discretization strategy, and the number of discrete nodes in LPCP is much less than that in SCP, which indicates that LGL discretization strategy is conducive to reducing the scale of optimization problems and reducing the amount of computation to a certain extent. At the same time, the objective function value of LPCP (10.868 km) and the objective function value of SCP (10.853 km) are very close. Compared with LPCP and GPOPS, LGL and LGR pseudospectral discretization methods are used respectively. Although the objective

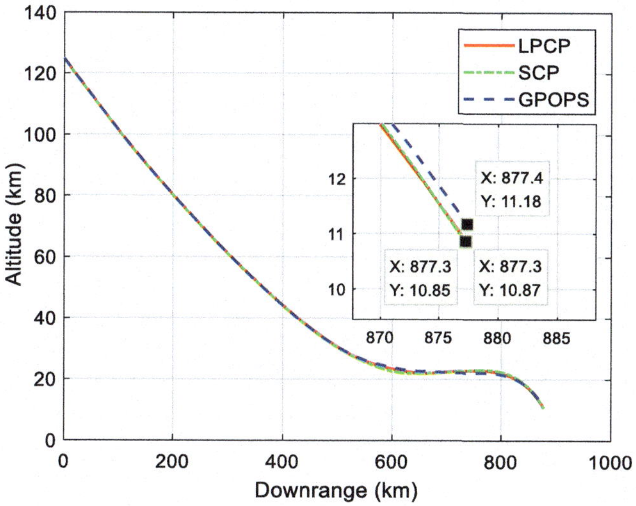

(a) Profiles of downrange versus altitude

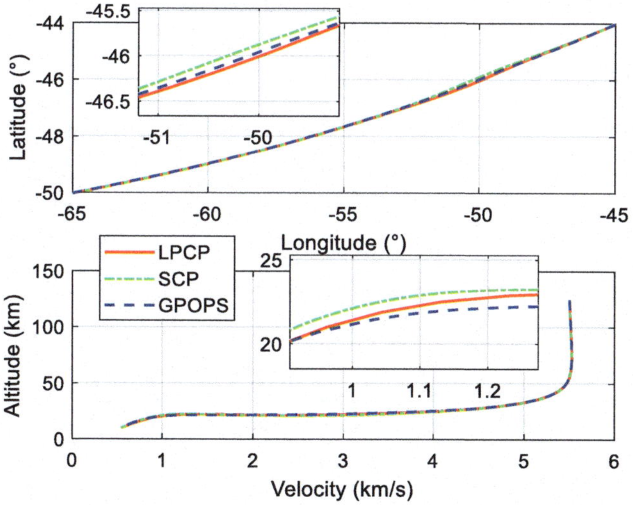

(b) Profiles of longtitude versus latitude and velocity versus altitude

Fig. 4.5 The results of the Mars entry trajectory optimization with maximum terminal altitude obtained by three methods

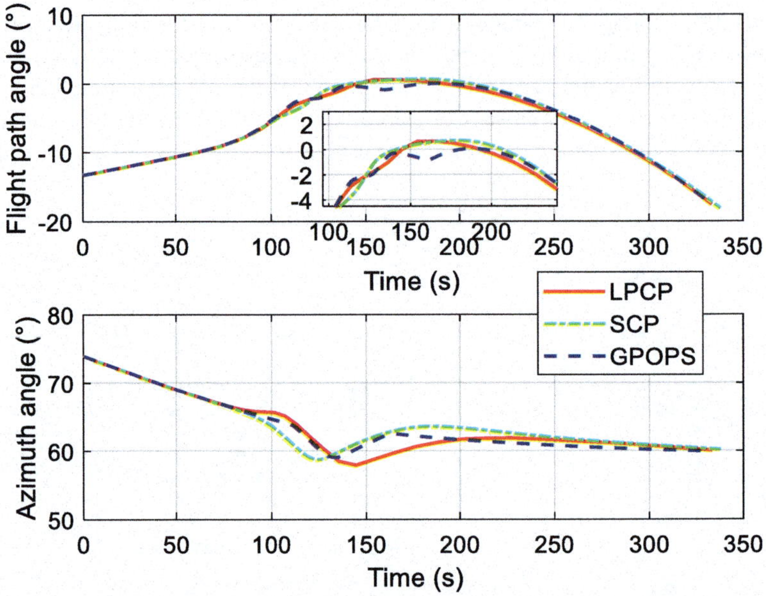

(c) Profiles of flight path angle and azimuth angle

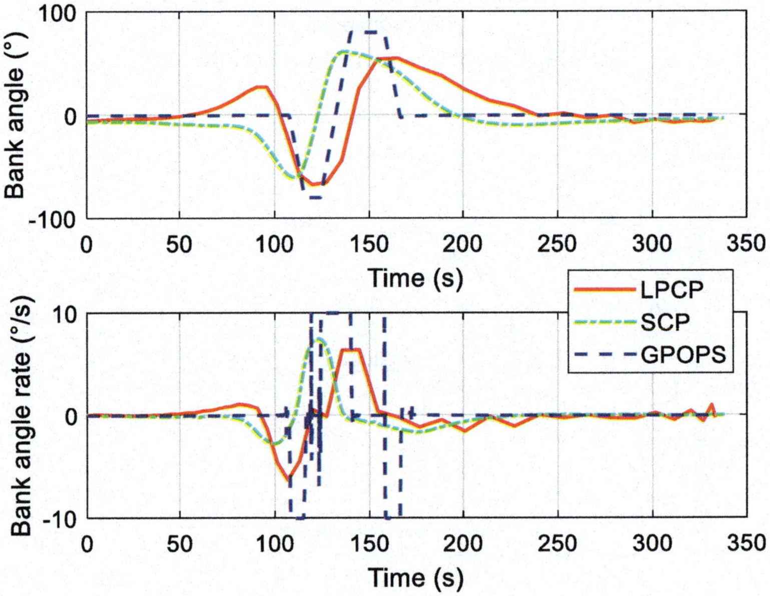

(d) Profiles of bank angle and its rate

Fig. 4.5 (continued)

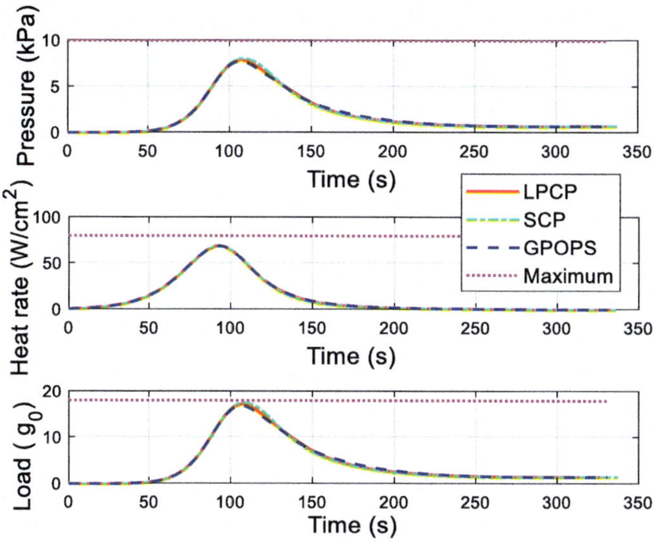

(e) Profiles of path constraints

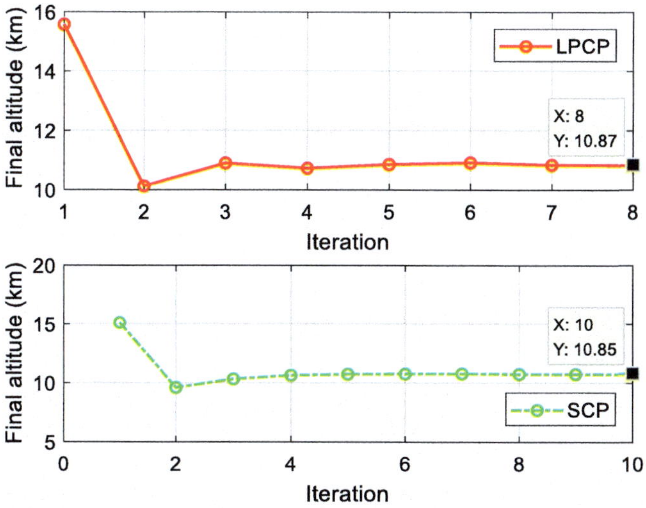

(f) Profiles of the objective function (altitude) of LPCP and SCP

Fig. 4.5 (continued)

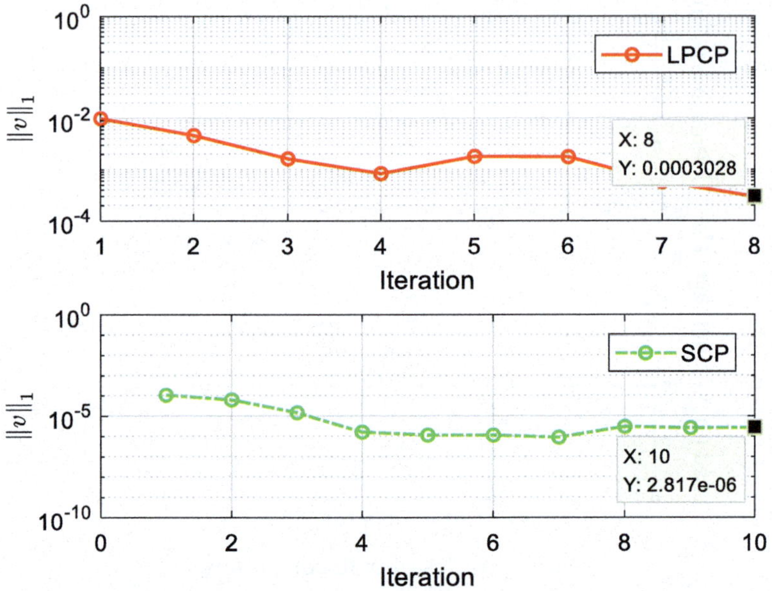

(g) Profiles of the virtual control of LPCP and SCP

Fig. 4.5 (continued)

function value of GPOPS (11.17 km) is larger than that of LPCP (10.868 km), the difference is only 300 m. Moreover, the CPU time of LPCP is smaller than that of GPOPS, while the objective function value is slightly smaller.

Table 4.9 Quantitative comparison of three algorithms

Parameters	LPCP	SCP	GPOPS	Parameters	LPCP	SCP	GPOPS
h_f(km)	10.868	10.853	11.17	t_f(s)	333.07	337.46	331.13
θ_f(°)	−45	−45	−45	σ_f(°)	−3.803	−3.273	−0.017
ϕ_f(°)	−44	−44	−44	$\lvert\dot{\sigma}\rvert_{max}$/(°/s)	6.446	7.512	10
V_f(m/s)	550	550	550	$\dot{Q}_{max}$(W/cm^2)	68.90	69.48	69.17
γ_f(°)	−17.48	−18.08	−17.21	q_{max}(kPa)	7.970	8.074	7.762
ψ_f(°)	60.13	60.27	59.98	$a_{max}(g_0)$	17.33	17.56	16.88
R_f(km)	877.26	877.26	877.38	CPU time(s)	6.195	10.92	48.66

4.4 Discussions

In this chapter, we propose a refinement of the Mars entry dynamics by utilizing the downrange angle as the independent variable. This transformation enables us to convert the optimal control problem, which previously had a free final time or energy, into a fixed downrange problem. Simultaneously, the r-SCP method enhances the time-optimal objective function by introducing a new state and relaxing the inequality constraint associated with the downrange angle rate. This modification effectively achieves convexification of the Mars entry time-optimal problem. These two enhancements alleviate the challenges faced by the SCP method when solving the Mars optimal entry problem. Additionally, the LPCP approach incorporates the LGL pseudospectral method and the convex optimization method to discretize the optimal control problem, resulting in notable improvements in the computational efficiency of the SCP method.

In summary, the developed r-SCP and LPCP methods exhibit superior computational efficiency compared to traditional SCP algorithms. However, there is a slight reduction in the optimal objective function value. Nonetheless, the optimal objective function values obtained using r-SCP and LPCP are comparable to those achieved with GPOPS, but with reduced computational requirements.

Chapter 5
Pseudospectral Model Predictive Convex Programming for Mars Entry Trajectory Planning

5.1 Generic Theory of PMPCP

5.1.1 Review of the MPCP Method

Consider a general nonlinear dynamic system defined in the time domain $[t_0, t_f]$ [132, 133]:

$$\dot{\boldsymbol{x}}(t) = \boldsymbol{f}(\boldsymbol{x}(t), \boldsymbol{u}(t)) \tag{5.1}$$

where the state $\boldsymbol{x} \in \mathbb{R}^{n_s}$ and control $\boldsymbol{u} \in \mathbb{R}^{n_c}$, n_s and n_c denote the dimensions of the state and control, respectively. The linearized error system along the reference trajectory $\boldsymbol{x}^*(t)$ and solution $\boldsymbol{u}^*(t)$ is given as:

$$\delta\dot{\boldsymbol{x}}(t) = \boldsymbol{A}^*\delta\boldsymbol{x}(t) + \boldsymbol{B}^*\delta\boldsymbol{u}(t) \tag{5.2}$$

$$\boldsymbol{A}^* = \left.\frac{\partial \boldsymbol{f}}{\partial \boldsymbol{x}}\right|_{\boldsymbol{x}^*(t),\boldsymbol{u}^*(t)}, \boldsymbol{B}^* = \left.\frac{\partial \boldsymbol{f}}{\partial \boldsymbol{u}}\right|_{\boldsymbol{x}^*(t),\boldsymbol{u}^*(t)} \tag{5.3}$$

$$\delta\boldsymbol{x}(t) = \boldsymbol{x}(t) - \boldsymbol{x}^*(t), \delta\boldsymbol{u}(t) = \boldsymbol{u}(t) - \boldsymbol{u}^*(t) \tag{5.4}$$

The error system (5.2) can be discretized at N time series $t_k = t_0 + (k-1)\Delta t$, $\Delta t = (t_f - t_0)/(N-1)$, $k = 1, 2, \ldots, N$, and the Euler rule is adopted to propagate the error system for $k = 1, 2, \ldots, N-1$:

$$\delta\boldsymbol{x}_{k+1} = (\boldsymbol{I} + \boldsymbol{A}_k^*\Delta t)\delta\boldsymbol{x}_k + \boldsymbol{B}_k^*\delta\boldsymbol{u}_k\Delta t \tag{5.5}$$

S. Li et al., *Trajectory Optimization and Guidance Methods for Mars Entry*,
https://doi.org/10.1007/978-981-99-6282-2_5

where $\boldsymbol{A}_k^* \triangleq \boldsymbol{A}^*(t_k)$, $\boldsymbol{B}_k^* \triangleq \boldsymbol{B}^*(t_k)$, $\delta\boldsymbol{x}_k \triangleq \delta\boldsymbol{x}(t_k)$, $\delta\boldsymbol{u}_k \triangleq \delta\boldsymbol{u}(t_k)$, $\boldsymbol{x}_k \triangleq \boldsymbol{x}(t_k)$, $\boldsymbol{u}_k \triangleq \boldsymbol{u}(t_k)$, $\boldsymbol{x}_k^* \triangleq \boldsymbol{x}^*(t_k)$, $\boldsymbol{u}_k^* \triangleq \boldsymbol{u}^*(t_k)$. Expanding $\delta\boldsymbol{x}_{k+1}$ recursively in the form of (5.5) for $i = k, k-1, \ldots, 1$, one can establish the sensitivity relation:

$$\begin{aligned}\delta\boldsymbol{x}_{k+1} &= (\boldsymbol{I} + \boldsymbol{A}_k^*\Delta t)\delta\boldsymbol{x}_k + \boldsymbol{B}_k^*\delta\boldsymbol{u}_k\Delta t \\ &= (\boldsymbol{I} + \boldsymbol{A}_k^*\Delta t)(\boldsymbol{I} + \boldsymbol{A}_{k-1}^*\Delta t)\delta\boldsymbol{x}_{k-1} + (\boldsymbol{I} + \boldsymbol{A}_k^*\Delta t)\boldsymbol{B}_{k-1}^*\delta\boldsymbol{u}_{k-1}\Delta t + \boldsymbol{B}_k^*\delta\boldsymbol{u}_k\Delta t \\ &= \boldsymbol{A}_k\delta\boldsymbol{x}_1 + \boldsymbol{B}_{k,1}\delta\boldsymbol{u}_1 + \boldsymbol{B}_{k,2}\delta\boldsymbol{u}_2 + \cdots + \boldsymbol{B}_{k,k}\delta\boldsymbol{u}_k \end{aligned} \tag{5.6}$$

$$\boldsymbol{A}_k = (\boldsymbol{I} + \boldsymbol{A}_k^*\Delta t)(\boldsymbol{I} + \boldsymbol{A}_{k-1}^*\Delta t)\cdots(\boldsymbol{I} + \boldsymbol{A}_1^*\Delta t) \tag{5.7}$$

$$\boldsymbol{B}_{k,i} = \frac{\boldsymbol{A}_k}{\boldsymbol{A}_i}\boldsymbol{B}_i^*\Delta t, i = k, k-1, \ldots, 1 \tag{5.8}$$

Note that the initial state increment $\delta\boldsymbol{x}_1$ can be determined via state estimation. In [133]:, $\delta\boldsymbol{x}_1$ is assumed zero since $\boldsymbol{x}_1$ is regarded as a known constant. Then (5.6) is rewritten as:

$$\delta\boldsymbol{x}_{k+1} = \boldsymbol{B}_{k,1}\delta\boldsymbol{u}_1 + \boldsymbol{B}_{k,2}\delta\boldsymbol{u}_2 + \cdots + \boldsymbol{B}_{k,k}\delta\boldsymbol{u}_k \tag{5.9}$$

In this way, the dynamic characteristics of the system can be approximately represented by the error system rather than the state-space equation.

Accordingly, the optimal control problem (OCP), denoted as P5-1, can be defined in the model predictive control framework:

$$\text{P5}-1: \quad \min_{\delta\boldsymbol{u}_1,\ldots,\delta\boldsymbol{u}_{N-1}} \sum_{k=1}^{N}\delta\boldsymbol{x}_k^T\boldsymbol{Q}\delta\boldsymbol{x}_k + \sum_{k=1}^{N-1}\delta\boldsymbol{u}_k^T\boldsymbol{R}\delta\boldsymbol{u}_k + \sum_{k=1}^{N-2}(\boldsymbol{u}_{k+1}^T - \boldsymbol{u}_k^T)\boldsymbol{R}_d(\boldsymbol{u}_{k+1} - \boldsymbol{u}_k) \tag{5.10}$$

$$\text{subject to } \delta\boldsymbol{x}_1 = 0,\ (5-9),\ \boldsymbol{g}(\boldsymbol{x}_k, \boldsymbol{u}_k) \le 0,\ \boldsymbol{h}(\boldsymbol{x}_k, \boldsymbol{u}_k) = 0, k = 1, 2, \ldots, N \tag{5.11}$$

where $\boldsymbol{x}_k = \boldsymbol{x}_k^* + \delta\boldsymbol{x}_k$, $\boldsymbol{u}_k = \boldsymbol{u}_k^* + \delta\boldsymbol{u}_k$. . In (5.11), the first two equalities denote the initial boundary condition and the approximate dynamics constraint. It should be noted that the terminal boundary condition is not involved, because it is treated as a convergence criterion after relaxation [136]. This point will be covered in a subsequent summary of the Model Predictive Control Programming (MPCP) [132] algorithm. Nonlinear functions $\boldsymbol{g}(\boldsymbol{x}, \boldsymbol{u})$ and $\boldsymbol{h}(\boldsymbol{x}, \boldsymbol{u})$ represent the inequality and equality path constraints on the state and control, respectively. In general, the two types of constraints may be convexified in MPCP or preserved in Model Predictive Static Programming (MPSP) [133]. Nevertheless, P5-1 can be solved as a SQP problem whether these constraints are convex or not, because the performance index is a quadratic function. In this performance index (5.10), the first and second terms minimize the state increments and control corrections, respectively; and the third term is imposed to smooth the control corrections. Obviously, the three quadratic terms with given positive semidefinite weighting matrices $\boldsymbol{Q}$, $\boldsymbol{R}$, and $\boldsymbol{R}_d$ are convex once the control correction and/or the state increment, which can also be expressed linearly with the control correction according to the linear sensitivity relation (5.6) or the sensitivity relation obtained later in PMPCP, are adopted as the decision variables. Therefore, the performance index is also convex as it is a linear function of convex terms. Note that the terminal control correction $\delta \boldsymbol{u}_N$ is not involved as it makes no difference to the terminal state. The MPCP algorithm is summarized in Algorithm 5.1.

Algorithm 5.1 Procedure of the MPCP

Step	Procedure
1	Initialize with the initial guess solution $\boldsymbol{u}_k^*$ $(k = 1, 2, \ldots, N-1)$, and propagate the system (5.1) using the control input $\boldsymbol{u}_k^*$ to get the reference trajectory $\boldsymbol{x}_k^*$ $(k = 1, 2, \ldots, N)$
2	Compute coefficients $\boldsymbol{A}_k$ and $\boldsymbol{B}_{k,i}$ using (5.7) and (5.8), and then solve problem P5-1 with the interior point method to obtain the control correction $\delta \boldsymbol{u}_k$ $(k = 1, 2, \ldots, N-1)$ and the state increment $\delta \boldsymbol{x}_k$, then $\boldsymbol{x}_k := \boldsymbol{x}_k^* + \delta \boldsymbol{x}_k$
3	If $\boldsymbol{x}_k$ satisfies the terminal boundary condition with the expected accuracy, i.e. $\left\|\boldsymbol{x}_N - \boldsymbol{x}_F^d\right\| \le p$ and $\sup\limits_{k=1,\ldots,N} \|\delta \boldsymbol{x}_k\|_\infty \le \varepsilon$, where $\boldsymbol{x}_F^d$ is the desired terminal state, and $p, \varepsilon > 0$ are the user-defined tolerances [136, 140], then go to Step 5, else go to Step 4
4	Let $\boldsymbol{u}_k := \boldsymbol{u}_k^* + \delta \boldsymbol{u}_k$, , propagate the system (5.1) using $\boldsymbol{u}_k$ to update $\boldsymbol{x}_k$, then $\boldsymbol{u}_k^* := \boldsymbol{u}_k$, $\boldsymbol{x}_k^* := \boldsymbol{x}_k$, go to Step 2
Step 5	Let $\boldsymbol{u}_k := \boldsymbol{u}_k^* + \delta \boldsymbol{u}_k$, then return $\boldsymbol{x}_k^* := \boldsymbol{x}_k$, $\boldsymbol{u}_k^* := \boldsymbol{u}_k$ as the numerical optimal solution and exit

5.1.2 *Pseudospectral Model Predictive Convex Programming, PMPCP*

Although the MPCP method successfully solves the constrained optimal control problem, it has at least two limitations. First, a small integration step is required to reduce the numerical integration error if the Euler rule is used to construct the sensitivity relation at evenly discretized time points. However, this strategy results in a large number of discretization nodes and a large-scale optimization problem that considerably increases the computation cost. For instance, 163 nodes are used for discretization over a 28-s flight in [132]. Second, the MPCP is initialized with the solution generated by existing toolboxes, but this step requires first converting the original problem into the form that is compatible with the toolboxes. To circumvent these limitations, the PMPCP is developed in this section to enhance the MPCP from the following aspects.

(1) **Pseudospetral Transcription**

For the PMPCP, the pseudospectral transcription technique is introduced to replace the even discretization strategy used in MPCP. Accordingly, the Euler rule is replaced by the Gaussian quadrature. Here, the Legendre–Gauss (LG) and Legendre–Gauss–Radau (LGR) pseudospectral transcriptions are adopted owing to their outstanding performance in terms of numerical accuracy. Nevertheless, only the LG and LGR transcriptions are detailed here, and other pseudospectral transcriptions can be performed analogously [134].

The LG collocation points $\{\tau_1, \tau_2, \ldots, \tau_N\}$ on the interval $(-1, 1)$, where $-1 < \tau_1$ and $\tau_N < 1$, are zeros of the N-th order Legendre polynomial $P_N(\tau)$:

$$P_N(\tau) = \frac{1}{2^N N!}\frac{d^N}{d\tau^N}[(\tau^2 - 1)^N] \tag{5.12}$$

Let the initial and terminal points $\tau_0 = -1$ and $\tau_{N+1} = 1$, then the continuous state increment $\delta\boldsymbol{x}(\tau)$ and control correction $\delta\boldsymbol{u}(\tau)$ can be approximated by the N-th order interpolating polynomials $\delta\boldsymbol{X}(\tau) \in \mathbb{R}^{n_s}$ and the (N-1)-th order interpolating polynomials $\delta\boldsymbol{U}(\tau) \in \mathbb{R}^{n_c}$, respectively:

$$\delta\boldsymbol{x}(\tau) \approx \delta\boldsymbol{X}(\tau) = \sum_{i=0}^{N} \delta\boldsymbol{X}(\tau_i)L_i(\tau),\ L_i(\tau) = \prod_{j=0, j\neq i}^{N} \frac{\tau - \tau_j}{\tau_i - \tau_j} \tag{5.13}$$

$$\delta\boldsymbol{u}(\tau) \approx \delta\boldsymbol{U}(\tau) = \sum_{i=1}^{N} \delta\boldsymbol{U}(\tau_i)\tilde{L}_i(\tau),\ \tilde{L}_i(\tau) = \prod_{j=1, j\neq i}^{N} \frac{\tau - \tau_j}{\tau_i - \tau_j} \tag{5.14}$$

where $L_i(\tau)$ and $\tilde{L}_i(\tau)$ are the bases of the Lagrange interpolating polynomials $\delta\boldsymbol{X}(\tau)$ and $\delta\boldsymbol{U}(\tau)$, respectively. They yield the following properties:

$$L_i(\tau_k),\ \tilde{L}_i(\tau_k) = \begin{cases} 1\, k = i \\ 0\, k \neq i \end{cases} \tag{5.15}$$

The point series $\{\tau_0, \tau_1, \ldots, \tau_N\}$ and $\{\tau_1, \tau_2, \ldots, \tau_N\}$ are the interpolation points for the state increment and the control correction, respectively. Note that equalities $\boldsymbol{X}(\tau_i) = \boldsymbol{x}(\tau_i)$ and $\boldsymbol{U}(\tau_i) = \boldsymbol{u}(\tau_i)$ hold at these interpolation points.

Differentiating the state increment in (5.13), one obtains:

$$\delta\dot{x}(\tau) \approx \delta\dot{X}(\tau) = \sum_{i=0}^{N} \delta X(\tau_i)\dot{L}_i(\tau) \tag{5.16}$$

where $i = 0, 1, \ldots, N$, and the derivative of the Lagrange interpolating polynomials at the LG collocation points can be determined offline:

$$D_{ki} \triangleq \dot{L}_i(\tau_k) = \sum_{l=0,l\neq i}^{N} \frac{\prod_{j=0,j\neq i,j\neq l}^{N} \tau_k - \tau_j}{\prod_{j=0,j\neq i}^{N} \tau_i - \tau_j} \tag{5.17}$$

where $k = 1, 2, \ldots, N$. Thus, the Gauss pseudospectral differentiation matrix $\boldsymbol{D}$ formed by D_{ki} is a $N \times (N+1)$ matrix.

Substituting Eq. (5.2) into Eq. (5.16) after time interval transformation, the differential equation is transcribed into algebraic equality constraints for $k = 1, 2, \ldots, N$ via the differentiation matrix:

$$\tau = \frac{2t}{t_f - t_0} - \frac{t_f + t_0}{t_f - t_0} \tag{5.18}$$

$$\sum_{i=0}^{N} \delta X(\tau_i) D_{ki} = \frac{t_f - t_0}{2}[A_k^* \delta X(\tau_k) + B_k^* \delta U(\tau_k)] \tag{5.19}$$

Let $\delta X_i \triangleq \delta X(\tau_i)$, $\delta\overline{X} = [\delta X_1; \delta X_2; \cdots; \delta X_N] \in \mathbb{R}^{Nn_s}$, $\delta U_i \triangleq \delta U(\tau_i)$, $\delta\overline{U} = [\delta U_1; \delta U_2; \cdots; \delta U_N] \in \mathbb{R}^{Nn_c}$, then these algebraic equalities in Eq. (5.19) can be rewritten as a vector equality:

$$\boldsymbol{D}_0 \delta X_0 + \boldsymbol{D}_{1:N} \delta\overline{X} = (\overline{A}^* \delta\overline{X} + \overline{B}^* \delta\overline{U}) \tag{5.20}$$

where $\boldsymbol{D}_i = [D_{1i}\boldsymbol{I}; D_{2i}\boldsymbol{I}; \cdots; D_{Ni}\boldsymbol{I}] \in \mathbb{R}^{Nn_s \times n_s}$, $\boldsymbol{I} \in \mathbb{R}^{n_s \times n_s}$ is an identity matrix, $\boldsymbol{D}_{1:N} = [\boldsymbol{D}_1, \boldsymbol{D}_2, \ldots, \boldsymbol{D}_N] \in \mathbb{R}^{Nn_s \times Nn_s}$, and

$$\overline{A}^* = \frac{t_f - t_0}{2} diag([A_1^*, \ldots, A_N^*]) \in \mathbb{R}^{Nn_s \times Nn_s} \tag{5.21}$$

$$\overline{B}^* = \frac{t_f - t_0}{2} diag([B_1^*, \ldots, B_N^*]) \in \mathbb{R}^{Nn_s \times Nn_c} \tag{5.22}$$

As a result, the state increment at LG collocation points is given by:

$$\delta\overline{X} = (\boldsymbol{D}_{1:N} - \overline{A}^*)^{-1}(\overline{B}^* \delta\overline{U} - \boldsymbol{D}_0 \delta X_0) \tag{5.23}$$

In this way, all state increments are expressed by the lumped control correction $\delta\overline{U}$ except the terminal state increment at the point τ_{N+1}, but it can be approximated via LG quadrature:

$$\begin{aligned}
\delta X_{N+1} =& \delta x_{N+1} = \delta x_0 + \int_{t_0}^{t_f} \delta\dot{x} dt \\
=& \delta x_0 + \int_{t_0}^{t_f} A^* \delta x + B^* \delta u d\tau \\
\approx& \delta X_0 + \frac{t_f - t_0}{2} \sum_{k=1}^{N} w_k (A_k^* \delta X_k + B_k^* \delta U_k) \\
=& \delta X_0 + W\overline{A}^* \delta\overline{X} + W\overline{B}^* \delta\overline{U}
\end{aligned} \tag{5.24}$$

$$W = [w_1 I, \ldots, w_N I] \in \mathbb{R}^{n_s \times N n_s},\ w_k = \frac{2}{(1 - \tau_k^2)\dot{P}_N^2(\tau_k)} \tag{5.25}$$

where $k = 1, 2, \ldots, N$, w_k is the quadrature weights associated with τ_k [134]. Substituting Eq. (5.23) into Eq. (5.24), the sensitivity relation of the terminal state increment and control corrections in PMPCP is obtained:

$$\begin{aligned}
\delta X_{N+1} =& \delta X_0 + W\overline{A}^* (D_{1:N} - \overline{A}^*)^{-1} (\overline{B}^* \delta\overline{U} - D_0 \delta X_0) + W\overline{B}^* \delta\overline{U} \\
=& [I - W\overline{A}^* (D_{1:N} - \overline{A}^*)^{-1} D_0] \delta X_0 + W[\overline{B}^* + \overline{A}^* (D_{1:N} - \overline{A}^*)^{-1} \overline{B}^*] \delta\overline{U}
\end{aligned} \tag{5.26}$$

As shown in Eq. (5.26), the sensitivity relation in PMPCP is linear because of the constant coefficient of $\delta\overline{U}$ and the known constant δX_0. Consequently, the convexity of problem P5-1 still holds if the constraint defined in Eq. (5.9) is replaced by Eq. (5.26). As such, we can analogously define the OCP (denoted as P5-2) of PMPCP:

$$\text{P5} - 2 : \min_{\delta\overline{U}} \sum_{k=1}^{N} \left(\delta X_k^T Q \delta X_k + \delta U_k^T R \delta U_k \right) + \sum_{k=1}^{N-1} (U_{k+1}^T - U_k^T) R_d (U_{k+1} - U_k) \tag{5.27}$$

$$\begin{aligned}
\text{subject to } (5-23), (5-26), \delta X_0 = x_I^d - X_0^* \ g(X_k, U_k) \le 0 \\
\delta X_{N+1} = x_F^d - X_{N+1}^* \ h(X_k, U_k) = 0
\end{aligned} \tag{5.28}$$

where x_I^d and x_F^d are the desired initial and terminal states, X_0^* and X_{N+1}^* are the initial and terminal states of the reference trajectory. Like MPCP, $\delta\overline{U}$ is also the only decision variable, which will greatly reduce the scale of optimization. Moreover, since there are linear relations (5.23) and (5.26) between the state increment and control correction, the quadratic term of the state increment with non-negative definite weighting matrix is convex, which also indicates that the performance index obtained by adding three convex functions is still a convex function about the control correction (or decision variable). The initial and terminal control corrections $\delta\boldsymbol{U}_0$ and $\delta\boldsymbol{U}_{N+1}$ are not obtained in P5-2, while they can be evaluated by the Lagrange interpolation. Also, P5-2 can be approximately solved via convex optimization after transcribing physical constraints $\boldsymbol{g}(\boldsymbol{x}, \boldsymbol{u})$ and $\boldsymbol{h}(\boldsymbol{x}, \boldsymbol{u})$ via the first-order Taylor expansion or lossless convexification.

Similarly, the error system (5.2) can be discretized at N LGR collocation points $\{\tau_1, \tau_2, \ldots, \tau_N\}$ (i.e. the N roots of $P_N(\tau) + P_{N-1}(\tau) = 0$). The state increment and control corrections along with their derivatives at points $\{\tau_1, \ldots, \tau_N, \tau_{N+1}\}$, where $\tau_1 = -1$, $\tau_N < 1$, and $\tau_{N+1} = 1$, can be evaluated as well. Since the principle using LGR collocation points is similar to that of LG points, only some key equations are presented here:

$$\sum_{i=1}^{N+1} \delta\boldsymbol{X}(\tau_i) D_{ki} = \frac{t_f - t_0}{2}[\boldsymbol{A}_k^* \delta\boldsymbol{X}(\tau_k) + \boldsymbol{B}_k^* \delta\boldsymbol{U}(\tau_k)] \tag{5.29}$$

$$\delta\boldsymbol{X}_{N+1} = \boldsymbol{D}_{N+1}^{-1}[(\overline{\boldsymbol{A}}^* - \boldsymbol{D}_{1:N})\delta\overline{\boldsymbol{X}} + \overline{\boldsymbol{B}}^* \delta\overline{\boldsymbol{U}}] \tag{5.30}$$

$$\mathrm{P5-3:} \min_{\delta\overline{\boldsymbol{X}}, \delta\overline{\boldsymbol{U}}} \sum_{k=1}^{N+1} \delta\boldsymbol{X}_k^T \boldsymbol{Q} \delta\boldsymbol{X}_k + \sum_{k=1}^{N} \delta\boldsymbol{U}_k^T \boldsymbol{R} \delta\boldsymbol{U}_k + \sum_{k=1}^{N-1} (\boldsymbol{U}_{k+1}^T - \boldsymbol{U}_k^T)\boldsymbol{R}_d(\boldsymbol{U}_{k+1} - \boldsymbol{U}_k) \tag{5.31}$$

$$\begin{aligned} \text{subject to } (5-29),\ \delta\boldsymbol{X}_0 &= \boldsymbol{x}_I^d - \boldsymbol{X}_0^*\ \boldsymbol{g}(\boldsymbol{X}_k, \boldsymbol{U}_k) \le 0 \\ \delta\boldsymbol{X}_{N+1} &= \boldsymbol{x}_F^d - \boldsymbol{X}_{N+1}^*\ \boldsymbol{h}(\boldsymbol{X}_k, \boldsymbol{U}_k) = 0 \end{aligned} \tag{5.32}$$

where the variables to be determined are $\delta\overline{\boldsymbol{X}} = [\delta\boldsymbol{X}_1; \delta\boldsymbol{X}_2; \cdots; \delta\boldsymbol{X}_N]$ and $\delta\overline{\boldsymbol{U}} = [\delta\boldsymbol{U}_1; \delta\boldsymbol{U}_2; \cdots; \delta\boldsymbol{U}_N]$, which is different from the problem P5-2. The reason is that the variables $\delta\overline{\boldsymbol{X}}$ and $\delta\boldsymbol{X}_{N+1}$ can be expressed by $\delta\overline{\boldsymbol{U}}$ in LG-PMPCP according to (5.23) and (5.26), while $\delta\overline{\boldsymbol{X}}$, $\delta\boldsymbol{X}_{N+1}$, and $\delta\overline{\boldsymbol{U}}$ are coupled in LGR-PMPCP as shown in (5.29) and (5.30). Nevertheless, the convexity of the performance index (5.31) still holds as it is essentially a non-negative quadratic function about the lumped decision variable $[\delta\overline{\boldsymbol{X}}; \delta\overline{\boldsymbol{U}}]$. Furthermore, it should be noted that not all pseudospectral methods can obtain the sensitivity relation in MPSP and MPCP. In other words, only pseudospectral methods such as LGL, CGL, and LG pseudospectral methods whose state variables can be completely parameterized by control variables can establish sensitivity relations like MPSP or MPCP.

(2) **Reference Trajectory Update**

It is important to determine the initial guess trajectory in initialization and the reference trajectory in subsequent optimizations for a sequential programming algorithm. In MPCP, the initial reference trajectory for the first iteration is generated by using the Proportional Navigation law or the open-source software *Falcon.m* [132]. However, this strategy requires the preliminary calculation that adds additional computation cost. In fact, it is feasible and reliable to initialize the MPCP or PMPCP with the trajectory generated by a constant control input owing to the insensitivity of the sequential convex optimization to the initial guess trajectory. Moreover, this sensitivity can be further reduced due to the introduction of the LG or LGR pseudospectral discretization technique, which preserves the continuity of dynamics as much as possible by reducing the discretization error of the sensitivity relation. Benefiting from the above advantage, the PMPCP can even be initialized with a linear interpolation trajectory between the desirable initial and the terminal states. Even so, the trajectory generated by integrating the nonlinear system with a constant control input is adopted here as the initial reference trajectory to avoid the unnecessary iteration steps in optimization.

There are also differences in the reference trajectory (i.e. the reference state) update for the sequential iteration process due to the different discretization strategies in the PMCP and PMPCP. In SCP and PSCP, the reference trajectory is directly updated by adopting the optimized state and control in the last iteration, but this strategy accumulates the discretization error in iterations and may result in the failure of algorithm [135]. In MPCP, the reference trajectory is updated by forward propagating the original nonlinear dynamic system (5.1) with the optimized control input and Euler rule, while in PMPCP, the interpolation procedure is performed at the non-uniform pseudospectral collocation points to improve the convergence. Intuitively, the reference trajectory for next iteration can be obtained through interpolating the state series generated by propagating the original nonlinear dynamic system with the current control input series and the Runge–Kutta rule. However, the integration or propagation accuracy cannot be guaranteed due to the sparsity of collocation points in the middle of integral interval. On the other hand, if using a large number of collocation points in Lagrange interpolation to avoid the above sparsity issue, it will increase computation complexity for interpolation and also incur numerical inaccuracy or even instability. Therefore, a three-step procedure is introduced. First, the Lagrange interpolation is employed to transcribe these control input series at the pseudospectral collocation points. Then, generate an evenly distributed control input series from this interpolated continuous control, which have denser points than the pseudospectral collocation points, especially in the middle of the integral interval. Subsequently, the original nonlinear dynamic system (5.1) is propagated in the current time interval by using the classic Runge–Kutta method and the evenly distributed control input series to generate an evenly distributed state series. Finally, estimate these state series at the LG or LGR points (i.e. the reference trajectory) for the next iteration by interpolating these evenly distributed state series. In this way, the three-step update strategy has two advantages in facilitating the convergence. On the one hand, the Runge–Kutta method used in PMPCP is more precise than the Euler rule employed in MPCP. On the other hand, the sensitivity relation in MPCP is also established based on the Euler rule. This relation is then used to optimize the control input. This strategy suggests that the quality of the reference trajectory obtained by propagating the original dynamics using the optimized control input, particularly its nonlinearity, may degrade. However, the nonlinearity of the discretized dynamics is improved in PMPCP owing to the pseudospectral discretization. As a result, the optimized control input and propagated reference trajectory preserve the nonlinear characteristics of the original dynamics as much as possible.

(3) **Characteristics of PMPCP**

In view of the PMPCP originating from the MPSP and MPCP methods, the theoretical properties of the PMPCP and existing methods are presented and compared in Table 5.1 to highlight the similarities and differences among those approaches. These comparisons manifest two benefits of PMPCP. On the one hand, the pseudospectral discretization guarantees the accuracy of the numerical solution and the efficiency of optimization. Convex optimization, on the other hand, makes it simple to cope with the path constraint.

5.1.3 Implementation of PMPCP

The procedure for implementing the PMPCP is summarized in Table 5.2.

5.1.4 Numerical Simulations

In order to verify the advantages of the LG-PMPCP and LGR-PMPCP methods in terms of computational efficiency and numerical accuracy, the two methods are assessed in this section. For the Mars entry trajectory optimization problem, the dimensionless dynamics is given by [91, 145]:

Table 5.1 Theoretical characteristics of compared methods [136–144]

Methods	Continuous or discrete	Quadrature rule	Programming approach	Parameterization of control	Path (or state) constraint	Control constraint
MPSP	Discrete	Euler	Static	Power series	Not mentioned or partial	Fully
QS-MPSP	Discrete	Euler	Static	Power series	Not mentioned or partial	Fully
C-QS-MPSP	Discrete	Euler	Static	Power series	Fully	Fully
G-QS-MPSP	Continuous	Gauss–Lobatto type	Static	Pseudospectral method (LGL)	Not mentioned	Fully
(M)-LPMPC	Discrete	Gauss type	Static	Pseudospectral method (LG)	Not mentioned	Fully
MPCP	Discrete	Euler	Convex	Power series, or trigonometric function	Fully	Fully
LG-PMPCP	Discrete	Gauss type	Convex	Pseudospectral method (LG)	Fully	Fully
LGR-PMPCP	Discrete	Gauss- Legendre type	Convex	Pseudospectral method (LGR)	Fully	Fully

Table 5.2 Procedure of the PMPCP

Step	Procedure
1	Initialize the system (5.1) with an initial state $\boldsymbol{X}_0$, then propagate this system using constant control inputs $\boldsymbol{U}_k^*$ ($k = 0, 1, \ldots, N$ for LG or $k = 1, 2, \ldots, N$ for LGR) to obtain the reference state $\boldsymbol{X}_k^*$ ($k = 0, 1, \ldots, N+1$ for LG or $k = 1, 2, \ldots, N+1$ for LGR)
2	Compute the coefficients using (5.21), (5.22) and (5.25) for LG-based error system (5.23) and (5.26). Solve problem P5-2 or P5-3 with the interior point method to obtain $\delta\boldsymbol{U}_k$ ($k = 1, 2, \ldots, N$ for LG or $k = 1, 2, \ldots, N+1$ for LGR) and $\delta\boldsymbol{X}_k$ ($k = 0, 1, \ldots, N$ for LG or $k = 1, 2, \ldots, N+1$ for LGR), then $\boldsymbol{X}_k := \boldsymbol{X}_k^* + \delta\boldsymbol{X}_k$
3	If $\sup\limits_{k=1,\ldots,N} \Vert\delta\boldsymbol{X}_k\Vert_\infty \le \varepsilon$ for LG or $\sup\limits_{k=1,\ldots,N+1} \Vert\delta\boldsymbol{X}_k\Vert_\infty \le \varepsilon$ for LGR, then go to Step 5, else go to Step 4
4	Let $\boldsymbol{U}_k := \boldsymbol{U}_k^* + \delta\boldsymbol{U}_k$, update $\boldsymbol{X}_k$ with $\boldsymbol{U}_k$ via the strategy discussed in Sect. 5.1.2, then $\boldsymbol{U}_k^* := \boldsymbol{U}_k$, $\boldsymbol{X}_k^* := \boldsymbol{X}_k$, go to Step 2
Step 5	Let $\boldsymbol{U}_k := \boldsymbol{U}_k^* + \delta\boldsymbol{U}_k$, return $\boldsymbol{X}_k^* := \boldsymbol{X}_k$ and $\boldsymbol{U}_k^* := \boldsymbol{U}_k$ as the numerical optimal solution and exit

$$
\begin{aligned}
\dot{r} &= V \sin\gamma \\
\dot{\theta} &= V \cos\gamma \sin\psi / r\cos\phi \\
\dot{\phi} &= V \cos\gamma \cos\psi / r \\
\dot{V} &= -D - \sin\gamma / r^2 \\
\dot{\gamma} &= L\cos\sigma / V + (V/r - 1/r^2 V)\cos\gamma \\
\dot{\psi} &= L\sin\sigma / V\cos\gamma + V\cos\gamma\sin\psi\tan\phi / r \\
\dot{\sigma} &= u
\end{aligned}
\tag{5.33}
$$

If the state $x = [r, \theta, \phi, V, \gamma, \psi, \sigma]^{\mathrm{T}}$, is defined, the dimensionless dynamics model can be rewritten in the form of a general dynamical system as

$$
\dot{\boldsymbol{x}} = \boldsymbol{f}(\boldsymbol{x}, u, t) \tag{5.34}
$$

At this point, the matrix $\boldsymbol{B} = \partial\boldsymbol{f}/\partial u = [0, 0, 0, 0, 0, 0, 1]^{\mathrm{T}}$ and

$$
\boldsymbol{A} = \partial\boldsymbol{f}/\partial\boldsymbol{x} \tag{5.35}
$$

$$
\boldsymbol{A} = \frac{\partial \boldsymbol{f}}{\partial \boldsymbol{x}} = \begin{bmatrix} 0 & 0 & 0 & a_{14} & a_{15} & 0 & 0 \\ a_{21} & 0 & a_{23} & a_{24} & a_{25} & a_{26} & 0 \\ a_{31} & 0 & 0 & a_{34} & a_{35} & a_{36} & 0 \\ a_{41} & 0 & 0 & a_{44} & a_{45} & 0 & 0 \\ a_{51} & 0 & 0 & a_{54} & a_{55} & 0 & 0 \\ a_{61} & 0 & a_{63} & a_{64} & a_{65} & a_{66} & 0 \\ 0 & 0 & 0 & 0 & 0 & 0 & 0 \end{bmatrix}
$$

$$
a_{14} = \sin\gamma,\ a_{15} = V\cos\gamma,\ a_{21} = -\frac{V\cos\gamma\sin\psi}{r^2\cos\phi},\ a_{23} = \frac{V\cos\gamma\sin\psi\sin\phi}{r\cos^2\phi},
$$

$$
a_{24} = \frac{\cos\gamma\sin\psi}{r\cos\phi},\ a_{25} = -\frac{V\sin\gamma\sin\psi}{r\cos\phi},\ a_{26} = \frac{V\cos\gamma\cos\psi}{r\cos\phi},
$$

$$
a_{31} = -\frac{V\cos\gamma\cos\psi}{r^2},\ a_{34} = \frac{\cos\gamma\cos\psi}{r},\ a_{35} = -\frac{V\sin\gamma\cos\psi}{r},
$$

$$
a_{36} = -\frac{V\cos\gamma\sin\psi}{r},\ a_{41} = -D_r + \frac{2\sin\gamma}{r^3},\ a_{44} = -D_V,
$$

$$
a_{45} = -\frac{\cos\gamma}{r^2},\ a_{51} = \frac{L_r\cos\sigma}{V} + \frac{2\cos\gamma}{r^3 V} - \frac{V\cos\gamma}{r^2},
$$

$$
a_{54} = \frac{L_V\cos\sigma}{V} - \frac{L\cos\sigma}{V^2} + \frac{\cos\gamma}{r^2V^2} + \frac{\cos\gamma}{r},\ a_{55} = \frac{\sin\gamma}{r^2V} - \frac{V\sin\gamma}{r},
$$

$$
a_{61} = \frac{L_r\sin\sigma}{V\cos\gamma} - \frac{V\cos\gamma\sin\psi\tan\phi}{r^2},\ a_{63} = \frac{V\cos\gamma\sin\psi}{r\cos^2\phi},
$$

$$
a_{64} = \frac{L_V\sin\sigma}{V\cos\gamma} - \frac{L\sin\sigma}{V^2\cos\gamma} + \frac{\cos\gamma\sin\psi\tan\phi}{r},\ a_{65} = \frac{L\sin\sigma\sin\gamma}{V\cos^2\gamma} - \frac{V\sin\gamma\sin\psi\tan\phi}{r},
$$

$$
a_{66} = \frac{V\cos\gamma\cos\psi\tan\phi}{r},\ D_r = -\frac{R_0}{h_s}D,\ D_V = \frac{R_0\rho_0 e^{-R_0(r-1)/h_s} V S_r C_D}{m},
$$

$$
L_r = -\frac{R_0}{h_s}L,\ L_V = \frac{R_0\rho_0 e^{-R_0(r-1)/h_s} V S_r C_L}{m}.
$$

The YALMIP toolbox [130] and MOSEK solver [131] are used to formulate and solve the convex optimization problem, respectively. Since the PMPCP method uses the state error system, it can only optimize the state increment and the control correction, and cannot directly handle the state or control variables. Therefore, the simulation in this section considers the control correction or control energy optimization problem, and the final time is fixed to 360 s.

In simulations, the parameters are given as follows: $R_0 = 3397.2$ km, $g_0 = 3.711$ m/s^2, $\rho_0 = 0.0158$ kg/m^3, $h_s = 9354.5$ m, $m = 2804$ kg, $S_r = 15.9$ m^2, $k_Q = 1.9027 \times 10^{-4}$, $R_n = 6.476$, $C_L = 0.36$, $C_D = 1.45$, $\dot{Q}_{\max} = 70$ W/cm^2, $q_{\max} = 8.5$ kPa, $a_{\max} = 18.5\ g_0$. The initial values and allowable variation ranges of the remaining state variables are shown in Table 4.1.

In this section, a comparison is made between the results obtained using the PMPCP method and those obtained using the Model Predictive Convex Programming (MPCP) method and the SCP method. Given the nonlinearity of the Mars entry problem, the virtual control is incorporated into the PMPCP, SCP, and MPCP methods to mitigate the effects of linearization errors and enhance the convergence performance of the algorithm. Consequently, a penalty term in the form of a 1-norm is added to the objective function. Additionally, details of the SCP method can be found in Sect. 4.2. Moreover, due to the direct manipulation of state and control variables in the SCP method, the control energy optimal control problem is formulated.

$$For\ MPCP\ and\ PMPCP : \underset{\delta\overline{U},\overline{v}}{minnize}\ w_v \|\overline{v}\|_1 + \sum_{k=1}^{N} \delta U_k^T \delta U_k \tag{5.36}$$

$$For\ SCP : \underset{\delta\overline{X},\delta\overline{U},\overline{v}}{mininize}\ w_v \|\overline{\boldsymbol{v}}\|_1 + \sum_{k=1}^{N} \boldsymbol{U}_k^T \boldsymbol{U}_k \tag{5.37}$$

where w_v is set to 10^6 and 10^8 in the MPCP (or PMPCP) and SCP, respectively.

The initial trajectory for all methods is determined by integrating the original nonlinear system using a constant bank angle of $-30°$. Furthermore, the convergence criterion for all methods is defined in Eq. (4.41). It is important to note that although the virtual control is not considered in the convergence criterion, it does not affect the accuracy of the convergence solution. In the simulation, both MPCP and SCP use 181 nodes, whereas PMPCP employs 60 collocation points, resulting in 62 nodes for the LG-PMPCP method and 61 nodes for the LGR-PMPCP method. Additionally, in order to demonstrate the performance improvement achieved by the reference trajectory update strategy, two simulation cases are conducted in this section. The first case updates the reference trajectory using the results from the previous iteration in MPCP and SCP, while the second case considers the results of numerical integration as the reference trajectory in PMPCP.

(1) **Updating the Reference Trajectory Directly**

Figure 5.1 shows the simulation results if the reference trajectory is updated using the results from the last iteration.

From Fig. 5.1a, b, it is evident that the optimal flight paths obtained by the four algorithms exhibit remarkable similarity, with closely aligned profiles in terms of altitude, speed, flight path angle, and azimuth angle. Figure 5.1c reveals that the optimal bank angle and its rate profiles obtained through the MPCP and PMPCP methods are highly comparable, displaying a pattern resembling Bang-Bang control. However, due to differences in the objective functions defined in Eqs. (5.36) and (5.37) in contrast to the other two methods, the control profile of the SCP method exhibits a consistent trend but deviates somewhat from their results. Additionally, the bank angle rate of the PMPCP method displays a zigzag pattern, whereas the corresponding curves of the MPCP and SCP methods demonstrate smoother variations. This phenomenon can be attributed to the utilization of Lagrange interpolation in the pseudospectral method. Although this method offers high accuracy in approximating smooth functions, it tends to oscillate when dealing with constant value functions. As evident from Fig. 5.1c, the bank angle exhibits Bang-Bang control characteristics, necessitating a switch between two constant values. Consequently, the bank angle rate profile of the PMPCP method displays oscillatory behavior. Moreover, Fig. 5.1e illustrates the control energy profile. The results indicate that the total control energy required by MPCP and SCP is higher. However, this is primarily due to the usage of a large number of uniformly distributed discrete nodes in these two methods, leading to a significant accumulation in the objective functions. Regarding path constraints, all four methods successfully fulfill them.

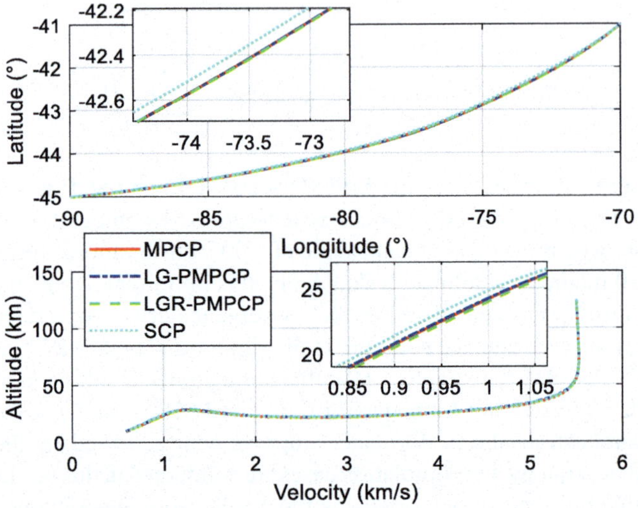

(a) Profiles of longitude versus latitude and velocity versus altitude

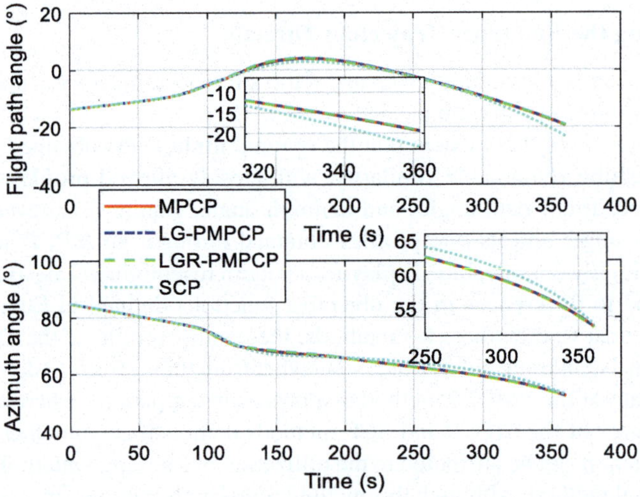

(b) Profiles of flight path angle and azimuth angle

Fig. 5.1 The Mars entry optimal solution obtained by updating the reference trajectory directly

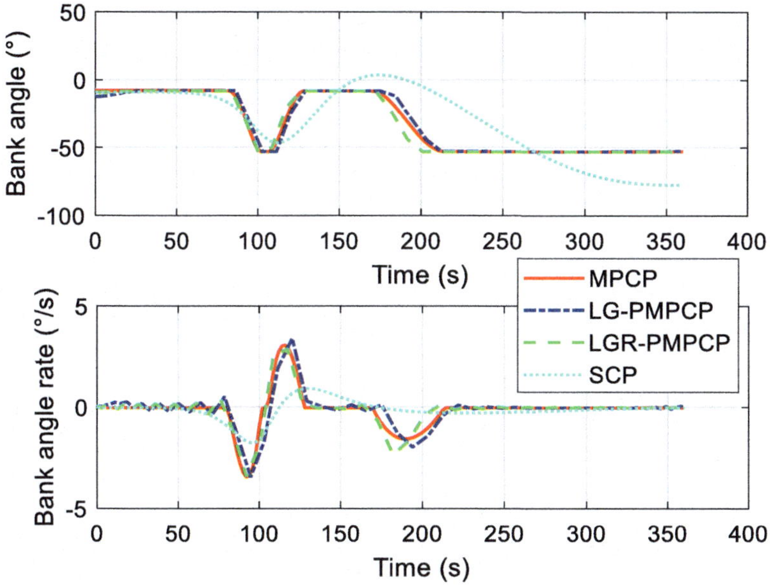

(c) Profiles of bank angle and its rate

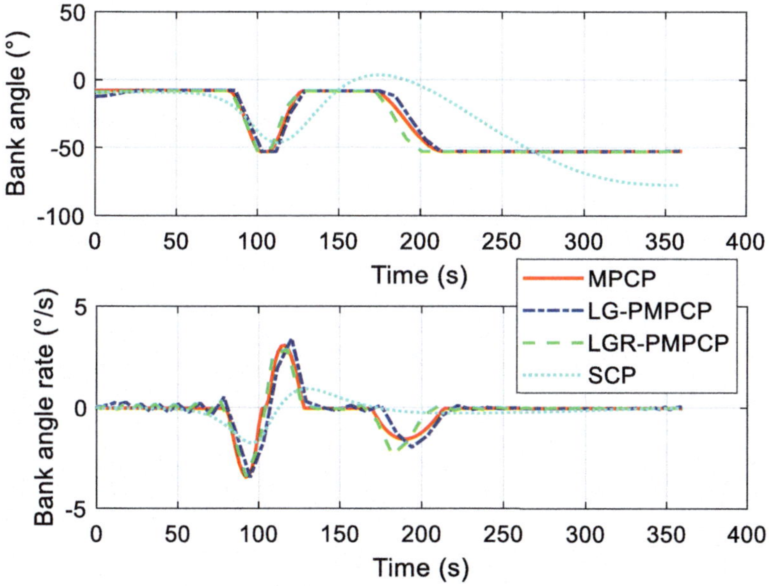

(d) Profiles of path constraints

Fig. 5.1 (continued)

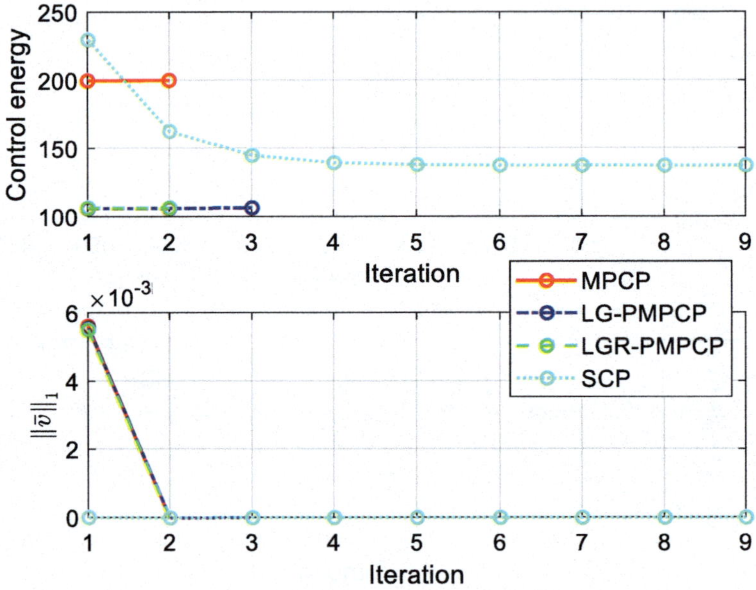

(e) Profiles of control energy and vritual control

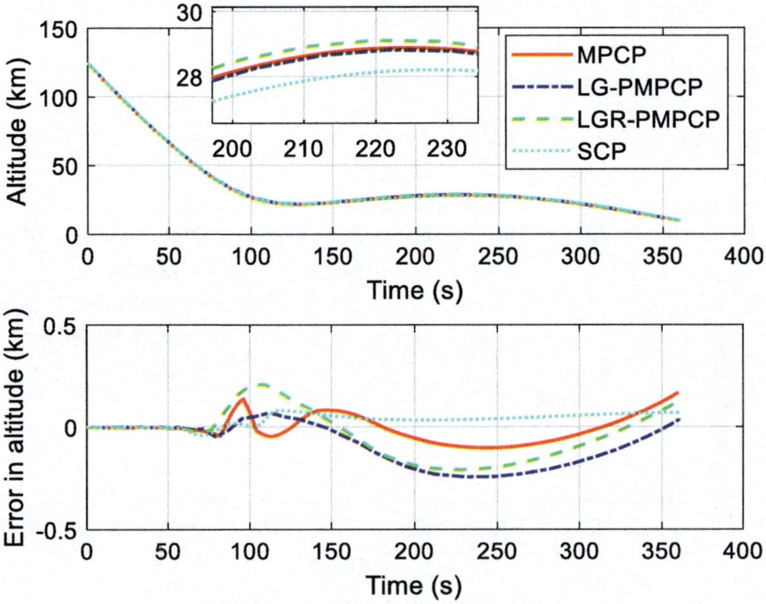

(f) Profiles of altitude and errors in altitude

Fig. 5.1 (continued)

Another aspect that requires a thorough explanation is the reliability of the optimization algorithm. Given that all four methods mentioned above rely on convex optimization, the resulting optimized control profile represents an approximate solution to the original problem. To assess the reliability of this approximate solution to the theoretical optimal solution, a commonly employed approach involves utilizing the optimized optimal control profile as the control input to integrate the original nonlinear system, and subsequently comparing the integrated trajectory with the optimized trajectory. If these trajectories exhibit similarity, it indicates that the optimized control profile is sufficiently close to the theoretical optimal solution. Hence, Fig. 5.1f illustrates the disparity between the optimal trajectories obtained through the four methods and the optimal integrated trajectory specifically in the altitude channel. It is evident that there exists minimal deviation between MPCP and PMPCP in terms of the maximum height error $\Delta|h|_{\max}$, thereby failing to adequately demonstrate the advantages of the pseudospectral discretization scheme. Conversely, SCP exhibits the smallest height error; however, achieving a solution below the convergence threshold requires multiple optimization iterations for SCP, resulting in a calculation time that is more than twice as long as the other methods.

To provide a more intuitive assessment of the performance of the aforementioned four algorithms, the corresponding key parameters are presented in Table 5.3. The number of iterations required for convergence is small for the four algorithms. However, the utilization of pseudospectral method reduces the size of the optimization problems, thereby reducing the computational time of the PMPCP method. Furthermore, it is noteworthy that the magnitude of virtual control during convergence is negligible for all four methods. This observation suggests that the linearized dynamics of the convex optimization problem closely approximate the original nonlinear dynamics.

(2) **Updating the Reference Trajectory by Numerical Integration**

When the reference trajectory is updated directly, the simulation solely demonstrates the enhanced computational efficiency of the PMPCP method, without any improvement in the numerical accuracy of the converged solution. However, if the reference trajectory is updated through numerical integration, the PMPCP method exhibits a substantial enhancement in the accuracy of the converged solution values, as illustrated in Fig. 5.2.

Table 5.3 Comparison of methods when updating the reference trajectory directly

Variables	MPCP	LG-PMPCP	LGR-PMPCP	SCP
h_f(km)	10	10	10	10
θ_f(°)	−70	−70	−70	−70
ϕ_f(°)	−41	−41	−41	−41
V_f(m/s)	588.3438	587.2247	590.8324	596.7975
γ_f(°)	−19.2368	−19.2801	−19.1656	−23.1555
ψ_f(°)	52.4264	52.3878	52.3425	52.7142
σ_f(°)	−52.2109	−51.8855	−52.2656	−77.0031
$\|\dot{\sigma}\|_{max}$/(°/s)	3.4114	3.4374	3.3873	1.3728
$\dot{Q}_{max}$(W/cm^2)	69.7770	68.5019	69.4345	69.5343
q_{max}(kPa)	8.0974	8.0974	8.0974	8.0974
a_{max}(g_0)	18.4856	18.4856	18.4856	18.4856
CPU time(s)	3.2741	3.1128	2.5561	8.4159
Iterations	2	3	2	9
$\Delta\|h\|_{max}$(km)	0.1722	0.2415	0.2109	0.0825
Final $\|\|\boldsymbol{v}\|\|_1$	2.4×10^{-12}	5.8×10^{-14}	3.8×10^{-14}	2.9×10^{-12}

Table 5.4 presents the key parameters for the four methods. Comparatively, the approach of updating the reference trajectory through numerical integration also yields a favorable trajectory optimization outcome. This includes meeting the path constraints and constraints on state and control variables. Importantly, the results obtained by the four methods are very close to those obtained by direct reference trajectory updating.

In the case of the SCP method, updating the reference trajectory through numerical integration does not yield a significant improvement in the numerical accuracy of the convergence solution. The maximum height error only decreases from 0.08309 km to 0.07954 km, the number of iterations reduces from 9 to 7, and the calculation time is 7.6805 s. In general, different methods for updating the reference trajectory have a tiny impact on the convergence solution of the SCP method. However, for the MPCP and PMPCP methods, updating the reference trajectory via numerical integration markedly enhances the accuracy of the convergence solution. The maximum height error of MPCP is reduced to 0.05327 km, whereas LG-PMPCP and LGR-PMPCP achieve maximum altitude errors of 0.03162 km and 0.009845 km, respectively. These values are significantly smaller than those obtained by directly updating the reference trajectory. Nevertheless, updating the reference trajectory through numerical integration does lead to an increase in the number of sequence optimization iterations for MPCP and PMPCP methods, resulting in a slight increase in computational requirements. However, the CPU time required for the PMPCP method remains lower than that of the MPCP method and SCP method.

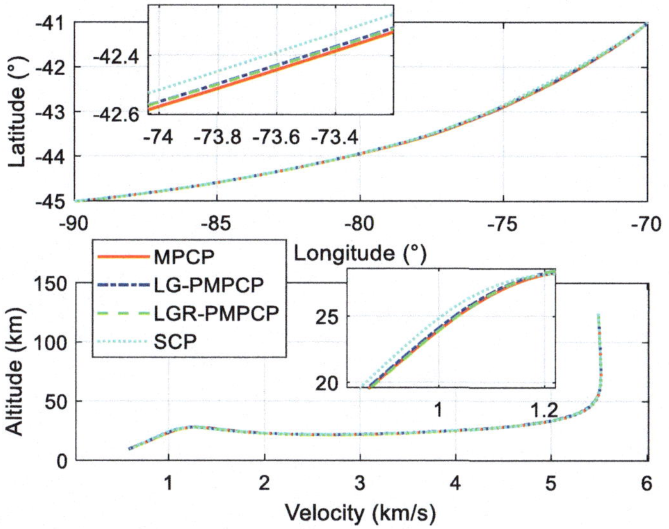

(a) Profiles of longitude versus latitude and velocity versus altitude

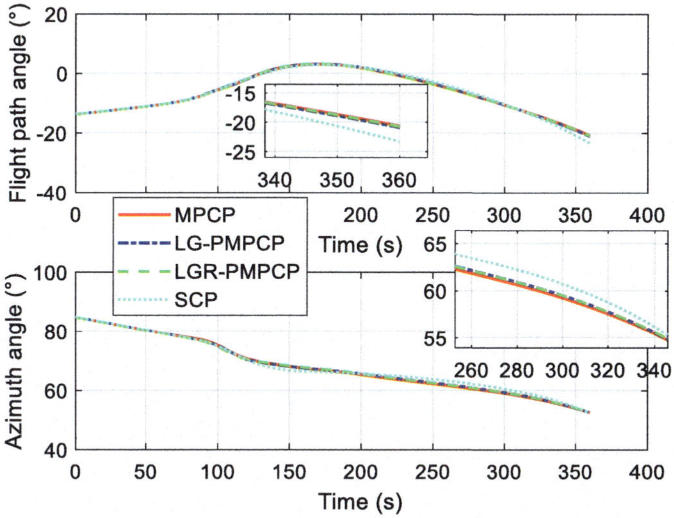

(b) Profiles of flight path angle and azimuth angle

Fig. 5.2 The Mars entry optimal solution obtained by updating the reference trajectory with the propagation solution

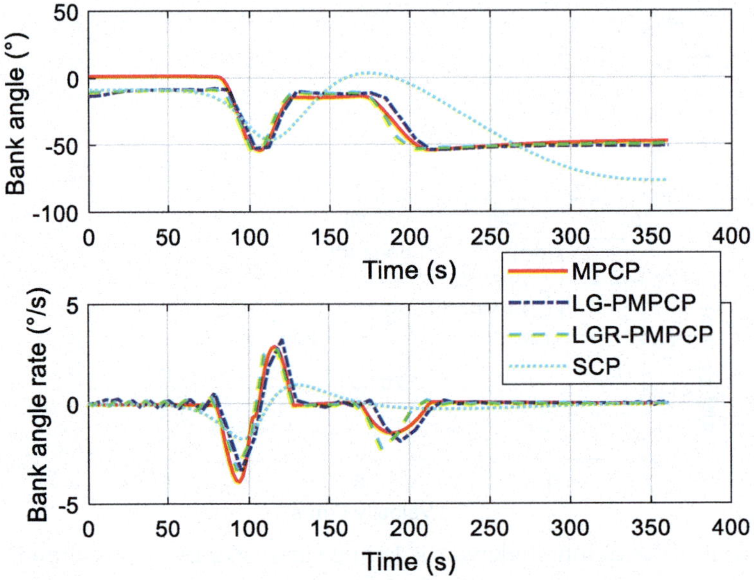

(c) Profiles of bank angle and its rate

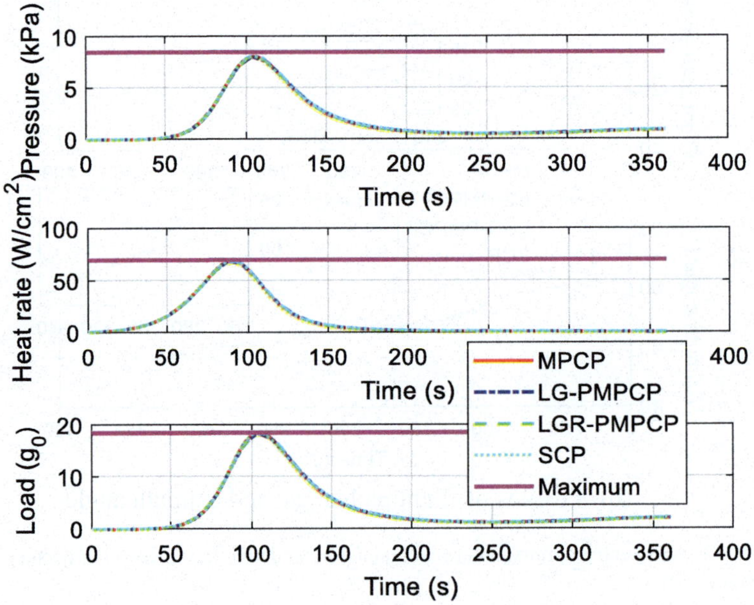

(d) Profiles of path constraints

Fig. 5.2 (continued)

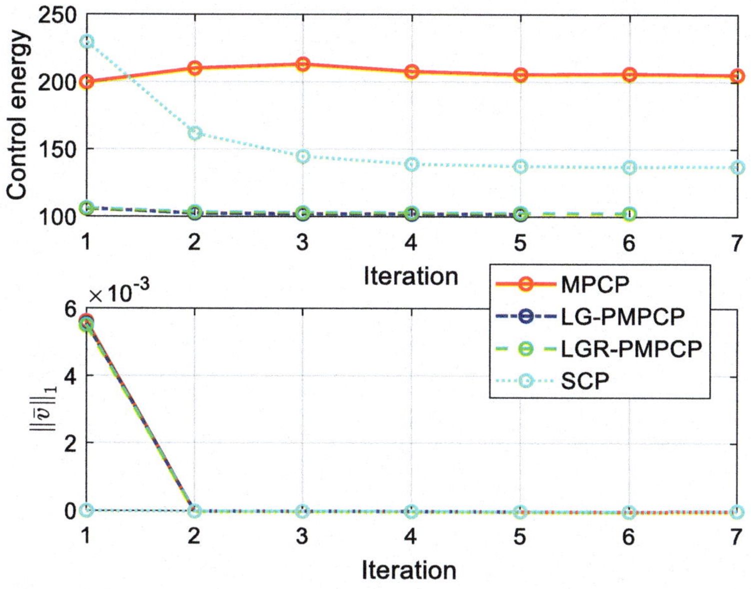

(e) Profiles of control energy and vritual control

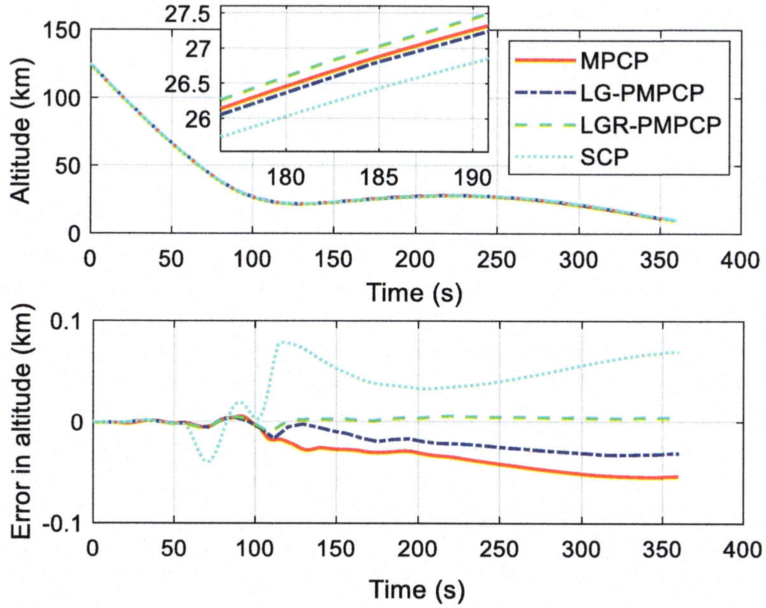

(f) Profiles of altitude and errors in altitude

Fig. 5.2 (continued)

Table 5.4 Comparison of methods when updating reference trajectory by numerical integration

Variables	MPCP	LG-PMPCP	LGR-PMPCP	SCP
h_f(km)	10	10	10	10
θ_f(°)	−70	−70	−70	−70
ϕ_f(°)	−41	−41	−41	−41
V_f(m/s)	581.4563	584.1781	584.5970	596.8115
γ_f(°)	−20.5131	−20.8223	−20.6040	−23.1220
ψ_f(°)	52.8139	52.7833	52.8806	52.6942
σ_f(°)	−46.9666	−50.55503	−48.7798	−76.5911
$\lvert\dot{\sigma}\rvert_{max}$/(°/s)	3.8848	3.3374	3.3265	1.3695
$\dot{Q}_{max}$(W/cm^2)	69.4625	68.4927	69.0461	69.5324
q_{max}(kPa)	8.0849	8.0445	7.9972	8.0974
$a_{max}(g_0)$	18.4572	18.3649	18.2569	18.4856
CPU time(s)	4.5051	4.0026	4.2380	7.6805
Iteration	7	5	6	7
$\Delta\lvert h\rvert_{max}$ (km)	0.05327	0.03162	0.009845	0.07954
Final $\lVert\boldsymbol{v}\rVert_1$	1.2×10^{-12}	3.3×10^{-11}	3.8×10^{-12}	2.5×10^{-12}

5.2 Mapped Chebyshev Pseudospectral Model Predictive Convex Programming

As described in Sect. 5.1, the Legendre pseudospectral model predictive convex optimization method enhance improves the existing model prediction convex optimization approach from two perspectives: the pseudospectral discretization and the reference trajectory updating strategy. Consequently, it improves both the convergence efficiency and the accuracy of the optimal solution. However, the pseudospectral method using Lagrange interpolation tends to exhibit oscillations when approximating a constant control variable. To alleviate this problem, this section proposes a predictive convex optimization method based on the mapped Chebyshev pseudospectral method [146, 147], which is referred to as MCGL-PMPCP in this section. By incorporating conformal mapping and barycenter rational Lagrange interpolation, the numerical stability of the Lagrange interpolation to the constant function is improved. This method effectively reduces the oscillation of the control while ensuring the accuracy of the solution.

Given the interpolation node $-1 = \tau_0 < \cdots < \tau_N = 1$ and the corresponding function value $f(\tau_0), \ldots, f(\tau_{N+1})$, the function $f(\tau)$ can be approximated by the Lagrange interpolation polynomial $f_N(\tau)$ of order N [148]:

$$f_N(\tau) = \sum_{i=0}^{N} f(\tau_i) L_i(\tau) \tag{5.38}$$

where the Lagrange interpolation basis function $L_i(\tau)$ satisfies:

$$L_i(\tau) = \prod_{j=0, j\neq i}^{N} \frac{\tau - \tau_j}{\tau_i - \tau_j}, \ L_i(\tau_k) = \begin{cases} 1\, k = i \\ 0\, k \neq i \end{cases} \tag{5.39}$$

At the same time, the interpolation condition $f_N(\tau_k) = f(\tau_k)$, $k = 0, 1, \ldots, N$ is satisfied.

Let $l(\tau) = (\tau - \tau_0)(\tau - \tau_1) \cdots (\tau - \tau_N)$ and define

$$w_i = \frac{1}{\prod_{i\neq j} (\tau_i - \tau_j)}, \ i = 1, 2, \ldots, N \tag{5.40}$$

then one has $w_i = 1\big/ l'(\tau_i)$. At this point, the Lagrange interpolation basis function $L_i(\tau)$ can be rewritten as:

$$L_i(\tau) = \prod_{j=0, j\neq i}^{N} \frac{\tau - \tau_j}{\tau_i - \tau_j} = l(\tau)\frac{w_i}{\tau - \tau_i} \tag{5.41}$$

In this case, the Eq. (5.38) can be further rewritten as:

$$f_N(\tau) = \sum_{i=0}^{N} f(\tau_i)L_i(\tau) = l(\tau)\sum_{i=0}^{N} f(\tau_i)\frac{w_i}{\tau - \tau_i} \tag{5.42}$$

which is the improved Lagrange interpolation formula.

If the Eq. (5.42) is interpolated with the constant function 1, that is, the function value $f(\tau_0), \ldots, f(\tau_{N+1}) \equiv 1$, then:

$$1 = \sum_{i=0}^{N} f(\tau_i)L_i(\tau) = l(\tau)\sum_{i=0}^{N} \frac{w_i}{\tau - \tau_i} \tag{5.43}$$

Divide both sides of Eq. (5.42) by each side of Eq. (5.43) to get:

$$f_N(\tau) = \frac{\sum_{i=0}^{N} f(\tau_i)\frac{w_i}{\tau - \tau_i}}{\sum_{i=0}^{N} \frac{w_i}{\tau - \tau_i}} \tag{5.44}$$

which is the barycentric Lagrange interpolation and has high numerical stability for pseudospectral discrete nodes.

5.2.1 Synthesized Algorithm

The standard pseudospectral method in CGL is known for its high numerical accuracy. However, when an excessive number of nodes are present, the pseudospectral differential matrix becomes susceptible to ill-conditioning. Consequently, significant errors arise at both ends of the interpolation interval, leading to a decrease in the precision of the optimized numerical solution. To tackle this problem, this section introduces a conformal mapping technique that transforms the non-uniform CGL nodes into uniformly mapped CGL nodes. This technique expands the analytical region of the function to be approximated, thereby improving the ill-conditioned nature of the CGL differential matrix.

Similar to the standard CGL pseudospectral method, the mapped CGL pseudospectral method also requires the specification of interpolation nodes and interpolation weights. The key distinction lies in the selection of interpolation nodes. The mapped CGL pseudospectral method employs the Kosloff-Tal-Eaer conformal mapping method [149] to transform the coordinates of the second type of CGL nodes, thereby obtaining the desired interpolation nodes:

$$\lambda_k = \frac{\arcsin(\alpha \tau_k)}{\arcsin(\alpha)}, \tau_k = \cos\left(\frac{\pi k}{N}\right), k = 0, 1, \ldots, N \tag{5.45}$$

where $0 \leq \alpha < 1$ represents conformal mapping parameters and characterizes the uniformity of mapping CGL collocation points. Figure 5.3 shows the mapped CGL collocation points corresponding to different α. It can be seen that when the value of α is closer to 1, the distribution of mapping CGL coordination points is more uniform.

Kosloff and Tal-Ezer provide the following selection procedure for the conformal mapping parameter α in order to guarantee the numerical correctness of the mapping CGL pseudo-spectral approach:

$$\alpha = 2\big/\left(t + t^{-1}\right), t = \varepsilon^{-1/N} \tag{5.46}$$

where ε is the expected numerical precision of calculation, N is the number of collocation points. The Eq. (5.46) can also be expressed as:

$$\alpha = \operatorname{sech}(|\ln \varepsilon|\big/N) \tag{5.47}$$

Further, the mapping CGL pseudospectral method uses the barycenter Lagrange interpolation method to approximate the state increment and control correction, which can reduce the oscillation of the interpolation solution at both ends and ensure the approximate accuracy of the variables:

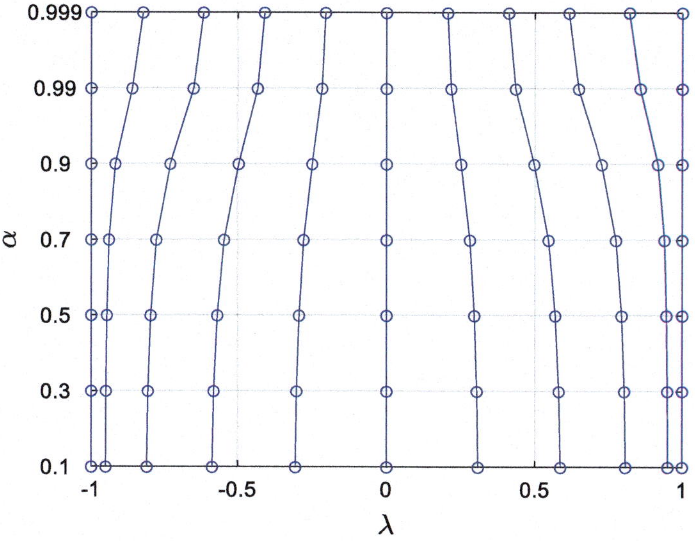

Fig. 5.3 The locations of CGL collocations under different α

$$\delta \boldsymbol{x}(\lambda) \approx \delta \boldsymbol{X}(\lambda) = \frac{\sum_{i=0}^{N} \delta \boldsymbol{x}(\lambda_i) \frac{w_i}{\lambda - \lambda_i}}{\sum_{i=0}^{N} \frac{w_i}{\lambda - \lambda_i}}, \delta \boldsymbol{u}(\lambda) \approx \delta \boldsymbol{U}(\lambda) = \frac{\sum_{i=0}^{N} \delta \boldsymbol{u}(\lambda_i) \frac{w_i}{\lambda - \lambda_i}}{\sum_{i=0}^{N} \frac{w_i}{\lambda - \lambda_i}} \tag{5.48}$$

where the weights of CGL collocation points are given as:

$$w_0 = 0.5, w_N = (-0.5)^N, w_k = (-1)^k, k = 1, 2, \ldots, N - 1 \tag{5.49}$$

The derivatives of state increment and control correction at mapped CGL collocation points are:

$$\delta \dot{\boldsymbol{x}}(\lambda_k) \approx \delta \dot{\boldsymbol{X}}(\lambda_k) = \sum_{i=0}^{N} \delta \boldsymbol{x}(\lambda_i) D_{ki}, \delta \dot{\boldsymbol{u}}(\lambda_k) \approx \delta \dot{\boldsymbol{U}}(\lambda_k) = \sum_{i=0}^{N} \delta \boldsymbol{u}(\lambda_i) D_{ki} \tag{5.50}$$

where D_{ki} is the element of the k–th row and i–th column of the first-order differential matrix mapped CGL pseudospectral method, which satisfies:

$$\begin{cases} D_{ki} = \frac{w_i / w_k}{\lambda_k - \lambda_i} & k \neq i \\ D_{kk} = -\sum\limits_{j=0, j \neq k}^{N} D_{kj} & k = i \end{cases} \tag{5.51}$$

In comparison to the first-order differential matrix of the standard CGL pseudospectral method, the first-order differential matrix of the mapped CGL pseudospectral method exhibits the ability to mitigate the ill-conditioned properties of the differential matrix. Figure 5.4 illustrates the element values of both the standard and mapped CGL pseudospectral methods when 21 CGL nodes are employed. It is evident that the first-order differential matrix of the mapped CGL pseudospectral method, which utilizes conformal mapping and barycentric Lagrange interpolation technology, demonstrates smaller element values at both ends of the diagonal. As a result, the ill-conditioned characteristics of the matrix are improved, thereby preventing the function to be approximated from diverging at the interval boundaries.

Like the LG- and LGR- PMPCM, the mapped Chebyshev pseudospectral method based MPCP also defines a pseudospectral sensitivity relation. Initially, original system is converted into a set of algebraic equality constraints at the mapped CGL node.

$$\sum_{i=0}^{N} \delta \boldsymbol{X}(\lambda_i) D_{ki} = \frac{t_f - t_0}{2}[\boldsymbol{A}_k^* \delta \boldsymbol{X}(\lambda_k) + \boldsymbol{B}_k^* \delta \boldsymbol{U}(\lambda_k)], k = 0, 1, \ldots, N \tag{5.52}$$

Let $\delta \boldsymbol{X}_i \triangleq \delta \boldsymbol{X}(\lambda_i)$,$\delta \boldsymbol{U}_i \triangleq \delta \boldsymbol{U}(\lambda_i)$,$\delta \overline{\boldsymbol{X}} = [\delta \boldsymbol{X}_0; \cdots ; \delta \boldsymbol{X}_N] \in \mathbb{R}^{(N+1)n_s}$,$\delta \overline{\boldsymbol{U}} = [\delta \boldsymbol{U}_0; \cdots ; \delta \boldsymbol{U}_N] \in \mathbb{R}^{(N+1)n_c}$, Then the Eq. (5.52) can be rewritten as a linear matrix equation:

$$\boldsymbol{D}_{0:N} \delta \overline{\boldsymbol{X}} = \overline{\boldsymbol{A}}^* \delta \overline{\boldsymbol{X}} + \overline{\boldsymbol{B}}^* \delta \overline{\boldsymbol{U}} \tag{5.53}$$

where $\boldsymbol{D}_i = [D_{0i} \boldsymbol{I}; \cdots ; D_{Ni} \boldsymbol{I}] \in \mathbb{R}^{(N+1)n_s \times n_s}$,$\boldsymbol{D}_{0:N} = [\boldsymbol{D}_0, \ldots, \boldsymbol{D}_N] \in \mathbb{R}^{(N+1)n_s \times (N+1)n_s}$, $\boldsymbol{I} \in \mathbb{R}^{n_s \times n_s}$ is the identity matrix, and

$$\overline{\boldsymbol{A}}^* = \frac{t_f - t_0}{2} diag([\boldsymbol{A}_0^*, \ldots, \boldsymbol{A}_N^*]) \in \mathbb{R}^{(N+1)n_s \times (N+1)n_s} \tag{5.54}$$

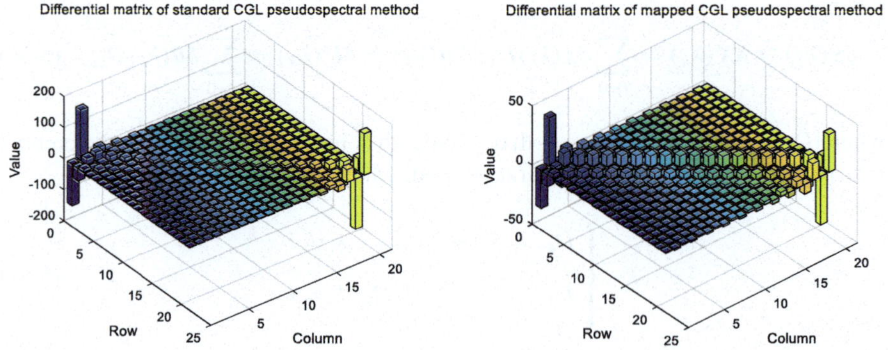

Fig. 5.4 First order differential matrix element values for standard CGL and mapping CGL pseudospectral methods

$$\overline{\boldsymbol{B}}^{*}=\frac{t_f-t_0}{2}diag([\boldsymbol{B}_0^{*},\ldots,\boldsymbol{B}_N^{*}])\in\mathbb{R}^{(N+1)n_s\times(N+1)n_c} \tag{5.55}$$

Equation (5.53) can be further rewritten as:

$$\delta\overline{\boldsymbol{X}}=(\boldsymbol{D}_{0:N}-\overline{\boldsymbol{A}}^{*})^{-1}\overline{\boldsymbol{B}}^{*}\delta\overline{\boldsymbol{U}} \tag{5.56}$$

which is the CGL pseudospectral sensitivity relation.

Similarly, the optimal control problem corresponding to the mapped Chebyshev pseudospectral predictive convex programming (MCGL-PMPCP) can be defined as:

$$P5-4:\sum_{k=0}^{N}\left(\delta\boldsymbol{X}_k^T\boldsymbol{Q}\delta\boldsymbol{X}_k+\delta\boldsymbol{U}_k^TR\delta\boldsymbol{U}_k\right)+\sum_{k=0}^{N-1}(\boldsymbol{U}_{k+1}-\boldsymbol{U}_k)^T\boldsymbol{R}_d(\boldsymbol{U}_{k+1}-\boldsymbol{U}_k) \tag{5.57}$$

$$\text{subject to }(5-56),\delta\boldsymbol{X}_0=\boldsymbol{x}_I^d-\boldsymbol{X}_0^{*},\delta\boldsymbol{X}_{N+1}=\boldsymbol{x}_F^d-\boldsymbol{X}_{N+1}^{*} \tag{5.58}$$

$$\boldsymbol{g}(\boldsymbol{X}_k,\boldsymbol{U}_k)\le 0,\boldsymbol{h}(\boldsymbol{X}_k,\boldsymbol{U}_k)=0 \tag{5.59}$$

where $\boldsymbol{x}_I^d$ and $\boldsymbol{x}_F^d$ show the desired initial and terminal states, $\boldsymbol{X}_0^{*}$ and $\boldsymbol{X}_{N+1}^{*}$ represent the initial and final state variables of the reference trajectory, $\boldsymbol{g}(\boldsymbol{X}_k,\boldsymbol{U}_k)$ and $\boldsymbol{h}(\boldsymbol{X}_k,\boldsymbol{U}_k)$ respectively represent the remaining inequality and equality constraints, which can be convexified using the first-order Taylor expansion.

In summary, the algorithm flow of MCGL-PMPCP is shown in Table 5.5.

Table 5.5 The flowchart of MCGL-PMPCP algorithm

Step	Procedure
1	Propagate the nonlinear system (5.1) with the initial state $\boldsymbol{X}_0$ and initial guess control $\boldsymbol{U}_k^{*}$, $k=0,1,\ldots,N$ to obtain the initial guess trajectory $\boldsymbol{X}_k^{*}$, $k=0,1,\ldots,N+1$
2	Calculate $\overline{\boldsymbol{A}}^{*}$ and $\overline{\boldsymbol{B}}^{*}$ according to Eqs. (5.54) and (5.55), then solve P4-4 via IPM to generate the control corrections $\delta\boldsymbol{U}_k$ and state increments $\delta\boldsymbol{X}_k$, and let $\boldsymbol{X}_k:=\boldsymbol{X}_k^{*}+\delta\boldsymbol{X}_k$
3	If $\sup\lVert\delta\boldsymbol{X}_k\rVert_\infty\le\varepsilon$, $k=0,1,\ldots,N$, where $\varepsilon>0$ is a customized constant, then go to Step 5; otherwise go to Step 4
4	Let $\boldsymbol{U}_k:=\boldsymbol{U}_k^{*}+\delta\boldsymbol{U}_k$, then propagate the nonlinear system (5.1) with $\boldsymbol{U}_k$, and update the reference trajectory $\boldsymbol{X}_k$ with the strategy described in Sect. 5.1.2; let $\boldsymbol{U}_k^{*}:=\boldsymbol{U}_k$, $\boldsymbol{X}_k^{*}:=\boldsymbol{X}_k$ and go to Step 2
5	Let $\boldsymbol{U}_k:=\boldsymbol{U}_k^{*}+\delta\boldsymbol{U}_k$, and output the optimal solution $\boldsymbol{X}_k^{*}:=\boldsymbol{X}_k$, $\boldsymbol{U}_k^{*}:=\boldsymbol{U}_k$, then the algorithm terminates

5.2.2 Numerical Simulations

To assess the effectiveness of the proposed MCGL-PMPCP method in mitigating constant control profile oscillations, the simulation scenario outlined in Sect. 5.1 is utilized to address the Mars entry with maximum final altitude. The MCGL-PMPCP method incorporates a virtual control and augments the penalty term associated with the virtual control within the objective function, where the penalty weight is set at 10^6. Additionally, the conformal mapping parameter α is assigned a value of 0.999. In the simulation, the performance of LG-PMPCP, LGR-PMPCP, CGL-PMPCP, and MCGL-PMPCP methods is compared by updating the reference trajectory through numerical integration, and the results are depicted in Fig. 5.5.

The simulation results clearly indicate that both the CGL-PMPCP and MCGL-PMPCP methods closely resemble the outcomes obtained from LG-PMPCP and LGR-PMPCP. This observation demonstrates that all four methods have the ability to converge towards the optimal solution. However, the key differentiating factors lie in the computational efficiency of each method and the smoothness of the control profile. Notably, in Fig. 5.5c, it is evident that only the bank angle rate profile of MCGL-PMPCP remains relatively stable, while the profiles of the other three methods exhibit a zigzag pattern. This distinction can be attributed to the utilization of Kosloff-Tal-Eaer conformal mapping in the MCGL-PMPCP method, which enhances the uniformity of the non-uniform CGL collocation points, followed by the adoption of barycentric Lagrange interpolation as a replacement for the Lagrange interpolation method. This improvement effectively mitigates oscillations at both ends of the constant profile, which are caused by the ill-conditioning of the standard CGL pseudospectral differential matrix. Furthermore, it enhances the stability of algorithm. The simulation results demonstrate this enhancement, as the initial segment (0–80 s) and middle segment (130–170 s) of the bank angle rate profile exhibit smooth curves rather than zigzag patterns. Regarding the bank angle rate profile beyond 200 s, since the bank angle actually exhibits an upward trend, the bank angle rate is not a constant value, resulting in oscillations for all four methods. However, concerning computational efficiency, Fig. 5.5e indicates that the CGL-PMPCP method requires 21 iterations, whereas MCGL-PMPCP achieves convergence in just 6 iterations. Additionally, the results presented in Table 5.6 demonstrate that the computation time for the CGL-PMPCP method exceeds 10 s, whereas the calculation time for MCGL-PMPCP and other Legendre pseudo-spectral models is comparable. It is worth noting, however, that due to the utilization of conformal mapping to enhance the uniformity of pseudo-spectral matching points in the MCGL-PMPCP method, there is a reduction in calculation accuracy, leading to a decrease in the precision of the converged solution. This aspect is further validated in Fig. 5.5f, where the height error of the MCGL-PMPCP method is the largest. Table 5.6 also corroborates this finding, with the corresponding height error measuring 0.1277 km, the highest among the four methods. In summary, the MCGL-PMPCP method effectively suppresses the oscillation phenomenon of the constant control quantity, albeit with a certain reduction in the precision of the numerical solution.

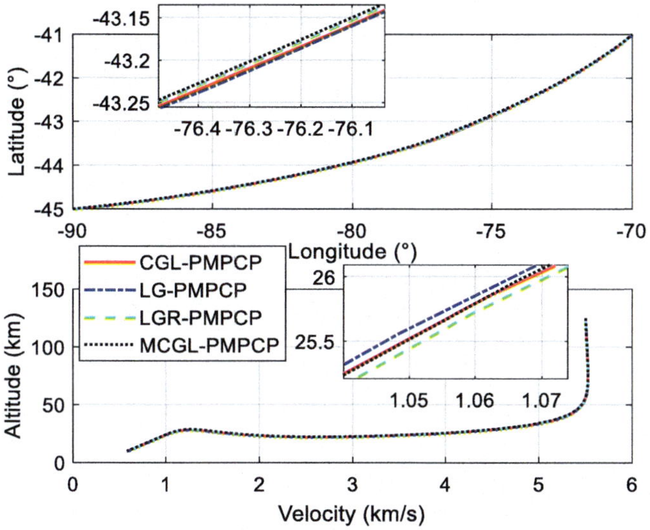

(a) Profiles of longitude versus latitude and velocity versus altitud

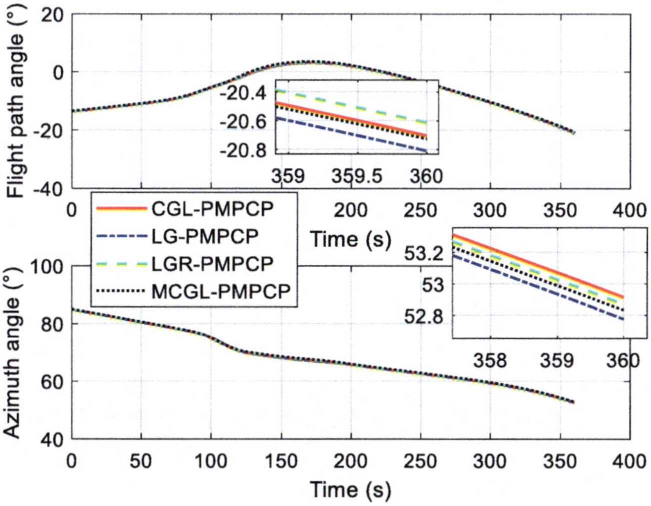

(b) Profiles of flight path angle and azimuth angle

Fig. 5.5 The Mars entry optimal solutions of four methods

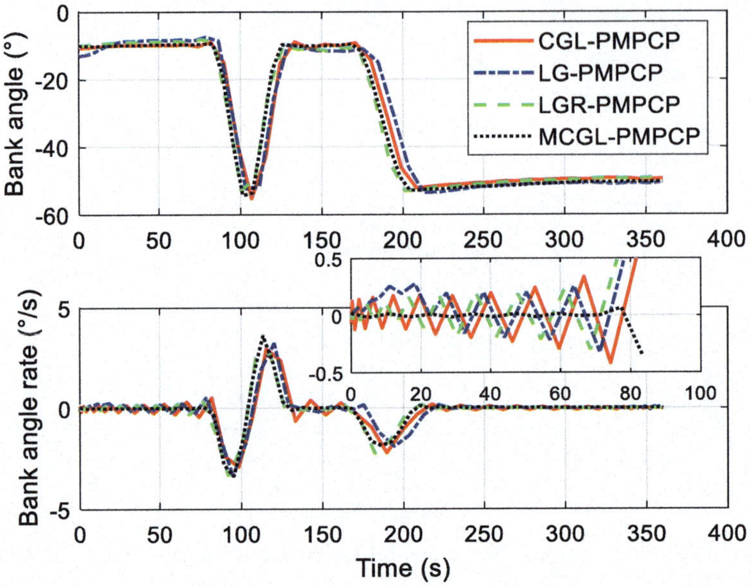

(c) Profiles of bank angle and its rate

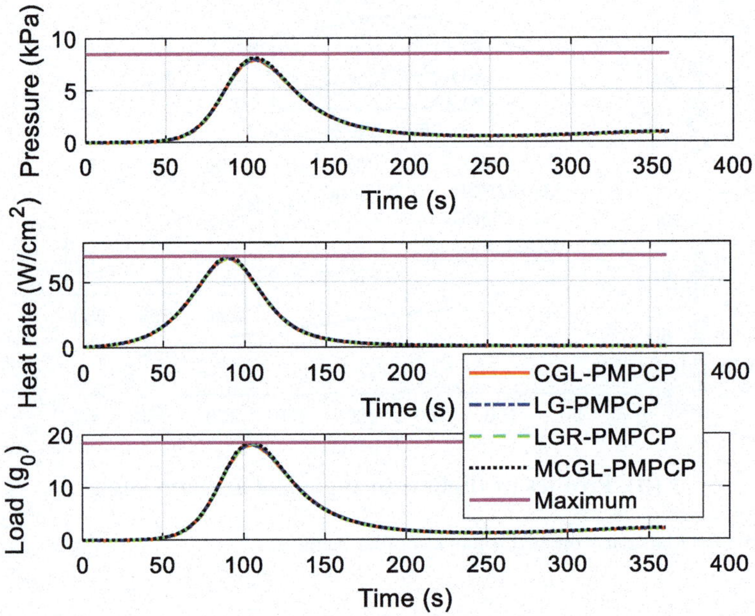

(d) Profiles of path constraints

Fig. 5.5 (continued)

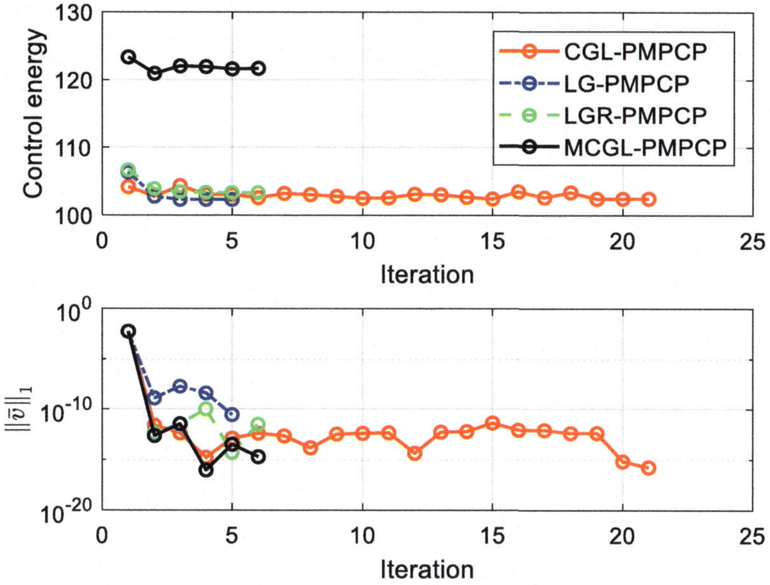

(e) Profiles of control energy and vritual control

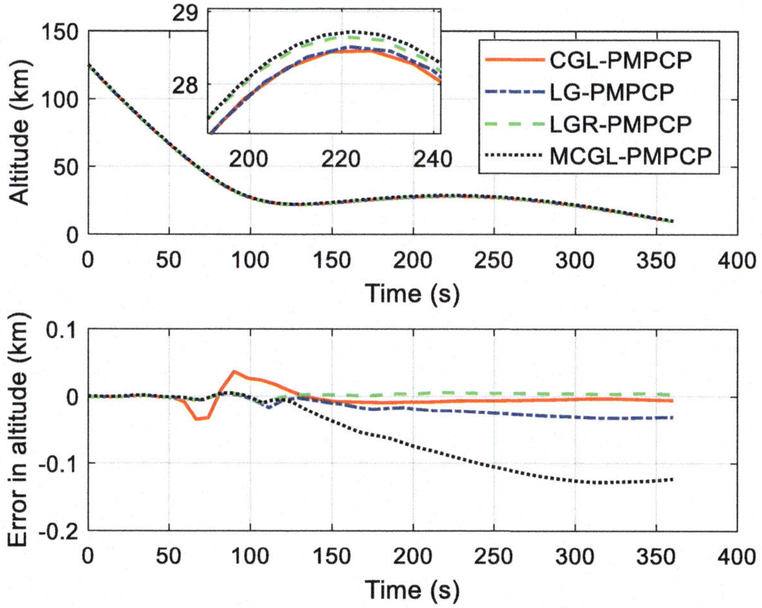

(f) Profiles of altitude and errors in altitude

Fig. 5.5 (continued)

Table 5.6 Comparison of four methods

Parameters	CGL-PMPCP	LG-PMPCP	LGR-PMPCP	MCGL-PMPCP
h_f(km)	10	10	10	10
θ_f(°)	−70	−70	−70	−70
ϕ_f(°)	−41	−41	−41	−41
V_f(m/s)	582.6519	584.1781	584.5970	586.4176
γ_f(°)	−20.7330	−20.8223	−20.6040	−20.7560
ψ_f(°)	52.8734	52.7833	52.8806	52.8074
σ_f(°)	−50.1394	−50.55503	−48.7798	−50.6697
$\lvert\dot{\sigma}\rvert_{max}$/(°/s)	2.9810	3.3374	3.3265	3.6152
$\dot{Q}_{max}$(W/cm^2)	69.5142	68.4927	69.0461	69.4905
q_{max}(kPa)	8.0030	8.0445	7.9972	8.0821
$a_{max}(g_0)$	18.2701	18.3649	18.2569	18.4507
CPU time(s)	10.3929	4.0026	4.2380	4.7950
Iteration	21	5	6	7
$\Delta\lvert h\rvert_{max}$ (km)	0.0378	0.03162	0.009845	0.1277
Final $\lVert \boldsymbol{v} \rVert_1$	2.0×10^{-16}	3.3×10^{-11}	3.8×10^{-12}	1.9×10^{-15}

5.3 Conclusions

In this chapter, the PMPCP method incorporates the Legendre pseudo-spectral method as a replacement for the uniform discrete method utilized in the MPCP method. This integration allows for the establishment of a pseudospectral sensitivity relation, which significantly reduces the computational load of the Mars entry trajectory problem. Additionally, the reference trajectory updating strategy of the MPCP method is enhanced, resulting in improved accuracy of the converged solution.

Furthermore, the MCGl-PMPCP method introduces Kosloff-Tal-Eaer conformal mapping and barycentric Lagrange interpolation technologies. This approach facilitates the adjustment of non-uniform CGL collocation points to nearly uniform MCGL collocation points, ensuring both computational efficiency and numerical accuracy. Consequently, the oscillation phenomenon observed in the constant control quantity within the LG- and LGR- PMPCP method is effectively suppressed.

Chapter 6
Indirect Sequential Convex Programming for Mars Entry Trajectory Planning

This chapter proposes indirect sequential convex programming (ISCP) to maximize the Mars entry terminal altitude. The main contributions of this chapter are threefold. (1) The ISCP improves the Unified Trigonometrization Method (UTM) [150, 152] by relaxing the optimal control options to the solution with a continuous and unique expression, thereby avoiding examination of the PMP in the classical UTM. (2) The SCP is used to solve the associated TPBVP of the Terminal Altitude Maximization (TAM) problem resulted by improved UTM. In this way, the TPBVP can be solved in polynomial computation time by the SCP instead of the time-consuming numerical continuation method. In addition, the costate variable of the constrained OCP can be estimated while solving the TPBVP, which is difficult to obtain in the traditional SCP method. (3) The extension of the hybrid strategy combining the indirect technique and convex optimization is another contribution of this chapter. Although this hybrid approach to solving OCPs has been investigated from one aspect lately [153], only the unconstrained OCP with a continuous non-singular solution was considered therein. As a step forward, the introduction of the UTM in this chapter can broaden the application of this hybrid paradigm to a more general class of OCPs. In other words, the ISCP can solve a class of constrained OCPs without considering the bang-bang or singular control structure as long as the control is affine.

The rest of this chapter is organized as follows. Section 6.1 describes the TAM problem. Sections 6.2 and 6.3 present the improved UTM and the ISCP, respectively. The numerical simulations are shown in Sect. 6.4. Section 6.5 concludes this chapter.

S. Li et al., *Trajectory Optimization and Guidance Methods for Mars Entry*,
https://doi.org/10.1007/978-981-99-6282-2_6

6.1 Mars Entry Terminal Altitude Maximization Problem

For the Mars entry terminal altitude maximization problem, the non-dimensional longitudinal translational dynamic is considered in view of the fact that the lateral dynamic is decoupled from the longitudinal one and governs the horizontal longitude and latitude.

$$\begin{cases} dr/d\tau = V\sin\gamma \\ dV/d\tau = -D - \sin\gamma/r^2 \\ d\gamma/d\tau = L\cos\sigma/V + V\cos\gamma/r - \cos\gamma/(r^2V) \\ ds/d\tau = V\cos\gamma/r \end{cases} \tag{6.1}$$

in which the dimensionless time τ is normalized by $t_s = \sqrt{R_0/g_0}$, R_0 is the Mars equatorial radius, g_0 is the gravitational acceleration at R_0, r scaled by R_0 is the radial distance from the Martian center to the mass center of the vehicle, V is the dimensionless velocity of the vehicle relative to the Mars, γ is the flight path angle, σ is the bank angle constrained to its feasible interval $0 < \sigma_{\min} \le \sigma \le \sigma_{\max} < \pi$, and s is the downrange angle or the dimensionless downrange $S = sR_0$. The non-dimensional lift and drag accelerations are:

$$L = R_0\rho V^2 S_r C_L/(2m),\quad D = R_0\rho V^2 S_r C_D/(2m) \tag{6.2}$$

where m, S_r, C_L and C_D are the mass, reference area, lift and drag coefficients of the vehicle, respectively. The atmospheric density ρ is modeled by the exponential function:

$$\rho = \rho_0 e^{-R_0(r-1)/h_s} \tag{6.3}$$

where ρ_0 is the atmospheric density at the Mars surface, and h_s denotes the scale height. It should be noted that the dynamics does not take into account the Martian rotation, which is small and has a negligible effect on this problem.

Define the state vector $\boldsymbol{x} = [r, V, \gamma, s]^T$ and the control variable $u = \cos\sigma$, then the longitudinal dynamics and control constraint are given by:

$$\dot{x} = \boldsymbol{f}(\boldsymbol{x}) + \boldsymbol{g}(\boldsymbol{x})u \tag{6.4}$$

$$u_{\min} \le u \le u_{\max} \tag{6.5}$$

where $u_{\min} = \cos\sigma_{\max}$, $u_{\max} = \cos\sigma_{\min}$, and $\boldsymbol{f}(\boldsymbol{x})$, $\boldsymbol{g}(\boldsymbol{x})$ are the right-hand-side terms in Eq. (6.1). The reason for choosing this control variable is that the sign of the bank angle has no effect on the longitudinal dynamics. Nevertheless, both $u = \cos\sigma$ and $u = \sigma$ are widely used.

The path constraints on dynamic pressure q, heat flux $\dot{Q}$, and normal load n are typically expressed as:

$$q = \rho (V V_s)^2/2 \le q_{\max}, \quad \dot{Q} = k_Q \sqrt{\rho / R_n} (V V_s)^3 \le \dot{Q}_{\max},$$
$$n = \sqrt{L^2 + D^2} \le n_{\max} \tag{6.6}$$

where $V_s = \sqrt{R_0 g_0}$, k_Q is a constant, R_n is the vehicle's nose radius, $q_{\max}$, $\dot{Q}_{\max}$, and $n_{\max}$ are the admissible maxima of the corresponding path constraints.

The boundary conditions are given by:

$$\boldsymbol{x}(\tau_0) = [r_0, V_0, \gamma_0, s_0]^T, \quad V(\tau_f) = V_f \tag{6.7}$$

where the subscripts '0' and 'f' denote the initial and terminal values of the corresponding variables, respectively.

The Mars entry terminal altitude maximization problem, denoted as P6.1, is formulated as follows:

P6.1: Minimize $J = -r(\tau_f)$.

subject to Eqs. (6.4), (6.5), (6.6), and (6.7).

6.2 Improvements to Unified Trigonometrization Method

6.2.1 Review of the UTM

For the problem P6.1, the classical optimal control theory employs the calculus of variation and PMP to reformulate the original problem into a TPBVP that can be solved via numerical methods. However, the process of reformulation is laborious. On the one hand, the inequality constraints lead to the Karush–Kuhn–Tucker (KKT) condition, which complicates the analysis of optimal control profiles. On the other hand, the possible bang-bang and singular arcs of the optimal control profile may result in numerical issues and non-convergence. To circumvent these issues, the UTM parameterizes the control as a function of trigonometric terms and introduces the regularization term to the objective function [150]. Thus, switching and singular profiles can be avoided by approximating the original control with a continuous and smooth function of the costate variables. Although the UTM can be applied to both control-affine and non-control-affine systems, only the methodology for the control-affine system is studied here because the longitudinal entry dynamics expressed in Eq. (6.4) is control-affine.

For the problem P6.1, the control input is first rewritten via the following nonlinear transformation in the UTM:

$$u = c_0 + c_1 \sin \overline{u} \tag{6.8}$$

in which $c_0 = (u_{\max} + u_{\min})/2$, $c_1 = (u_{\max} - u_{\min})/2$, and the new control $\overline{u}$ is unconstrained. Also, the objective function is redefined by introducing the regularization term:

$$J = -r(\tau_f) + \int_0^{\tau_f} [\varepsilon_c \cos \overline{u} + \varepsilon_q \sec(a_q \pi / 2) + \varepsilon_{\dot{Q}} \sec(a_{\dot{Q}} \pi / 2) + \varepsilon_n \sec(a_n \pi / 2)] d\tau \tag{6.9}$$

$$\begin{aligned} &a_q = (2q - q_{\max} - q_{\min}) / q_{\max}, \quad a_{\dot{Q}} = (2\dot{Q} - \dot{Q}_{\max} - \dot{Q}_{\min}) / \dot{Q}_{\max}, \\ &a_n = (2n - n_{\max} - n_{\min}) / n_{\max} \end{aligned} \tag{6.10}$$

where $q_{\min} = \dot{Q}_{\min} = n_{\min} = 0$, and $\varepsilon_c, \varepsilon_q, \varepsilon_{\dot{Q}}, \varepsilon_n > 0$ are the user-defined continuation parameters of the UTM. When these continuation parameters are very small, the objective function in Eq. (6.9) is almost the same as the original objective function. The cosine term in Eq. (6.9) is an error-control term that represents the degree of the approximation or smoothness to the original control profile. As shown in Fig. 6.1, the red solid line denotes the normalized extremum of the original optimal control, the approximate solution $\sin \overline{u}$ is represented by the blue dashed line, and the black dotted line orthogonal to the approximate solution is $\cos \overline{u}$. When the original optimal control exhibits the bang-bang feature, the cosine term is introduced to the neighborhood of the discontinuous point to approximate or smooth the control, and ε_c is used to control the smoothness of the solution with respect to the original solution. In this way, the jump phenomenon does not occur because the discontinuous control profile is approximated by the smooth differentiable sine function. Meanwhile, the other secant terms serve as penalties for violation of the three path constraints described in Eq. (6.6). It should be emphasized that parameters $a_q, a_{\dot{Q}}, a_n$ in Eq. (6.10) should be controlled to stay in the interval $[-1, 1]$ to guarantee that the arguments in these secant functions are within the interval $[-\pi/2, \pi/2]$, otherwise the penalty on the path constraint is invalid once the secant term does not approach infinity as presented in Fig. 6.2.

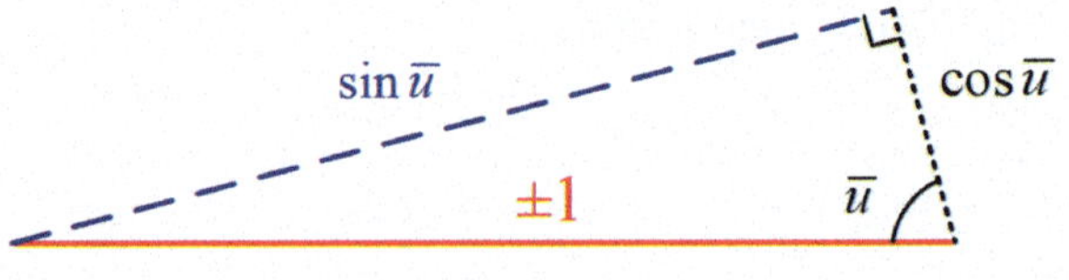

Fig. 6.1 Smoothing of the bang-bang control in the UTM

Fig. 6.2 The secant function in the UTM serving as the penalty term

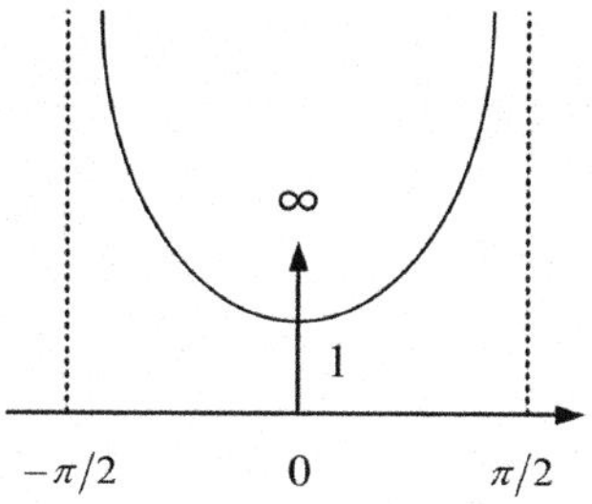

With the above formulations, the constrained OCP P6.1 is converted to the unconstrained one. The Hamiltonian of the unconstrained problem is thus defined as:

$$\begin{aligned} H &= \boldsymbol{\lambda}^T[\boldsymbol{f}(\boldsymbol{x}) + \boldsymbol{g}(\boldsymbol{x})u] + \varepsilon_c \cos\overline{u} + \varepsilon_q \sec(a_q\pi/2) \\ &\quad + \varepsilon_{\dot{Q}} \sec(a_{\dot{Q}}\pi/2) + \varepsilon_n \sec(a_n\pi/2) \\ &= \lambda_r V \sin\gamma + \lambda_V(-D - \sin\gamma/r^2) \\ &\quad + \lambda_\gamma[L(c_0 + c_1 \sin\overline{u})/V + V\cos\gamma/r - \cos\gamma/(r^2V)] + \lambda_s V\cos\gamma/r \\ &\quad + \varepsilon_c \cos\overline{u} + \varepsilon_q \sec(a_q\pi/2) + \varepsilon_{\dot{Q}} \sec(a_{\dot{Q}}\pi/2) + \varepsilon_n \sec(a_n\pi/2) \end{aligned} \tag{6.11}$$

in which $\boldsymbol{\lambda} = [\lambda_r, \lambda_V, \lambda_\gamma, \lambda_s]^T$ is the costate vector.

According to the first-order necessary condition, the costate equations and transversality conditions are given by:

$$\dot{\boldsymbol{\lambda}} = -\frac{\partial H}{\partial x} \tag{6.12}$$

$$\boldsymbol{\lambda}(\tau_f) = \frac{\partial\phi(\boldsymbol{x}(\tau_f), \tau_f)}{\partial\boldsymbol{x}(\tau_f)} + \boldsymbol{\mu}^T\frac{\partial\boldsymbol{\psi}(\boldsymbol{x}(\tau_f), \tau_f)}{\partial\boldsymbol{x}(\tau_f)} \tag{6.13}$$

$$H(\tau_f) = -\frac{\partial\phi(\boldsymbol{x}(\tau_f), \tau_f)}{\partial\tau_f} - \boldsymbol{\mu}^T\frac{\partial\boldsymbol{\psi}(\boldsymbol{x}(\tau_f), \tau_f)}{\partial\tau_f} \tag{6.14}$$

where the terminal cost is $\phi(\boldsymbol{x}(\tau_f), \tau_f) = -r(\tau_f)$, the terminal equality constraint is $\boldsymbol{\psi}(\boldsymbol{x}(\tau_f), \tau_f) = V(\tau_f) - V_f$, and $\boldsymbol{\mu}$ is the Lagrange multiplier associated with the terminal constraint. Then the costate equations and Hamiltonian become:

$$
\begin{aligned}
\dot{\lambda}_r &= -\frac{\partial H}{\partial r} = -\lambda_V\left(-D_r + \frac{2\sin\gamma}{r^3}\right) \\
&\quad - \lambda_\gamma\left(\frac{L_r(c_0 + c_1\sin\overline{u})}{V} - \frac{V\cos\gamma}{r^2} + \frac{2\cos\gamma}{r^3 V}\right) \\
&\quad + \lambda_s\frac{V\cos\gamma}{r^2} - \varepsilon_q\frac{\pi}{2}a_q^r\sec\left(\frac{\pi}{2}a_q\right)\tan\left(\frac{\pi}{2}a_q\right) \\
&\quad - \varepsilon_{\dot{Q}}\frac{\pi}{2}a_{\dot{Q}}^r\sec\left(\frac{\pi}{2}a_{\dot{Q}}\right)\tan\left(\frac{\pi}{2}a_{\dot{Q}}\right) \\
&\quad - \varepsilon_n\frac{\pi}{2}a_n^r\sec\left(\frac{\pi}{2}a_n\right)\tan\left(\frac{\pi}{2}a_n\right) \\
\dot{\lambda}_V &= -\frac{\partial H}{\partial V} = -\lambda_r\sin\gamma + \lambda_V D_V \\
&\quad - \lambda_\gamma\left(\frac{L_V(c_0 + c_1\sin\overline{u})}{V} - \frac{L(c_0 + c_1\sin\overline{u})}{V^2} + \frac{\cos\gamma}{r} + \frac{\cos\gamma}{r^2V^2}\right) \\
&\quad - \lambda_s\frac{\cos\gamma}{r} - \varepsilon_q\frac{\pi}{2}a_q^V\sec\left(\frac{\pi}{2}a_q\right)\tan\left(\frac{\pi}{2}a_q\right) \\
&\quad - \varepsilon_{\dot{Q}}\frac{\pi}{2}a_{\dot{Q}}^V\sec\left(\frac{\pi}{2}a_{\dot{Q}}\right)\tan\left(\frac{\pi}{2}a_{\dot{Q}}\right) - \varepsilon_n\frac{\pi}{2}a_n^V\sec\left(\frac{\pi}{2}a_n\right)\tan\left(\frac{\pi}{2}a_n\right) \\
\dot{\lambda}_\gamma &= -\frac{\partial H}{\partial\gamma} = -\lambda_r V\cos\gamma + \lambda_V\frac{\cos\gamma}{r^2} \\
&\quad + \lambda_\gamma\left(\frac{V\sin\gamma}{r} - \frac{\sin\gamma}{r^2V}\right) + \lambda_s\frac{V\sin\gamma}{r} \\
\dot{\lambda}_s &= 0
\end{aligned} \tag{6.15}
$$

$$
\lambda_r(\tau_f) = -1,\ \ \lambda_V(\tau_f) = \text{Free},\ \ \lambda_\gamma(\tau_f) = 0,\ \ \lambda_s(\tau_f) = 0 \tag{6.16}
$$

$$
H(\tau_f) = 0 \tag{6.17}
$$

in which $L_r = \partial L/\partial r = -LR_0/h_s$, $L_V = R_0\rho VS_rC_L/m$, $D_r = \partial D/\partial r = -DR_0/h_s$, $D_V = R_0\rho VS_rC_D/m$, and a_i^j denotes the partial derivative of the variable $a_i \in \{a_q, a_{\dot{Q}}, a_n\}$ with respect to the state variable $j \in \{r, V\}$:

$$
\begin{aligned}
a_q^r &= 2q_r/q_{\max},\ \ a_q^V = 2q_V/q_{\max} \\
a_{\dot{Q}}^r &= 2\dot{Q}_r/q_{\max},\ \ a_{\dot{Q}}^V = 2\dot{Q}_V/q_{\max} \\
a_n^r &= 2n_r/q_{\max},\ \ a_n^V = 2n_V/q_{\max}
\end{aligned} \tag{6.18}
$$

$$
q_r = \partial q/\partial r = -R_0\rho(VV_s)^2/(2h_s),\ \ q_V = \partial q/\partial V = \rho VV_s^2 \tag{6.19}
$$

$$\dot{Q}_r = \partial \dot{Q} / \partial r = -R_0 k_Q (V V_s)^3 \sqrt{\rho / R_n} \Big/ (2h_s),$$
$$\dot{Q}_V = \partial \dot{Q} / \partial V = 3k_Q V_s^3 V^2 \sqrt{\rho / R_n} \tag{6.20}$$

$$n_r = \partial n / \partial r = -R_0^2 \rho V^2 S_r (\sqrt{C_L^2 + C_D^2}) \Big/ (2mh_s),$$
$$n_V = \partial n / \partial V = R_0 \rho V S_r (\sqrt{C_L^2 + C_D^2}) \Big/ m \tag{6.21}$$

The optimal control options can be obtained by using the stationary condition:

$$\partial H / \partial \overline{u} = \lambda_\gamma L c_1 \cos \overline{u} / V - \varepsilon_c \sin \overline{u} = 0 \tag{6.22}$$

Then, the PMP is used to select the optimal control from the following options:

$$\overline{u}^* = \begin{cases} \arctan(\lambda_\gamma L c_1 / (V \varepsilon_c)) \\ \arctan(\lambda_\gamma L c_1 / (V \varepsilon_c)) + \pi \end{cases} \tag{6.23}$$

It should be noted that only the first option can be obtained by Eq. (6.22), and the source of the second option is described in the next subsection. Thus, Eqs. (6.4), (6.7), (6.15), (6.16), (6.17), and (6.23) form the TPBVP to be solved.

For the resulted TPBVP, its convergence is sensitive to the initial costate. In the meanwhile, the initial costate is difficult to set because it has no clear physical significance. To mitigate this issue, the UTM solves the TPBVP by using the MATLAB built-in function *bvp4c* through the numerical continuation. In the continuation process, a simple TPBVP is first defined and solved without considering the physical significance and values of some critical parameters, and the simple solution serves as the initial guess for the subsequent complex TPBVPs with practical parameters. Thus, the choice of the parameters for numerical continuation, the value of parameters, and the continuation step size are all critical because these steps directly define auxiliary TPBVPs and affect convergence of the continuation process. For instance, parameters ε_c, ε_q, $\varepsilon_{\dot{Q}}$, and ε_n must be constants close to zero in the complex problem. However, what small values of these parameters can simplify the problem is uncertain. This situation is also faced for the remaining parameters, such as $q_{\max}$, $\dot{Q}_{\max}$, and $n_{\max}$. As a result, the preparatory work of determining a suitable combination of parameters for convergence is laborious. Furthermore, it takes intensive computations to solve these auxiliary TPBVPs. The detailed continuation sets of parameters are given in Sect. 6.4.

6.2.2 Improved Optimal Solution

There are several shortcomings in the UTM. First, there are two options for the optimal control profile, implying that the control profile is not unique, and the PMP should be evaluated accordingly to select the correct one, but this evaluation inevitably increases the computation time. Second, continuation sets (discussed in Sect. 6.2) are created by decreasing these continuation parameters ε_c, ε_q, $\varepsilon_{\dot{Q}}$, and ε_n one by one in order to approximate the original optimal solution. However, it is a challenge and trade-off to determine the step size of the continuation parameter because a small step size ensures the convergence but increases the computation time. Third, the values of a_q, $a_{\dot{Q}}$, a_n must be bounded in the interval $[-1, 1]$, otherwise the UTM is not able to implement state path constraints using these secant penalty terms. To cope with this problem, the tolerable maxima $q_{\max}$, $\dot{Q}_{\max}$, $n_{\max}$ of path constraints are considered as the continuation parameters in the UTM to constrain the values of a_q, $a_{\dot{Q}}$, a_n. However, as previously stated, it is difficult to select the appropriate continuation parameter.

In fact, by analyzing the Hamiltonian function, the optimal control profile can be estimated and approximated uniquely. Here, the function H in Eq. (6.11) is first rewritten as a periodic one of the 2π period:

$$
\begin{aligned}
H = & \sqrt{a^2 + b^2}\cos(\overline{u} - \theta) + \lambda_r V \sin\gamma + \lambda_V(-D - \sin\gamma / r^2) \\
& + \lambda_\gamma[Lc_0 / V + V\cos\gamma / r - \cos\gamma / (r^2 V)] + \lambda_s V \cos\gamma / r \\
& + \varepsilon_q \sec(a_q \pi / 2) + \varepsilon_{\dot{Q}} \sec(a_{\dot{Q}} \pi / 2) + \varepsilon_n \sec(a_n \pi / 2)
\end{aligned}
\tag{6.24}
$$

where $a = \lambda_\gamma Lc_1 / V$, $b = \varepsilon_c$, $\cos\theta = b / \sqrt{a^2 + b^2}$, $\sin\theta = a / \sqrt{a^2 + b^2}$, $\theta = \arctan(a / b)$. Obviously, the Hamiltonian H holds the minimum when the optimal control $\overline{u}^*$ satisfies:

$$
\overline{u}^* = 2k\pi + \pi + \theta, \quad k = 0, \pm 1, \pm 2, \cdots \tag{6.25}
$$

Note that this solution is exactly the second option in Eq. (6.23) if the periodic term $2k\pi$ is omitted, which shows that the first term in Eq. (6.23) is actually not the optimum, thus explaining why UTM needs to use PMP to determine the optimal option in Eq. (6.23).

Furthermore, since $\varepsilon_c \to 0^+$, then $\cos\theta > 0$ and.

$$
\theta = \begin{cases} |\theta| \to (\pi / 2)^- = \pi / 2 - \mu & a > 0 \\ 0 & a = 0, \quad \mu \to 0^+ \\ -|\theta| \to (-\pi / 2)^+ = \mu - \pi / 2 & a < 0 \end{cases} \tag{6.26}
$$

where $\mu \to 0^+$. Then, substituting Eq. (6.26) into Eq. (6.25), the optimal control is rewritten as:

$$\overline{u}^* = \begin{cases} 2k\pi + \pi + |\theta| & a > 0 \\ 2k\pi + \pi & a = 0 \\ 2k\pi + \pi - |\theta| & a < 0 \end{cases} \tag{6.27}$$

Two facts are indicated by Eq. (6.27). First, the bang-bang profile can be automatically determined if the sign of a reverses in the interval $2k\pi + [\pi - |\theta|, \pi + |\theta|]$, and this reversal occurs at the intermediate points $2k\pi + \pi$. Second, the optimal control in Eq. (6.27) can be further approximated as follows considering the small ε_c:

$$\overline{u}^* = \begin{cases} 2k\pi + \pi + |\theta| = 2k\pi + \pi + \pi/2 - \mu \approx 2k\pi + \pi + \pi/2 + \mu \\ \quad = 2(k+1)\pi - (\pi/2 - \mu) = 2(k+1)\pi + \arctan(-a/b) & a > 0 \\ 2k\pi + \pi & a = 0 \\ 2k\pi + \pi - |\theta| = 2k\pi + \pi/2 + \mu \approx 2k\pi + \pi/2 - \mu \\ \quad = 2k\pi - (\mu - \pi/2) = 2k\pi + \arctan(-a/b) & a < 0 \end{cases} \tag{6.28}$$

If $a \neq 0$, it is obvious that $\overline{u}^*$ can be simplified to $2k\pi + \arctan(-a/b)$ according to Eq. (6.28). In the case of $a = 0$, considering $\overline{u}^* = 2k\pi + \pi, \theta = \arctan(a/b) = \arctan(-a/b) = 0$ and Eq. (6.8), then

$$\begin{aligned} u = c_0 + c_1 \sin \overline{u}^* &= c_0 + c_1 \sin(2k\pi + \pi) \triangleq c_0 + c_1 \sin(2k\pi + 0) \\ &\triangleq c_0 + c_1 \sin(2k\pi + \theta) \triangleq c_0 + c_1 \sin[2k\pi + \arctan(-a/b)] \end{aligned} \tag{6.29}$$

Therefore, one has $\overline{u}^* = 2k\pi + \arctan(-a/b)$ when $a = 0$. Then, the expression of $\overline{u}^*$ can be written as follows without considering the value of a:

$$\overline{u}^* = 2k\pi + \arctan(-a/b) \tag{6.30}$$

The optimal control $\overline{u}^*$ in Eq. (6.27), the improved $\overline{u}^*$ in Eq. (6.30), and the θ in Eq. (6.26) are illustrated in Fig. 6.3:

where the black square, circle, and triangle denote the original optimal control $\overline{u}^*$; the red plus, star, and cross signs represent θ. In the axis of Eq. (6.27), the original optimal control $\overline{u}^*$ is evaluated when $k = -1$ and $k = 0$. Then, the original optimal

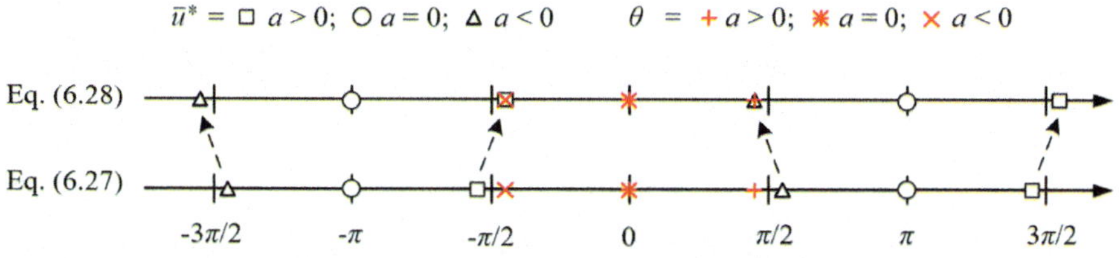

Fig. 6.3 Illustrations of the optimal control and its improved version

solution is approximated and replaced by the improved one in the axis of Eq. (6.28) according to the previous discussions.

Furthermore, the periodic term $2k\pi$ can be dropped considering Eq. (6.8), so the improved optimal control is given as:

$$\overline{u}^* = \arctan(-a/b) \tag{6.31}$$

At this point, the option of optimal control is unique, and there is no need to evaluate the PMP. Of course, a sufficiently small ε_c is the prerequisite for all the above formulations and approximations.

6.3 Indirect Sequential Convex Programming

For the UTM, there are considerable auxiliary TPBVPs to be solved during the homotopy or numerical continuation. It is not only difficult but also time-consuming to find appropriate continuation parameters for solving TPBVPs via the shooting or collocation methods. Note that the system of ordinary differential equations in TPBVPs can be approximated as a series of equality constraints by means of numerical integration rules. In this way, as long as a proper objective function is defined, the TPBVP can be reformulated as a constrained optimization problem. In this regard, sequential convex programming (SCP) can handle the optimization problem within a polynomial time. Therefore, the SCP is adopted here to solve the TPBVP. Nonetheless, some preliminary procedures including convexification and discretization are required before converting the TPBVP into the convex optimization problem, which will be detailed in this section. In addition, because a_q, $a_{\dot{Q}}$, a_n in the regularization terms must be bounded in the interval $[-1, 1]$, the indirect sequential convex programming (ISCP) directly relaxes the path constraints in Eq. (6.6) into the convex inequality constraints to ensure that the values of a_q, $a_{\dot{Q}}$, a_n are within $[-1, 1]$. Accordingly, parameters ε_q, $\varepsilon_{\dot{Q}}$, and ε_n can be assigned as zero in the objective function. These details of ISCP are given as follows.

6.3.1 Convexification and Discretization

Define the new state vector $\boldsymbol{y} = [r, V, \gamma, s, \lambda_r, \lambda_V, \lambda_\gamma, \lambda_s]^T$, then the differential equation of $\boldsymbol{y}$ is:

$$d\boldsymbol{y}/d\xi = \tau_f \boldsymbol{h}(\boldsymbol{y}) \tag{6.32}$$

where function $\boldsymbol{h}(\boldsymbol{y})$ is defined with respect to the scaled dimensionless time $\xi = \tau/\tau_f \in [0, 1]$, and τ_f is the non-dimensional terminal time to be optimized. The expression of $\boldsymbol{h}(\boldsymbol{y})$ is shown in Eqs. (6.1) and (6.15). To incorporate the TPBVP into the convex optimization framework, the first-order Taylor series approximation is used to convexify the nonlinear differential equation in Eq. (6.32), the transversality condition $H(\tau_f) = 0$ in Eq. (6.17), and the path constraints in Eq. (6.6):

$$
\begin{aligned}
d\boldsymbol{y}/d\xi &= \tau_f \boldsymbol{h}(\boldsymbol{y}) \\
&= \boldsymbol{h}(\boldsymbol{y}^*)(\tau_f - \tau_f^*) + \tau_f^*[\boldsymbol{h}(\boldsymbol{y}^*) + \boldsymbol{A}(\boldsymbol{y}^*)(\boldsymbol{y} - \boldsymbol{y}^*)] + \boldsymbol{v} \\
&= \boldsymbol{h}(\boldsymbol{y}^*)\tau_f + \tau_f^* \boldsymbol{A}(\boldsymbol{y}^*)\boldsymbol{y} - \tau_f^* \boldsymbol{A}(\boldsymbol{y}^*)\boldsymbol{y}^* + \boldsymbol{v}
\end{aligned} \tag{6.33}
$$

$$
\tau_f H(\boldsymbol{y}_f) \approx H(\boldsymbol{y}_f^*)\tau_f + \tau_f^* \boldsymbol{B}(\boldsymbol{y}_f^*)(\boldsymbol{y}_f - \boldsymbol{y}_f^*) + \Delta = 0 \tag{6.34}
$$

$$
p_i \approx p_i(r^*, V^*) + p_i'(r^*, V^*)[r - r^*, V - V^*]^T \leq p_{i,\max} \tag{6.35}
$$

where $\boldsymbol{y}^*$ denote the reference solution for the first-order Taylor series expansion, $\boldsymbol{y}_f^* = \boldsymbol{y}^*(\tau_f)$ is the state vector at the final time τ_f, $\boldsymbol{A}(\boldsymbol{y}^*) = \partial \boldsymbol{h}/\partial \boldsymbol{y}\big|_{\boldsymbol{y}=\boldsymbol{y}^*}$, $\boldsymbol{B}(\boldsymbol{y}_f^*) = \partial H/\partial \boldsymbol{y}_f\big|_{\boldsymbol{y}_f=\boldsymbol{y}_f^*}$, p_i' is the partial derivative of the variable $p_i \in \{q, \dot{Q}, n\}$ with respect to r and V, the expression of p_i' is given in Eqs. (6.19)–(6.21), and $p_{i,\max} \in \{q_{\max}, \dot{Q}_{\max}, n_{\max}\}$. The virtual control $\boldsymbol{v}$ and virtual buffer Δ are introduced to compensate for the linearization error and mitigate the possibility of artificial infeasibility.

To apply the SCP method for solving the TPBVP, the adaptive trust-region technique is utilized to relax the state vector $\boldsymbol{y}$ and terminal time τ_f around the reference solution $\boldsymbol{y}^*$ and τ_f^*, which ensures the feasibility of the first-order Taylor series approximation.

$$
\left\| \boldsymbol{y} - \boldsymbol{y}^* \right\|_2 \leq \delta, \quad \left\| \tau_f - \tau_f^* \right\|_2 \leq \Gamma/t_s \tag{6.36}
$$

where δ is a variable to be determined in the optimization, and Γ is a user-defined value to ensure the bounded increment of the final flight time between two successive iterations.

The following objective function is defined to establish a mathematical programming problem (denoted as P6.2):

P6.2: minimize $J = w_v \|\boldsymbol{v}\|_2 + w_\delta \|\delta\|_2 + w_\Delta \|\Delta\|_2$.

subject to Eqs. (6.7), (6.16), (6.33), (6.34), (6.35), and (6.36)

where w_v, w_δ, and $w_\Delta \in \mathbb{R}^+$ are the weight coefficients prescribed by users.

So far, the original optimal control problem P6.1 has been transcribed to the convex optimization problem P6.2. To obtain the solution of P6.2 in the framework of convex optimization, the problem P6.2 is further discretized at a series of points $\{\xi_1, \xi_2, \cdots, \xi_{N+1}\}$ with the trapezoidal rule, where $\xi_{i+1} = \xi_i + 1/N$, $\xi_1 = 0$ and $\xi_{N+1} = 1$. Then, these equality and inequality constraints are rewritten into discrete forms:

$$\begin{aligned} \boldsymbol{y}(\xi_1) &= [r_0, V_0, \gamma_0, s_0, \text{Free}, \text{Free}, \text{Free}, \text{Free}]^T \\ \boldsymbol{y}(\xi_{N+1}) &= [\text{Free}, V_f, \text{Free}, \text{Free}, \lambda_{r,f}, \text{Free}, \lambda_{\gamma,f}, \lambda_{s,f}]^T \end{aligned} \tag{6.37}$$

$$\begin{aligned} \boldsymbol{y}_{i+1} = \boldsymbol{y}_i + [&\boldsymbol{h}(\boldsymbol{y}_i^k)\tau_f + \tau_f^k \boldsymbol{A}(\boldsymbol{y}_i^k)\boldsymbol{y}_i - \tau_f^k \boldsymbol{A}(\boldsymbol{y}_i^k)\boldsymbol{y}_i^k + \boldsymbol{v}_i \\ &+ \boldsymbol{h}(\boldsymbol{y}_{i+1}^k)\tau_f + \tau_f^k \boldsymbol{A}(\boldsymbol{y}_{i+1}^k)\boldsymbol{y}_{i+1} - \tau_f^k \boldsymbol{A}(\boldsymbol{y}_{i+1}^k)\boldsymbol{y}_{i+1}^k + \boldsymbol{v}_{i+1}]/(2N) \end{aligned} \tag{6.38}$$

$$H(\boldsymbol{y}_{N+1}^k)\tau_f + \tau_f^k \boldsymbol{B}(\boldsymbol{y}_{N+1}^k)(\boldsymbol{y}_{N+1} - \boldsymbol{y}_{N+1}^k) + \Delta = 0 \tag{6.39}$$

$$p_j(r_i^k, V_i^k) + p_j'(r_i^k, V_i^k)[r_i - r_i^k, V_i - V_i^k]^T \le p_{j,\max} \tag{6.40}$$

$$\left\| \boldsymbol{y}_i - \boldsymbol{y}_i^k \right\|_2 \le \delta_i, \ \left\| \tau_f - \tau_f^k \right\|_2 \le \Gamma / t_s \tag{6.41}$$

where $\lambda_{r,f} = -1$, $\lambda_{\gamma,f} = 0$, $\lambda_{s,f} = 0$, $i = 1, \cdots, N+1$, $j = 1, 2, 3$, and k is the index of iterations during the sequential convex programming, variables with the superscript 'k' represent the solution obtained in the k-th iteration and act as the reference trajectory for the subsequent iteration.

Moreover, the objective function of P6.2 is redefined as follows:

$$J = w_v \|\overline{v}\|_2 + w_\delta \left\|\overline{\delta}\right\|_2 + w_\Delta \|\Delta\|_2 \tag{6.42}$$

in which $\overline{v} = [\boldsymbol{v}_1; \cdots; \boldsymbol{v}_{N+1}]$, $\overline{\delta} = [\delta_1, \cdots, \delta_{N+1}]$, $\boldsymbol{y}_i$, and Δ are four decision variables of the SCP.

The ISCP algorithm is summarized in Table 6.1.

Table 6.1 The flowchart of Indirect Sequential Convex Programming

Step	Procedure
1	Let the iteration index $k = 0$, propagate the nonlinear dynamics Eq. (6.32) with the initial state vector $\boldsymbol{x}_0$, costate vector $\boldsymbol{\lambda}_0$, and terminal time τ_f^0 to generate the initial guess trajectory $\{\boldsymbol{y}_1^0, \cdots, \boldsymbol{y}_{N+1}^0\}$
2	For $k \ge 1$, solve the convex problem P6.2 defined as follows to find the optimal solution $\{\boldsymbol{y}_1^k, \cdots, \boldsymbol{y}_{N+1}^k\}$ and variables $\overline{v}$, $\overline{\delta}$, Δ by taking the previous solution $\{\boldsymbol{y}_1^{k-1}, \cdots, \boldsymbol{y}_{N+1}^{k-1}\}$ as the reference trajectory:
P6.2: Minimize Eq. (6.42) Subject to Eqs. (6.37)–(6.41)	
3	If $\sup\|\boldsymbol{v}_i\|_2 \le v_{\max}$, $\sup\|\delta_i\|_2 \le \delta_{\max}$, then go to Step 4, where $v_{\max}$ and $\delta_{\max}$ are the customized convergence thresholds; otherwise let $k = k + 1$ and go to step 2
4	Algorithm stops, the numerical optimal trajectory for Mars entry with maximum terminal altitude is found to be $\{\boldsymbol{y}_1^k, \cdots, \boldsymbol{y}_{N+1}^k\}$, and the optimal control profile can be obtained through Eq. (6.31)

6.3.2 *Further Discussions of UTM, SCP, and ISCP*

The SCP method is used to solve the TPBVP generated by the modified UTM method. Hence, this method is a hybrid of the indirect and direct methods. To illustrate the connection and difference between the above types of methods, Fig. 6.4 is given as below:

As presented in Fig. 6.4, the first path colored in blue represents the process of UTM. It transcribes the original Mars entry terminal altitude maximization problem into a TPBVP and then solves the resulted boundary value problem with the MATLAB function bvp4c in the framework of homotopy. Although UTM satisfies the first-order necessary conditions, it has two drawbacks. First, the PMP must be examined repeatedly due to the two optimal options in Eq. (6.23). Second, it is a challenge to determine the proper continuation sets of parameters, and solving these auxiliary problems is also time-consuming. The traditional SCP method is represented by the No. 2 green path. This approach can generate a solution in polynomial time that is at least feasible. However, it is typically difficult to acquire the costate information. The third path in red denotes the proposed ISCP method. This method first improves the UTM from two optimal solution options into a unique and continuous one, then solves the corresponding TPBVP by formulating it as a convex optimization problem. Thus, the TPBVP can be solved in polynomial time using only finite-time convex optimizations rather than repeatedly solving a large number of auxiliary TPBVPs. Furthermore, if the continuation sets are poorly defined, the UTM cannot converge, whereas ISCP at least guarantees a feasible solution.

6.4 Numerical Demonstration

This section verifies the effectiveness of the ISCP in the Mars entry problem by comparing it with UTM and the general pseudospectral optimal problem software (GPOPS). All simulations are run in the MATLAB R2017b environment. The parameters for simulation are given as follows: $R_0 = 3397$ km, $g_0 = 3.712$ m/s^2, $\rho_0 = 0.0158$ kg/m^3, $h_s = 9354.5$ m, $m = 3300$ kg, $S_r = 15.9$ m^2, $C_L = 0.348$, $C_D = 1.45$,

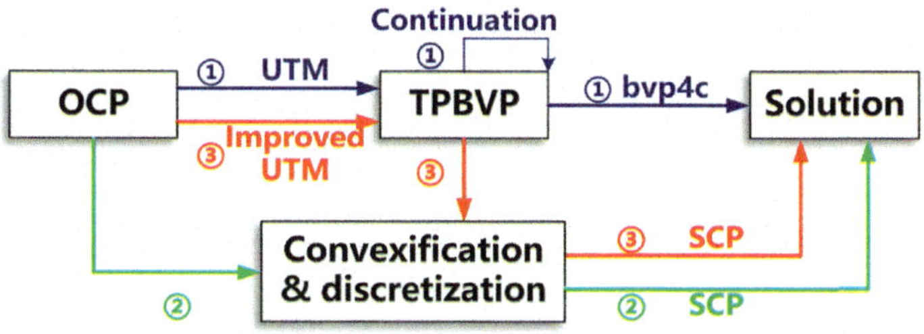

Fig. 6.4 Schematic diagram of UTM, SCP, and ISCP

$R_n = 0.6$ m, $\sigma_{\min} = 30°$, $\sigma_{\max} = 120°$, $k_Q = 1.9027 \times 10^{-4}$, $q_{\max} = 10$ kPa, $\dot{Q}_{\max} = 70$ W/cm^2, $n_{\max} = 5\ g_E$, $g_E = 9.8$ m/s^2, $r_0 = 3522$ km, $V_0 = 6000$ m/s, $\gamma_0 = -11.5°$, $s_0 = 0$ km, $V_f = 540$ m/s, $\varepsilon_c = 1 \times 10^{-10}$, $w_v = 10$, $w_\delta = 1$, $w_\Delta = 30$, $v_{\max} = 1 \times 10^{-3}$, $\delta_{\max} = 1 \times 10^{-3}$, $\Gamma = 20$ s, $\boldsymbol{\lambda}_0 = [0, 0, 0, 0]^T$, and $\tau_f^0 = 280/t_s$. It is noted that the parameters ε_q, $\varepsilon_{\dot{Q}}$, ε_n are set as 0 in the ISCP for both cases with and without path constraints, and 100 discretization intervals are set for ISCP. ISCP uses the modeling toolbox YALMIP and solver MOSEK to solve the convex sub-problem. The GPOPS calls the default nonlinear programming optimizer IPOPT to find the solution with the prescribed 1×10^{-4} mesh tolerance. As for the UTM, the MATLAB's built-in function *bvp4c* is used to solve these TPBVPs generated by the numerical continuation logic. As stated in Sect. 6.2, the continuation process constructs numerous auxiliary TPBVPs by establishing a series of continuation sets to reduce the sensitivity to the initial costate. The parameters of continuation sets must be carefully specified in advance. Hence, Table 6.2 lists all the continuation parameters required in both Mars entry cases with and without path constraints. During the continuation process, a continuation set is generated by updating the value of one parameter step by step while leaving the values of other parameters unchanged. For example, the altitude increases from 50 to 125 km with a step size of 5 km in the case of without path constraints, while another parameter ε_c is equal to its initial value of 1. Then, the altitude holds at 125 km and the parameter ε_c decreases in intervals $[10, 1] \times 10^{-i}$ with a step size of -0.5×10^{-i}, where $i = 1, 2, \ldots, 7$. As for the case with path constraints, parameters $\dot{Q}_{\max}$, $n_{\max}$, and $q_{\max}$ should be initialized with large constants to ensure that the corresponding variables a_q, $a_{\dot{Q}}$, a_n are bounded in the interval. In addition, ε_c, ε_q, $\varepsilon_{\dot{Q}}$, and ε_n should also decrease to small values in order. In this way, there are total 143 and 1035 continuation problems in the cases without and with path constraints, respectively.

Table 6.2 Continuation sets for the cases without and with path constraints

Parameters	Values	
	Without path constraints	With path constraints
r (km)	$[50{:}5{:}125] + R_0$	$[50{:}5{:}125] + R_0$
$\dot{Q}_{\max}$ (W/cm^2)	Free	$[200{:}-1{:}70]$
$n_{\max}$ (g_E)	Free	$[50{:}-1{:}5]$
$q_{\max}$ (kPa)	Free	$[100{:}-10{:}10]$
$\varepsilon_{\dot{Q}}$	0	$[10{:}-0.5{:}1] \times 10^{-i}, i = 1, 2, \ldots, 7$
ε_n	0	$[10{:}-0.5{:}1] \times 10^{-i}, i = 1, 2, \ldots, 7$
ε_q	0	$[10{:}-0.5{:}1] \times 10^{-i}, i = 1, 2, \ldots, 7$
ε_c	$[10{:}-0.5{:}1] \times 10^{-i}, i = 1, 2, \ldots, 7$	$[10{:}-0.2{:}1] \times 10^{-i}, i = 1, 2, \ldots, 10$

6.4.1 Mars Entry Without Path Constraints

The results of Mars entry without path constraints are shown in Figs. 6.5, 6.6, 6.7, and 6.8, including the state, costate variables, control, and path constraints profiles generated via the ISCP, UTM, and GPOPS. It should be noted that the main contribution of this chapter is to improve the UTM and solve the corresponding TPBVP with SCP. Hence, the simulation mainly compares the results generated by the ISCP and UTM. In addition, the result of the well-established pseudospectral optimal control software GPOPS is presented as the benchmark. All results of the three methods are very close, implying that these methods converge to the optimal solution with high accuracy. The costate variables in the GPOPS are estimated by using the KKT multipliers of the LGR pseudospectral method. As depicted in Fig. 6.6, the switching times of the bang–bang control in the ISCP, UTM, and GPOPS are 135.25 s, 135.18 s, and 135.06 s, respectively. Note that the last spike of the optimal control profile in UTM and ISCP is caused by $\lambda_\gamma(\tau_f) = 0$, but this spike does not contribute to the state at the final time. Consequently, this phenomenon is not considered as a control switch. For the objective function, the terminal altitude attains about 11.3673 km for both UTM and GPOPS, whereas it is 11.2855 km for the ISCP, which is only 82 m (0.72%) lower. However, a clear difference is that the entry terminates at 270.50 s in ISCP, whereas the other two methods generate entry trajectories with the flight time of 281.00 s, and 281.38 s, respectively. This difference attributes to the introduced virtual control and buffer. For GPOPS, there are only four elements in the state vector. By contrast, eight elements are included in the lumped state vector $\boldsymbol{y} = [r, V, \gamma, s, \lambda_r, \lambda_V, \lambda_\gamma, \lambda_s]^T$ of ISCP. Obviously, a larger linearization error will be introduced due to the increase of the state vector dimension. Although the magnitude of the linearization error can be suppressed by using the virtual control and buffer, the linearization error affects the objective function of the original optimal control problem as shown in Eq. (6.42). Once the proportions of $\overline{\delta}$ and Δ are large, the optimality of the solution will be compromised. This effect is also shown in Fig. 6.7, where the profile of λ_γ is slightly different from the others. Another intriguing observation in Fig. 6.6 is the shift of the terminal bank angle from 30° to 79.45° for the ISCP and UTM. This is because the zero value of $\lambda_\gamma(\tau_f)$ in Eq. (6.16) leads to zero optimal control solution $\overline{u}$ from Eq. (6.31), then the terminal bank angle of 79.45° is obtained from $\cos\sigma = c_0 + c_1 \sin\overline{u}$. Nevertheless, the terminal bank angle has no effect on the trajectory solution. Also, the dynamic pressure, heat flux, and normal load related to the state variables with respect to their constraints are also consistent as presented in Fig. 6.8.

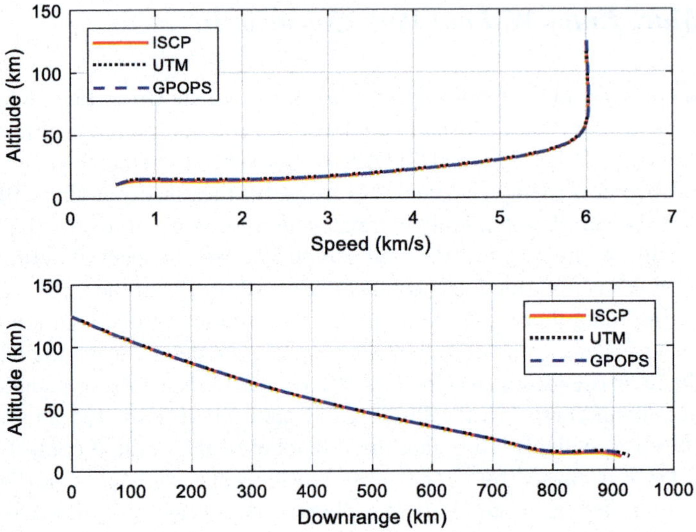

Fig. 6.5 Altitude profiles versus speed and downrange without path constraints

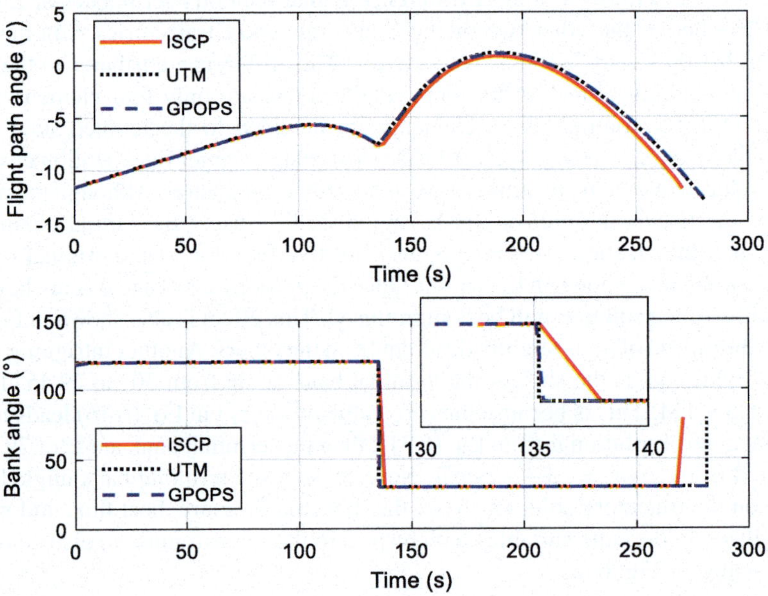

Fig. 6.6 Flight path angle and bank angle histories without path constraints

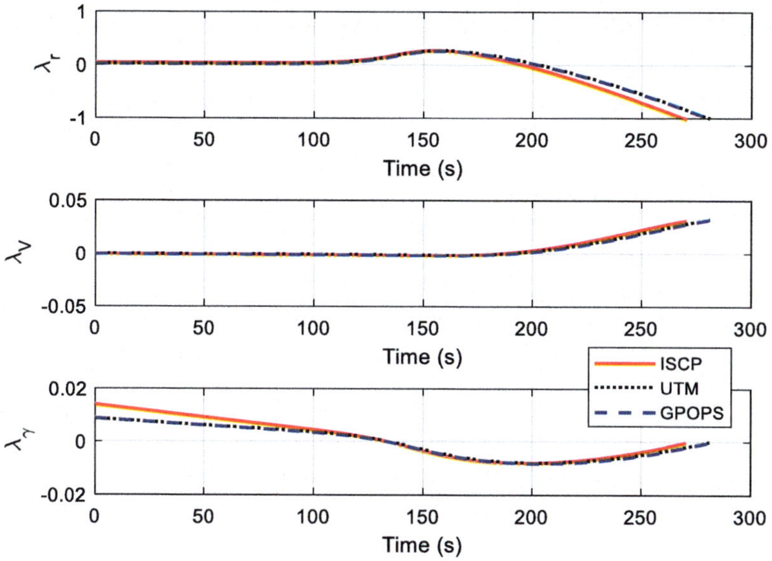

Fig. 6.7 Costate histories without path constraints

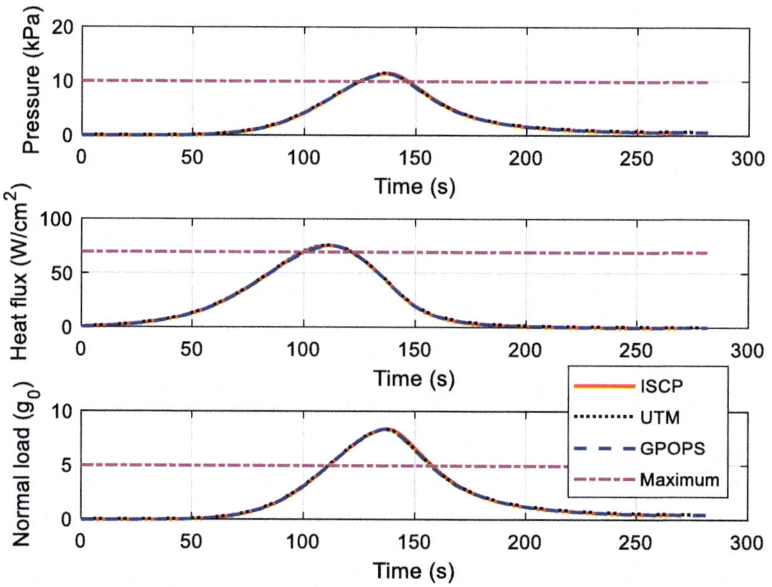

Fig. 6.8 Pressure, heat flux, and normal load histories without path constraints

6.4.2 Mars Entry with Path Constraints

The numerical simulation results of Mars entry with path constraints generated by the ISCP, UTM, and GPOPS are presented in Figs. 6.9, 6.10, 6.11, 6.12, and 6.13. In contrast to the previous case, the solution of ISCP differs from those of UTM and GPOPS, such as the downrange in Fig. 6.9 and flight path angle in Fig. 6.10 of the ISCP. Another most prominent difference is the control structure. All methods take two switches to generate the bang−bang control profile except for the ISCP. The control profile of UTM switches at 112.36 s and 167.91 s, while the control profile of GPOPS switches at 100.24 s and 160.70 s. However, the control of ISCP has only one switch at 166.22 s. Also, the sign of λ_γ in ISCP changes only once as depicted in Fig. 6.11, while the profile of λ_γ crosses zero twice in other methods as shown in Fig. 6.13. This phenomenon has three primary reasons. The first reason is the introduced cosine term in Eq. (6.9). The optimal solution of the original problem and the numerical solution of P6.2 are not completely identical, since the cosine term alters the objective function to some extent. Second, the way of addressing path constraints is different between UTM and ISCP. In ISCP, path constraints are directly considered in the SCP process, rather than being addressed by introducing three secant terms into the objective function as Eq. (6.9). In this case, TPBVPs obtained by improved UTM for unconstrained and constrained Mars entry scenarios are actually the same because the resulted Hamiltonian, the costate equations, and transversality conditions are identical according to the schematic diagram of ISCP described in Sect. 6.3. This difference is also clearly illustrated in Fig. 6.12, where the profiles of path constraints in ISCP is obviously different from that of UTM. Third, the virtual control and buffer relax the original TPBVP, which also has an impact on the optimal terminal altitude. The values of the terminal altitude with ISCP, UTM, and GPOPS are 10.8776 km, 10.4974 km, and 10.4895 km, respectively, and the ISCP achieves the highest altitude. In addition, the nature of the Mars entry problem may attribute to the optimal control profile. As mentioned in some existing works, multiple optimal bank profiles exist in the upper atmosphere region because the bank modulation is not relevant for the weak atmosphere. By contrast, the optimal bank profile is unique in the dense atmosphere region (< 60 km). This point can be seen from Fig. 6.10 and Fig. 6.13, where the three optimal bank profiles after 150 s are close. Besides, the terminal bank angle shifts to 79.45° in the ISCP as explained in the previous case study. As for the flight time, there is not much difference. The ISCP, UTM, and GPOPS terminate the entry at 313.62 s, 316.67 s, and 315.20 s, respectively.

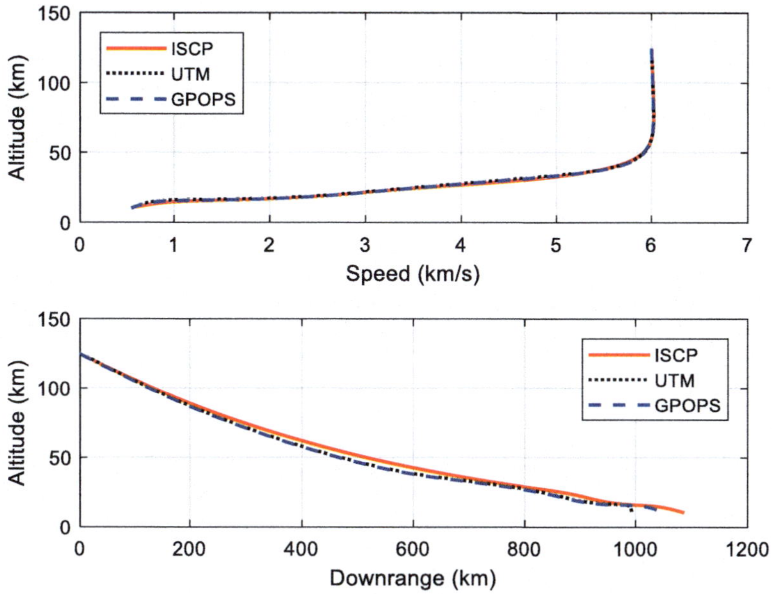

Fig. 6.9 Altitude profiles versus speed and downrange with path constraints

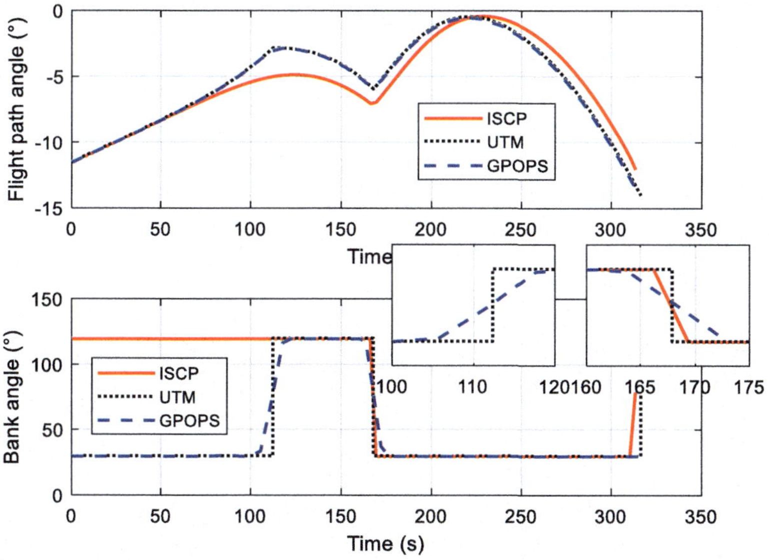

Fig. 6.10 Flight path angle and bank angle histories with path constraints

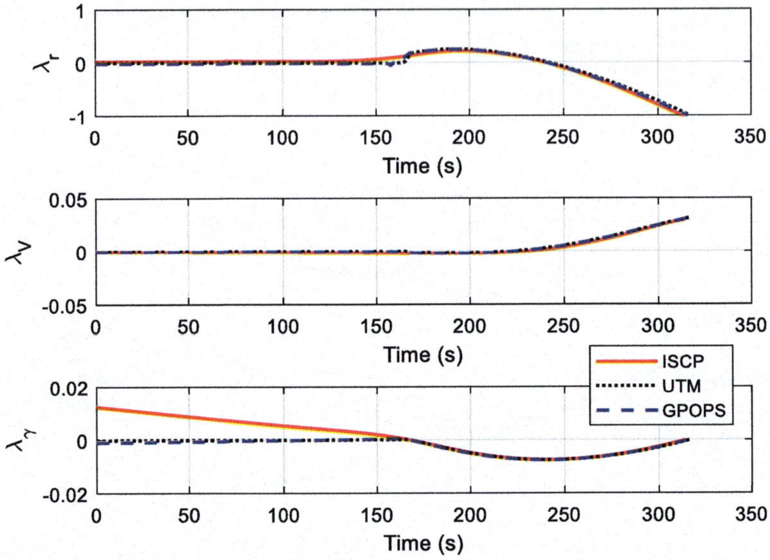

Fig. 6.11 Costate variable histories with path constraints

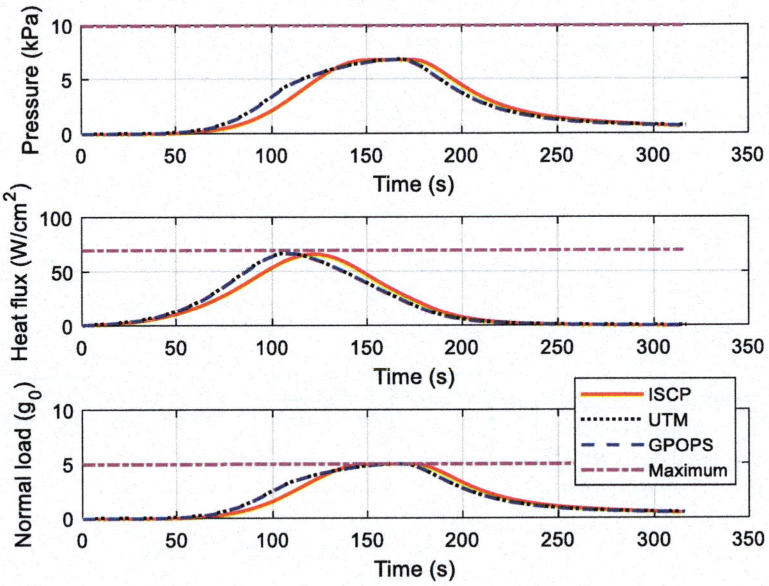

Fig. 6.12 Pressure, heat flux, and normal load histories with path constraints

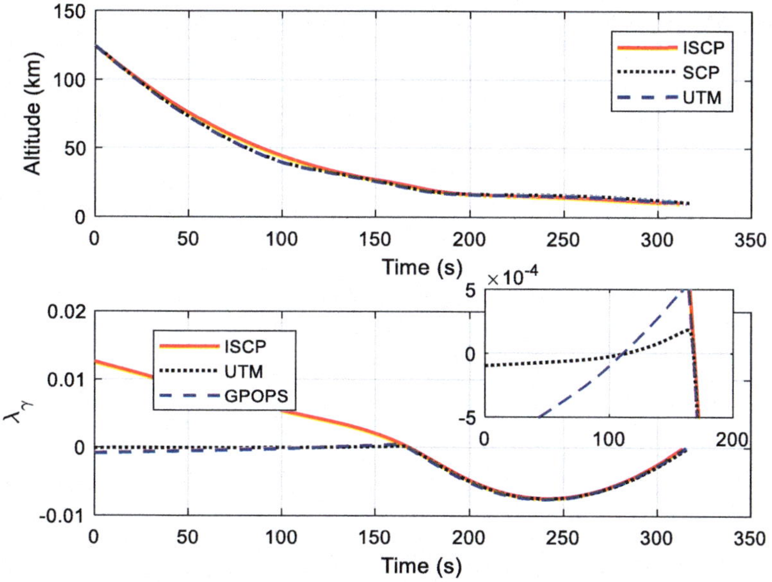

Fig. 6.13 Altitude profiles and costate variable of flight path angle profiles with path constraints

6.4.3 Further Comparison and Analyses

The ISCP provides a more efficient way to find the numerical optimal solution for the maximum terminal altitude problem of Mars entry. More comparison results are presented in Fig. 6.14 and Table 6.3. As shown in Fig. 6.14, ISCP can converge within several iterations. Table 1 shows that the CPU time of ISCP is even less than that of GPOPS with the default solver IPOPT. For UTM, it took hundreds of seconds to solve 143 (357.73 s) and 1035 (1072.78 s) continuation sets. It should be noted that the open-source UTM1.0 toolbox lacks some files for the Mars entry case, hence the two CPU times were measured using our own code rather than that toolbox. Although the UTM1.0 toolbox may improve computational performance, it still takes 36.79 s to solve 110 auxiliary TPBVPs and produce the solution to the Mars aerocapture problem that is similar to the Mars entry mission. However, 6800 auxiliary TPBVPs are configured for the Mars entry terminal altitude maximization problem in UTM1.0. Thus, the UTM obviously requires much more computations to obtain the solution of the Mars entry problem than other methods, even the UTM1.0 toolbox is used.

Nevertheless, it must be stated that the convergence of ISCP relies on the values of the customized weights w_v, w_δ, w_Δ in Eq. (6.42). They represent the proportion of different factors in the objective function, and different combinations of weights may lead to distinct solutions. For the Mars entry problem, the objective function

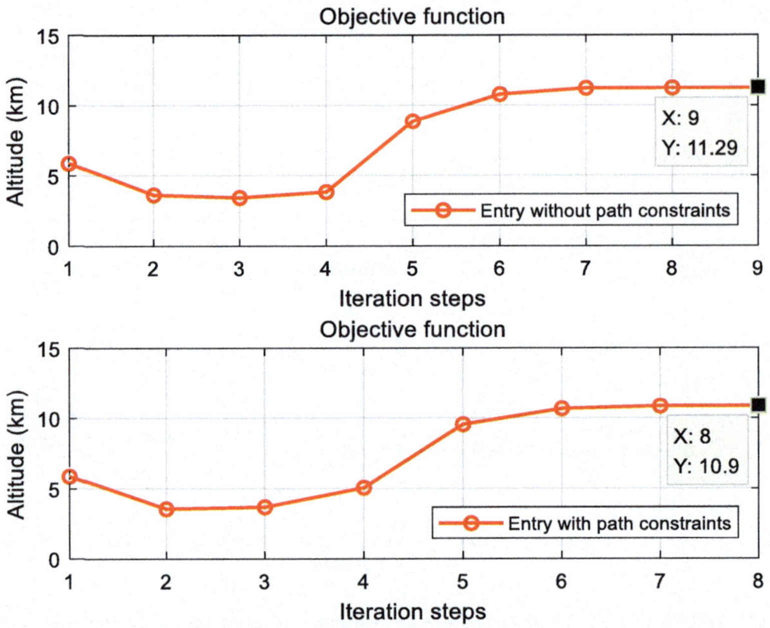

Fig. 6.14 The convergence histories of terminal altitude for ISCP

mainly depends on the virtual control and virtual buffer, which determine whether linearization works and whether the optimal solution is found, respectively. As shown in Table 6.3, the values of the virtual buffer for the cases with (−0.00230) and without (−0.00186) path constraints are a bit large. In fact, the virtual buffer relaxes the nonlinear equality constraint at the final flight time to a linear one, which inevitably has an impact on the optimality of the solution. Both convergence and optimality may be enhanced if the fixed-final-time problem is investigated, which will be a future consideration.

6.5 Discussions

The proposed ISCP algorithm combines the UTM and SCP to generate the bang–bang control profile for the Mars entry terminal altitude maximization problem. For the case without path constraints, the ISCP provides a numerical solution that is close to that of the well-established UTM and GPOPS. Although the solution obtained by the ISCP in the case with path constraints is not featured by the bang–bang structure, the optimal terminal altitude of ISCP is comparable to the results of the other methods. Quantitative analyses also show that the ISCP does not require the time-consuming continuation computations compared with the UTM. Meanwhile, the ISCP provides an alternative way of solving the TPBVP, namely the convex optimization approach.

Table 6.3 Metrics of the Mars entry without/with path constraints

Metrics	Without path constraints			With path constraints		
	ISCP	UTM	GPOPS	ISCP	UTM	GPOPS
Terminal altitude (km)	11.2855	11.3673	11.3673	10.8776	10.4974	10.4895
Terminal speed (m/s)	540	540	540	540	540	540
Terminal time (s)	270.50	281.00	281.38	313.62	316.67	315.20
Switching time (s)	135.25	135.18	135.06	166.22	112.36, 167.91	100.24, 160.70
Maximum pressure (kPa)	11.5127	11.4611	11.4720	6.8200	6.8204	6.8199
Maximum heat flux (W/cm^2)	76.1042	76.0776	76.0789	66.6920	67.0486	66.8362
Maximum normal load (g_E)	8.4404	8.3968	8.4106	4.9966	4.9969	5.0000
Maximum of $\|\boldsymbol{v}_i\|_2$	4.3×10^{-13}	–	–	0.001082	–	–
Maximum of $\|\delta_i\|_2$	2.47×10^{-6}	–	–	4.89×10^{-5}	–	–
Virtual buffer Δ	−0.00230	–	–	−0.00186	–	–
CPU Time (s)	2.7241	357.73	3.6734	5.722	1072.78	9.8672

This approach can not only solve the TAM problem, but also can be applied to solve a class of constrained OCPs without considering the bang–bang or singular control structure as long as the control is affine. At last, ISCP also generates the costate information like GPOPS, and this information has not been attained in the traditional SCP method.

Chapter 7
Mars Entry and Powered Descent Using Collaborative Optimization

7.1 Problem Formulation

Since the flight process of the high-mass mid lift-to-drag ratio rigid vehicle includes atmospheric entry and powered descent phases. In order to achieve propellant-optimal guidance, we have to coordinate the behaviors for entry phase and powered descent phase. More specifically, the handover from optimal entry guidance to the optimal powered descent guidance should be adjusted to occur at an appropriately vertical- and horizontal-range to the predefined landing site.

7.1.1 Optimal Mars Atmospheric Entry Problem

In this problem, one needs to find an optimal bank angle (σ^*) profile which is the required entry guidance.

1. **Mars Atmospheric Entry Dynamics**

The entry vehicle is modeled as an unpowered point mass flying in stationary Martian atmosphere. Entry vehicle is commanded to fly at a constant trim angle-of-attack. The three degree-of-freedom dynamical equations defined with respect to the Mars centered Mars-fixed coordinate system are given by:

$$\dot{r} = V \sin\gamma \tag{7.1}$$

$$\dot{\theta} = \frac{V \cos\gamma \sin\psi}{r \cos\lambda} \tag{7.2}$$

S. Li et al., *Trajectory Optimization and Guidance Methods for Mars Entry*,
https://doi.org/10.1007/978-981-99-6282-2_7

$$\dot{\lambda} = \frac{V}{r} \cos \gamma \cos \psi \tag{7.3}$$

$$\dot{V} = -D - g_m \sin \gamma \tag{7.4}$$

$$\dot{\gamma} = \left(\frac{V}{r} - \frac{g_m}{V} \right) \cos \gamma + \frac{L}{V} \cos \sigma \tag{7.5}$$

$$\dot{\psi} = \frac{V}{r} \sin \psi \cos \gamma \tan \lambda + \frac{L \sin \sigma}{V \cos \gamma} \tag{7.6}$$

where r is the radial distance from the center of Mars to the center of mass of the entry vehicle, V is the velocity of the entry vehicle, θ is the longitude and λ is the latitude, γ is the flight path angle. The heading (azimuth) angle is ψ, and σ is the bank angle defined as the angle about the velocity vector from the local vertical plane to the lift vector. Mars gravitational acceleration $g_m = \mu/r^2$, where μ denotes the gravitational constant of the Mars. L and D are aerodynamic lift and drag accelerations, respectively, defined by $D = \frac{1}{2}\rho v^2 C_D S/m$ and $L = \frac{1}{2}\rho v^2 C_L S/m$, and C_L and C_D are the aerodynamic lift and drag coefficients respectively, S represents vehicle's reference surface area, and m denotes the mass of the entry vehicle. Ballistic coefficient is defined as $B_c = C_D S/m$, and the Mars atmospheric density is defined as $\rho(h) = \rho_0 \exp[\beta(h)]$, where $h = r - r_m$, $\rho_0 = \beta_1 / \beta_2 T$, $T = 1.4 \times 10^{-13} h^3 - 8.85 \times 10^{-9} h^2 - 1.245 \times 10^{-3} h + 205.3645$, $\beta_1 = 559.351005946503$, $\beta_2 = 188.95110711075$, $\beta(h) = -0.000105h$, r_m is reference radius of the Mars.

2. **Process Constraints**

The entry vehicle's normal overload, dynamic pressure and aerodynamic heating rate should meet certain constraints during Mars atmospheric entry.

$$n_a = \sqrt{L^2 + D^2} \leq n_{\max} \tag{7.7}$$

$$q = \tfrac{1}{2}\rho V^2 \leq q_{\max} \tag{7.8}$$

$$\dot{Q} = k_Q \rho^{0.5} V^{3.15} \leq \dot{Q}_{\max} \tag{7.9}$$

where n_a, q, $\dot{Q}$ are the normal overload, dynamic pressure and aerodynamic heating rate, respectively; $n_{\max}$, $q_{\max}$, $\dot{Q}_{\max}$ are the tolerable maximum values, and k_Q is the heat flux coefficient.

The bank angle and its maneuver rate are constrained with prescribed bounds as follows

$$|\sigma| \leq \sigma_{\max} \tag{7.10}$$

$$|\dot{\sigma}| \le \dot{\sigma}_{\max} \tag{7.11}$$

where $\sigma_{\max}$ and $\dot{\sigma}_{\max}$ are the corresponding limit values.

3. **Boundary Conditions**

The entry trajectory of the vehicle starts from the initial conditions and ends at a point where the prescribed terminal conditions are met. The initial conditions are assumed to be specified:

$$\left[r(t_0), \theta(t_0), \lambda(t_0), V(t_0), \gamma(t_0), \psi(t_0)\right] = \left[r_0, \theta_0, \lambda_0, V_0, \gamma_0, \psi_0\right] \tag{7.12}$$

where t_0 denotes the initial time of entry phase, $r_0 = r_m + h_0$, and $h_0, \theta_0, \lambda_0, V_0, \gamma_0, \psi_0$ are the given initial altitude, longitude, latitude, velocity, flight path angle, and azimuth angle. The terminal conditions are assumed to be specified:

$$\left[r(t_f), \theta(t_f), \lambda(t_f)\right] = \left[r_f, \theta_f, \lambda_f\right] \tag{7.13}$$

where t_f denotes the terminal time of entry phase, $r_f = r_m + h_f$, and h_f, θ_f, λ_f are the desired terminal altitude, longitude, and latitude. Note that the desired terminal three-dimensional location will be optimized subsequently.

4. **Objective Function**

Unlike in a Mars robotic EDL mission employing parachute deceleration where it is desired to achieve as high as possible a final altitude of the entry phase, one has found that using a lower energy in a heavy-mass Mars landing mission with supersonic retro-propulsion system is very effective to produce a low-velocity powered descent initiation condition for propellant usage reduction in powered descent [154, 155]. Hence, the objective function is defined as minimizing the final energy of the entry phase, that is

$$\min \quad J_E = E(t_f) = \frac{[V(t_f)]^2}{2} - \frac{\mu}{r(t_f)} \tag{7.14}$$

7.1.2 *Optimal Mars Powered Descent Problem*

This problem is to find an optimal thrust ($\mathbf{T}^*$) profile which is the desired powered descent guidance.

1. **Mars Powered Descent Dynamics**

In formulating the guidance problem for Mars powered descent, we model the dynamics using the following three-degree-of-freedom dynamic equations with respect to a Cartesian coordinate system with the origin at the center of mass of the Mars. The major forces acting on the vehicle are gravitational force and thrust force. Other forces acting on the vehicle, such as aerodynamic disturbance and gravitational perturbation, are typically minimal and ignored here. Thus, the dynamical equations for powered descent can be expressed as follows [154, 155].

$$\dot{\mathbf{r}} = \mathbf{V} \tag{7.15}$$

$$\dot{\mathbf{V}} = -\frac{\mu}{\|\mathbf{r}\|^3}\mathbf{r} + \frac{\mathbf{T}}{m} \tag{7.16}$$

$$\dot{m} = -\frac{\|\mathbf{T}\|}{I_{sp} g_0} \tag{7.17}$$

where $\mathbf{r}$ and $\mathbf{V}$ are the position vector and velocity vector of the vehicle respectively, $\mathbf{T}$ is the commanded thrust vector with the thrust magnitude $\|\mathbf{T}\| = \sqrt{T_x^2 + T_y^2 + T_z^2}$, I_{sp} is the specific impulse of the retro-rocket, and Earth's sea-level gravitational acceleration $g_0 = 9.8\,\text{m/s}^2$.

2. **Process Constraints**

The available onboard fuel should be sufficient for the entire powered descent process, thus the mass of the vehicle satisfies

$$m_{dry} \le m(t) \le m_0 \tag{7.18}$$

where m_{dry} denotes the dry mass of the vehicle. The magnitude of the thrust generated by retro-rocket engine is limited by

$$0 \le \Gamma_{\min} \le \|\mathbf{T}\| \le \Gamma_{\max} \tag{7.19}$$

where $\Gamma_{\min}$ and $\Gamma_{\max}$ are the minimal and maximal values of the magnitude of thrust.

3. **Boundary Constraints**

Once the ignition opportunity of powered descent is determined, the initial state of the powered descent would be given as

$$\mathbf{r}(t_i) = \mathbf{r}_i,\ \ \mathbf{V}(t_i) = \mathbf{V}_i,\ \ m(t_i) = m_i \tag{7.20}$$

where t_i denotes the initial time of powered descent, $\mathbf{r}_i, \mathbf{V}_i, m_i$ are the initial position vector, velocity vector, and mass of the vehicle in the powered descent phase respectively. Note that the initial position vector is a handover condition that will be optimized subsequently.

To achieve pin-point soft landing, the terminal location should close to the target landing site $\mathbf{r}^* = \left[r_x^*, r_y^*, r_m\right]^{\mathrm{T}}$, and the terminal velocity should close to zero. Thus, the terminal conditions in this problem take the form as

$$\left[r_x\left(t_f\right) - r_x^*\right]^2 + \left[r_y\left(t_f\right) - r_y^*\right]^2 \leq \Delta r_{xy}^2, \quad r_z\left(t_f\right) - r_m \leq \Delta r_z \tag{7.21}$$

$$\begin{aligned} &-2\,\text{m/s} \leq V_z\left(t_f\right) \leq 0\,\text{m/s}, \left|V_x\left(t_f\right)\right| < V_z\left(t_f\right)\tan(\Delta_{angle}), \\ &\left|V_y\left(t_f\right)\right| < V_z\left(t_f\right)\tan(\Delta_{angle}) \end{aligned} \tag{7.22}$$

$$\left|T_x\left(t_f\right)/T_z\left(t_f\right)\right| < \tan(\Delta_{angle}), \left|T_y\left(t_f\right)/T_z\left(t_f\right)\right| < \tan(\Delta_{angle}) \tag{7.23}$$

where t_T denotes the terminal time of powered descent, and Δr_{xy} and Δr_z are the horizontal and vertical deviation limits of the final flameout position. Equations (7.22) and (7.23) constrain the vehicle's final touchdown close to the vertical attitude using Δ_{angle}.

4. **Objective Function**

Here, we only take the optimal propellant consumption into consideration. That is

$$\min J_P = \int_{t_i}^{t_T} \|\mathbf{T}(t)\|\,\mathrm{d}t \tag{7.24}$$

It signifies that the vehicle costs the least amount of propellant to terminate its powered descent precisely at the target landing site with zero velocity.

7.2 Integrated Guidance Strategy Design

For a given entry vehicle, entry condition, and landing site, the propellant usage obviously is affected by the condition at powered descent initiation and the powered descent guidance [155]. Since propellant-optimal guidance has already been adopted in the powered descent, the key problem for designing integrated guidance is how to determine an appropriate condition for ignition, i.e. the optimal handover condition from entry phase to the powered descent.

7.2.1 Optimal Handover

1. **Handover State Transformation**

To combine the entry phase with the powered descent phase, one should firstly clarify the transformation relationship of handover state between terminal state of the entry phase and the initial state of the powered descent phase. The transformation can be expressed as

$$\left[r_f, 0, 0\right]^{\mathrm{T}} = \boldsymbol{C}_1(t_f)\boldsymbol{r}_i \tag{7.25}$$

$$[V_f, 0, 0]^{\mathrm{T}} = \boldsymbol{C}_2(t_f)\boldsymbol{V}_i \tag{7.26}$$

$$\boldsymbol{C}_1 = \begin{bmatrix} \cos\lambda(t)\cos\theta(t) & \cos\lambda(t)\sin\theta(t) & \sin\lambda(t) \\ -\sin\theta(t) & \cos\theta(t) & 0 \\ -\sin\lambda(t)\cos\theta(t) & -\sin\lambda(t)\sin\theta(t) & \cos\lambda(t) \end{bmatrix} \tag{7.27}$$

$$\begin{aligned} \boldsymbol{C}_2 = & \begin{bmatrix} \cos\gamma(t)\cos\psi(t) & \cos\gamma(t)\sin\psi(t) & -\sin\gamma(t) \\ -\sin\psi(t) & \cos\psi(t) & 0 \\ \sin\gamma(t)\cos\psi(t) & \sin\gamma(t)\sin\psi(t) & \cos\gamma(t) \end{bmatrix} \begin{bmatrix} 0 & 0 & 1 \\ 0 & 1 & 0 \\ -1 & 0 & 0 \end{bmatrix} \\ & \begin{bmatrix} \cos\lambda(t)\cos\theta(t) & \cos\lambda(t)\sin\theta(t) & \sin\lambda(t) \\ -\sin\theta(t) & \cos\theta(t) & 0 \\ -\sin\lambda(t)\cos\theta(t) & -\sin\lambda(t)\sin\theta(t) & \cos\lambda(t) \end{bmatrix} \end{aligned} \tag{7.28}$$

where $\boldsymbol{C}_1$ and $\mathbf{C}_2$ are corresponding transformation matrices. Note that the three-dimensional position (Eq. (7.25)) for handover is determined through solving an optimal handover problem, while the three-dimensional velocity (Eq. (7.26)) for handover is based on the entry trajectory evolution under specific optimal entry guidance.

2. **Optimal Handover Problem**

For minimizing the propellant consumption, the most important contribution that the entry guidance can make is to offer an entry trajectory which permits a low mechanical energy for ignition at a specific handover location. Meanwhile, the powered descent requires sufficient space margin for optimal propellant consumption deceleration and realizes safely touchdown at a target landing site. Therefore, an optimal handover problem is formulated as:

Find the handover location $\mathbf{r}_i$, ($\mathbf{r}_i$ can be transcribed into $(r_f, \theta_f, \lambda_f)$ through Eqs. (7.25) and (7.27));

Minimize the performance index J_P defined in Eq. (7.24);

Subject to

(a) Mars atmospheric entry dynamics (Eqs. (7.1)–(7.6)), process constraints (Eqs. (7.7)–(7.11)), initial conditions (Eq. (7.12)), and final condition:

$$\begin{aligned} & r_m < r_f,\ \left(\theta_f - \theta^*\right)^2 + \left(\lambda_f - \lambda^*\right)^2 \le \Delta_{\theta\lambda}^2, \\ & \left[\theta^*, \lambda^*\right] = \arg\left\{\mathbf{C}_1(\theta, \lambda)\mathbf{r}^* = [r_m, 0, 0]^{\mathrm{T}}\right\} \end{aligned} \tag{7.29}$$

(b) Mars powered descent dynamics (Eqs. (7.15)–(7.17)), process constraints (Eqs. (7.18) and (7.19)), final conditions (Eqs. (7.21)–(7.23)), and initial conditions:

$$\mathbf{r}(t_i) = \mathbf{C}_1^{-1}\left[r(t_f), 0, 0\right]^{\mathrm{T}},\ \mathbf{V}(t_i) = \mathbf{C}_2^{-1}\left[v(t_f), 0, 0\right]^{\mathrm{T}},\ m(t_i) = m_i \tag{7.30}$$

where $\Delta_{\theta\lambda}$ is a specific range near the longitude and latitude of the target landing site, operator arg{·} represents the solution of the equation in brackets. Note that is set based on the design experience of reachable and controllable sets for planetary entry and landing [110], so as to improve computational efficiency.

7.2.2 Integrated Guidance Framework

A new integrated guidance framework for high-mass Mars entry and powered descent is designed as shown in Fig. 7.1. In the computational framework, optimal guidance behaviors for the entry phase and powered descent are produced through successively solving the optimal Mars atmospheric entry problem and optimal Mars powered descent problem. To appropriately connect these two phases and minimize propellant consumption, an optimal handover location is determined by solving the optimal handover problem. In order to obtain the optimal solution iteratively, large amount of trial solutions from both the optimal Mars atmospheric entry problem and optimal Mars powered descent problem are required.

Clearly, the optimal Mars atmospheric entry problem and optimal Mars powered descent problem are typical optimal control problems which can be solved numerically through direct methodology. In order to efficiently address these two problems with high dimension and high non-linearity, *hp*-adaptive pseudospectral method is employed in this study. Unlike typical pseudospectral methods, *hp*-adaptive pseudospectral method taps the advantages of both the finite element method and the global pseudospectral method to improve the accuracy of the solution in a computationally efficient manner [156].

For solving the optimal handover problem, large amount of trial data input from those two problems would result in great-scale computations. And those trial data

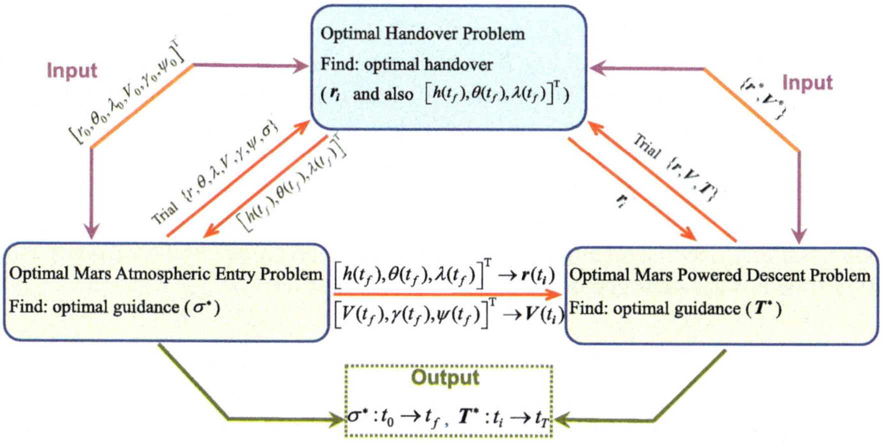

Fig. 7.1 Computational framework of integrated guidance

is stochastic for the optimal handover problem, thus causing inevitably convergence difficulty when using typical deterministic optimization algorithms. Unlike the classic optimal control framework where the system must be modeled as a series of deterministic differential equations, the system model formulated in reinforcement learning framework can be a statistical model learned from large-scale experimental data [157]. Therefore, to ensure efficient convergence to the optimal solution of the optimal handover problem, reinforcement learning method is used in this study. The details of the reinforcement learning and *hp*-adaptive pseudospectral methods are presented in Sect. 7.4.

7.3 Methods of Solution

7.3.1 Reinforcement Learning

The concepts used to define the reinforcement learning (RL) problem have clear counterparts to those used to define the typical optimal control problems: the controller becomes the policy, the control variable becomes the action, the dynamics becomes the state transition probabilities, and the cost function becomes the utility, whereas the state remains the state [158, 159]. RL concerned with how dose an agent (e.g. the vehicle) take appropriate actions under stochastic input conditions to maximize a cumulative reward (e.g. minimize fuel consumption).

1. **Markov Decision Process**

In the RL framework, the optimal handover problem is cast as a Markov decision process (MDP) where the transition between states for a given action is modeled using appropriate transition probability. The MDP is consisted of the following five factors [158, 159]: (1) state set $\boldsymbol{S}$, and $\boldsymbol{s}$ denotes one state in $\boldsymbol{S}$; (2) action set $\boldsymbol{A}$, and $\boldsymbol{a}$ denotes one action in $\boldsymbol{A}$; (3) reward function $R(\boldsymbol{s})$, which maps a state to real number set; (4) state transition probability P_s, which defines that the probability distribution over the new state $\mathbf{s}' \in \mathbf{S}$ is transitioned to the desired final/next state under the given action $\boldsymbol{a}$; (5) optional discount rate κ_d, which is typically used for infinite horizon problems.

In implementation, RL algorithm conventionally learns a policy $Pol(\mathbf{s}) : \mathbf{S} \mapsto \mathbf{A}$ that maps each state to an optimal action. Those actions, which can maximize the policy's utility over the resulting system state trajectory, are optimal for a given state trajectory. For a state trajectory I, the utility is defined as

$$U^{(i)}(Pol) = E\left[R\left(\mathbf{s}_0^{(i)}\right) + \kappa_d R\left(\mathbf{s}_1^{(i)}\right) + \kappa_d^2 R\left(\mathbf{s}_2^{(i)}\right) + \cdots\right] \tag{7.31}$$

where $E[\cdot]$ represents mathematical expectation.

Any possible starting location will define a unique trajectory over a certain scope. For the optimal handover problem and given the initial and target states, reinforcement learning means that a policy that will generate an optimal action (i.e. handover location) is learned through simulated experience of both the entry and powered descent problem. For this type of problem, one can estimate a policy's utility according to the sample trajectories (i.e. trial results of the optimal entry and powered descent problems). That is

$$U(Pol) = \min U^{(i)}(Pol) \tag{7.32}$$

Generally, RL determines the optimal policy using a value function. For a given policy, the value function can be expressed in the form of

$$Va^{Pol}(\mathbf{s}) = R(\mathbf{s}) + \kappa \sum_{\mathbf{s}' \in \mathbf{S}} P_s(\mathbf{s}') Va^{Pol}(\mathbf{s}') \tag{7.33}$$

The optimal value function $Va^*(\mathbf{s})$ can be easily found by dynamic programming algorithm [26]. Then, the optimal policy can be determined by

$$Pol^*(\mathbf{s}) = \arg \max_{\mathbf{a} \in \mathbf{A}} \sum_{\mathbf{s}' \in \mathbf{S}} P_s(\mathbf{s}') Va^*(\mathbf{s}') \tag{7.34}$$

2. **MDP State, Action, Reward and State Transition Probability Distribution**

The state vector is defined by $\mathbf{s} = \left[r_x, r_y, r_z, V_x, V_y, V_z\right]^{\mathrm{T}}$, which are the position and velocity of the vehicle. The action is defined by the vehicle's thrust vector $\mathbf{a} = \left[T_x, T_y, T_z\right]$. The reward is function of mass cost, which is defined as

$$R(\mathbf{s} = \mathbf{s}_f) = -\left(m_i - m_f\right) \tag{7.35}$$

where $\mathbf{s}_f$ denotes the final state, and m_i and m_f are the initial and final mass of the vehicle.

Generally, the reward function is chosen to be linear in the change in the vehicle's mass. The choice of the reward function is motivated by the fact that during the learning process the system is rewarded if the guidance algorithm chooses trajectories that tend to maximize the reward, i.e., powered descent trajectories that minimize the fuel consumption under the condition of given optimal entry trajectories. That is to say, the best integrated guidance will be found out when the optimal entry trajectory where final state matched with the initial state of the selected optimal powered descent.

To integrate the powered descent guidance with entry guidance, here, the MDP's state transition probability distribution comprises the initial state (i.e. stochastic handover state), which is modeled based on the given dynamics model mentioned in Sect. 7.2. Initial state element is modeled as [158].

$$s_{i,j} = \overline{s}_{i,j} + \Pr(\varepsilon) \tag{7.36}$$

where subscript I denotes initial, $j = 1, 2, ..., 6$, $\overline{s}_{i,j}$ is the mean of $s_{i,j}$, ε is the standard deviation, $\Pr(\varepsilon)$ represents the approximate probabilistic distribution. This equation is obtained through analyzing the probabilistic distribution of the final state set of entry trajectory family by MATLAB's *ksdensity* function [160].

3. **Policy Architecture Approach**

In the RL framework, the policy is determined through simulated experiences, where the agent (i.e. controller) learns the optimal handover that will cause the vehicle descent to the target position with minimum fuel. Clearly, the optimal policy must have the ability to map the current state into a specific action that generates state trajectories with the desired properties mentioned above. In this chapter, the policy is implemented using a multi-layer neural network [161]. A hyperbolic tangent activation function is employed for hidden neurons, and a linear activation function for output neurons. More precisely, the output of hidden layer node j is given by

$$a_j = \tanh\left(\sum_{i=1}^{m} w_{ij}\right) \tag{7.37}$$

where w_{ij} is the weight connecting i-th input to j-th hidden layer node, m denotes the total amount of the input. For the linear layer, the output of hidden layer node k is given by

$$a_k = \sum_{j=1}^{n} w_{jk} \tag{7.38}$$

where w_{jk} denotes the weight connecting hidden layer node j to output layer node k, and n denotes the total amount of the hidden layer node.

4. **Apprenticeship Learning**

This step is to optimize the policy through a series of simulated experiences that are generated using the MDP model above. Policy optimization requires weight initialization followed by an iterative process that tweaks the network weights until the utility is maximized. Here, the apprentice learning technique is implemented by computing the policy for specified number of fuel-optimal trajectories that drive the spacecraft to the target landing site with zero velocity. Optimal trajectories can be generated using MATLAB GPOPS software [119], and the RL framework can be found on https://github.com/pat-coady/.

7.3.2 Hp-Adaptive Pseudospectral Method

In this method, the h-method divides the problem into a series of mesh, while the p-method enlarges the degree of the polynomial to achieve convergence in a single mesh [156]. Thus, the accuracy of optimal solution can be improved through increasing adaptively the degree of the approximate polynomial in the mesh interval and/or refining the mesh.

1. **Adaptive Criteria**

Consider the time interval $t \in [t_0, t_f]$, let $[t_{k-1}, t_k]$ be the k-th interval on which the solution is collocated at N_k points. Let e be an accuracy tolerance for the discretized dynamics and path constraints. The k-th mesh interval is considered to be within the accuracy tolerance if the maximum violation in the constraints equations on $[t_{k-1}, t_k]$ is below e. The control in mesh intervals $1, \ldots, K-1$ can be approximated using N_k-th degree Lagrange polynomial. For $\forall k \in [1, K-1]$, the support points of the Lagrange basis $\hat{L}^{(k)}(\tau)$ are the N_k Legendre–Gauss-Radau (LGR) points [134] and the non-collocated point $\tau = 1$, i.e.

$$u^{(k)}(\tau) = \sum_{i=1}^{N_k+1} u_i^{(k)} \hat{L}_i^{(k)}(\tau),\ \hat{L}_i^{(k)}(\tau) = \prod_{i=1, i\neq j}^{N_k+1} \frac{\tau - \tau_i^{(k)}}{\tau_j - \tau_i^{(k)}} \tag{7.39}$$

Since the final time t_f is not collocated, the control in the final mesh interval K is approximated by use of the $(N_k - 1)$-th degree Lagrange polynomial, i.e.

$$u^{(k)}(\tau) = \sum_{i=1}^{N_k} u_i^{(k)} \tilde{L}_i^{(k)}(\tau),\ \tilde{L}_i^{(k)}(\tau) = \prod_{i=1, i\neq j}^{N_k} \frac{\tau - \tau_i^{(k)}}{\tau_j - \tau_i^{(k)}} \tag{7.40}$$

where the support points of the Lagrange basis $\tilde{L}^{(k)}(\tau)$ are just the N_k LGR points.

The error in the dynamics and constraints equations at M points $\left(\bar{t}_1^{(k)}, \cdots, \bar{t}_M^{(k)}\right) \in \left[t_{k-1}, t_k\right]$ in each mesh interval $k \in [1, \ldots, K]$ can be expressed as [19,20]

$$a_{li}^{(k)} = \left|\dot{x}_i^{(k)}\left(\bar{t}_l^{(k)}\right) - f_i^{(k)}\left(\mathbf{x}_l^{(k)}, u_l^{(k)}, \bar{t}_l^{(k)}\right)\right| \tag{7.41}$$

$$b_{lj}^{(k)} = \mathbf{g}_j^{(k)}\left(\mathbf{x}_l^{(k)}, u_l^{(k)}, \bar{t}_l^{(k)}\right) \tag{7.42}$$

where s; $i = 1, 2, \ldots, 6$; and $j = 1, 2, \ldots, 9$. If $a_{li}^{(k)}$ and $b_{lj}^{(k)}$ is less than e on the current mesh interval, then the specified error tolerance e has been satisfied on this interval. If any $a_{li}^{(k)}$ or $b_{lj}^{(k)}$ is larger than e, then either the mesh interval will be refined or the degree of the polynomial on the interval will be increased.

In order to determine if the mesh should be refined or if the degree of the approximate polynomial should be increased, the curvature of the m-th element of the state approximation in the k-th mesh interval is calculated firstly as criteria, i.e.

$$\kappa^{(k)}(\tau) = \frac{\left|\ddot{X}_m^{(k)}(\tau)\right|}{\left|\left[1 + \dot{X}_m^{(k)}(\tau)^2\right]^{3/2}\right|} \tag{7.43}$$

where $X_m^{(k)}(\tau)$ denotes the component of the state in the k-th mesh interval corresponding to the maximal value of either $a_{li}^{(k)}$. Furthermore, the ratio of the maximum to the mean curvature is defined as

$$r_k = \frac{\kappa_{\max}^{(k)}}{\overline{\kappa}^{(k)}} \tag{7.44}$$

where $\kappa_{\max}^{(k)}$ and $\overline{\kappa}^{(k)}$ are the maximum and mean value of $\kappa^{(k)}(\tau)$, respectively. If $r_k > r_{k\max}$, where $r_{k\max} > 0$ is a user-defined constant, then the mesh should be refined since there exists a large curvature relative to the rest of the mesh intervals. If $r_k < r_{k\max}$, then the degree of polynomial should be increased to obtain a better approximation on the k-th mesh interval.

2. **Increasing Polynomial Degree**

The degree N_k of the polynomial in the k-th mesh interval is determined by [156]

$$N_k = M_d + \text{ceil}\left[\lg\left(e_{\max}^{(k)} / e\right)\right] + A_d \tag{7.45}$$

where M_d denotes the initial degree of the polynomial in the mesh,$A_d > 0$ is an integer constant which controls the growth of the number of collocation points in the mesh, $e_{\max}^{(k)}$ is the maximal value of $a_{li}^{(k)}$ and $b_{lj}^{(k)}$ over l, I, and j, and ceil$(\cdot)$ is the operator that rounds to the next highest integer. In other words, the increase in the degree of the polynomial in the k-th mesh interval depends on the ratio of the maximum error to the specified error tolerance.

3. **Refining Mesh**

The new number of mesh intervals is determined by

$$n_k = \text{ceil}\left[c_e \lg\left(e_{\max}^{(k)} / e\right)\right] \tag{7.46}$$

where c_e is an integer constant which controls the growth of the number of mesh intervals. The locations of the new mesh points are determined using the integral of a curvature density function

$$p(\tau) = c_\tau \kappa(\tau)^{1/3} \tag{7.47}$$

where c_τ is a constant, such that

$$\int_{-1}^{1} p(\zeta)\mathrm{d}\zeta = 1 \tag{7.48}$$

The cumulative distribution function is

$$F(\tau) = \int_{-1}^{\tau} p(\zeta)\mathrm{d}\zeta \tag{7.49}$$

The n_k new mesh points are collocated such that

$$F(\tau_i) = \frac{i-1}{n_k}, \quad 1 \le i \le n_k + 1 \tag{7.50}$$

If $n_k = 1$, then no new sub-intervals are created. Therefore, the minimal value for n_k is two at least.

4. **Algorithm Flow**

The *hp*-adaptive pseudospectral algorithm begins by forming a coarse mesh, using approximate polynomials of a fixed degree M_d on each mesh interval, and solving the resulting NLP problem. For each interval in the mesh, based on the adaptive criteria, we determine if the mesh should be refined or if the degree of the polynomial should be increased. When we refine the mesh, we calculate the new number of mesh intervals via Eq. (7.46) and collocate new mesh points via Eq. (7.50). When we increase the degree of the polynomials, we determine the new degree based on Eq. (7.44). Then, we continue the refinement process (as listed in Table 7.1) until the specified error tolerance is satisfied.

Table 7.1 Refinement process

Step	Procedure
1	Solve the NLP using the current mesh Begin: For $k = 1, \ldots, K$,
2	If $e_{\max}^{(k)} \le e$, then continue the next k
3	If either $r_k \ge r_{k\max}$ or $N_k > M_d$, then refine the k-th mesh interval into n_k sub-intervals based on Eq. (7.47). Set the degree of the polynomials on each of the sub-intervals to be M_d and proceed to the next k
4	Otherwise, set the degree of the polynomials on the k-th sub-interval to be N_k based on Eq. (7.46) End: For $k = 1, \ldots, K$
5	Return to step 1

7.4 Numerical Simulations

7.4.1 Simulation Setup

All simulation computations are performed on a mobile workstation with Intel Core(TM)I7820HQ CPU at 2.9 GHz and 32 GB of memory and the MATLAB R2016b environment. The optimal solutions for Mars entry and powered descent problems are obtained using the open-source MATLAB software GPOPS (4.x version) with the adaptive pseudospectral method and the NLP solver SNOPT. The number of discrete points for state and control variables is 20. The design and target matrices (training set) for the neural network training phase are generated by sampling 1000 trajectories generated using GPOPS, with 15 samples per trajectory. The neural network is designed to have 32 hidden layer nodes. MATLAB's neural network toolbox is adopted to model the network and execute the training process via the Bayesian regularization algorithm [162]. To establish a pre-optimization baseline, in each case, the trained neural network was simulated for 1000 different handover conditions obtained from 1000 optimal entry trajectories. The MATLAB convex programming (CVX) software is adopted for the convex optimization steps [163]. The default configuration of the CVX is used, and the solver SDPT3 4.0 within CVX is called. Perfect navigation is assumed so that the capability of the guidance algorithms can be assessed. Those parameters of the vehicle and Mars, process constraints, and boundary conditions are set as listed in Table 7.2, Table 7.3, and Table 7.4, respectively.

Table 7.2 Physical parameters of the vehicle and Mars

Parameters	Symbols	Values	Units
Initial landing mass	m_0	56×10^3	kg
Dry mass	m_{dry}	43×10^3	kg
Lift-to-drag ratio	C_L/C_D	0.58	–
Reference surface area	S	160	m^2
Ballistic coefficient	B_c	397	kg/m^2
Specific impulse	I_{sp}	360	s
Mars reference radius	r_m	3397×10^3	m
Mars gravitational constant	μ	4.282829×10^{13}	m^3/s^2

Table 7.3 Parameters of process constraints

Parameters	Symbols	Values	Units
Maximum overload	$n_{\max}$	4	g_0
Maximum dynamic pressure	$q_{\max}$	7.7×10^3	Pa
Maximum heat rate	$\dot{Q}_{\max}$	100	W/m^2
Maximum bank angle magnitude	$\sigma_{\max}$	100	deg
Maximum bank angle rate	$\dot{\sigma}_{\max}$	15	deg/s
Maximum thrust	$\Gamma_{\max}$	800×10^3	N
Minimum thrust	$\Gamma_{\min}$	200×10^3	N

Table 7.4 Parameters of boundary conditions

Parameters	Symbols	Values	Units
Entry altitude	h_0	129	km
Entry velocity	V_0	4695	m/s
Initial longitude	θ_0	−1.63	deg
Initial latitude	λ_0	−22.18	deg
Initial flight path angle	γ_0	−10.6	deg
Initial azimuth angle	ψ_0	4.6	deg
Target landing site	(θ^*, λ^*)	(0,0)	deg
Final landing angle limit	Δ_{angle}	10	deg
Searching domain for handover	$\Delta_{\theta\lambda}$	0.6	deg
Final horizontal deviation upper limit	Δr_{xy}	30	m
Final vertical deviation upper limit	Δr_z	2	m

7.4.2 Simulation Results

Scenario 1. Integrated Guidance Using Different Predefined Handover Conditions

In fact, the selection of the location to transition from entry phase to powered descent is a crucial performance variable. In order to illustrate how dramatically the handover location affects the optimal entry trajectory and optimal powered descent trajectory, five different points are selected to be the initial point of the powered descent (also can be transformed into the terminal point of the atmospheric entry), as listed in Table 7.5. Thus, corresponding integrated guidance will be produced through successively solving the optimal Mars atmospheric entry problem and optimal Mars powered descent problem. The results of this scenario will also be used to compare with Scenario 2.

The five trajectories steered by optimal atmospheric entry guidance and optimal powered descent guidance are shown in Fig. 7.2. It is demonstrated that there exist

Table 7.5 Experimental handover locations

Location	Case 1	Case 2	Case 3	Case 4	Case 5	Unite
$\mathbf{r}_i$	[−4, −16,7.5]	[−3.5, −14,6.75]	[−3, −12,6]	[−2.5, −10,5.25]	[−2, −8,4.5]	km

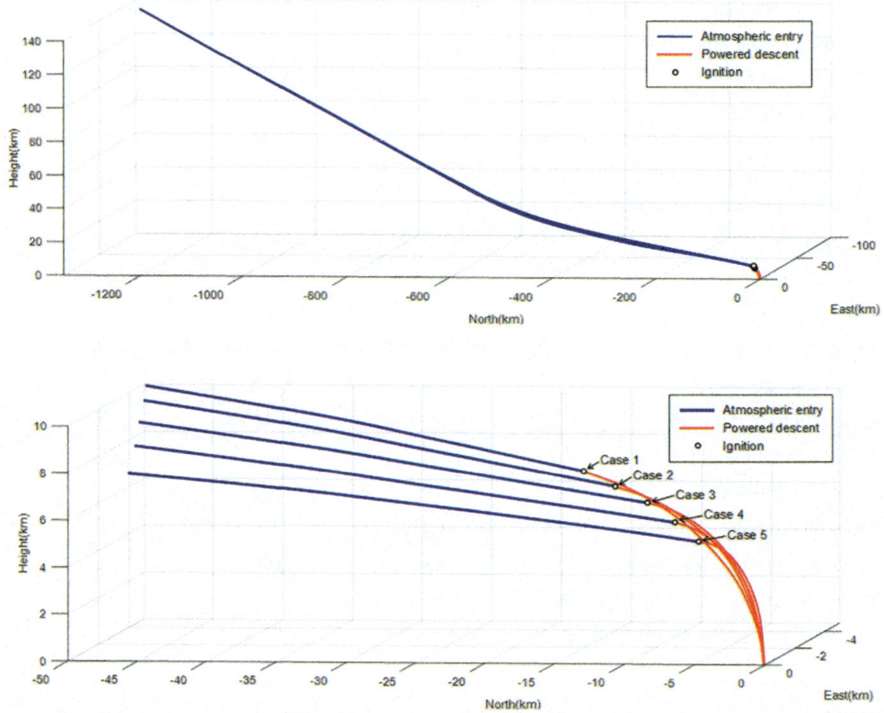

Fig. 7.2 Trajectories in Mars atmospheric entry and powered descent phases for five cases in Table 7.5. (The lower subplot is a local magnified view of the upper subplot.)

optimal atmospheric entry trajectories and corresponding optimal powered descent trajectories connected by the given five handover locations. The altitude-velocity profiles, bank angle profiles and thrust profiles are illustrated in Fig. 7.3, Fig. 7.4, and Fig. 7.5, respectively. The terminal position and velocity of entry trajectories match well with their corresponding initiation states of powered descent. Meanwhile, the optimal guidance can bring a nearly vertical attitude for the final touchdown. In general, the farther away the handover position (ignition location) is from the landing site, the longer the throttle-back phase lasts. This is the most influential factor in the fuel-optimal powered descent responsible for saving propellant. The total flight times among the powered descent for the five cases range from 55 s in case 4 to 82.5 s in case 1. And the total flight time s among the atmospheric entry for the five cases range from 517.5 s in case 1 to 538 s in case 5.

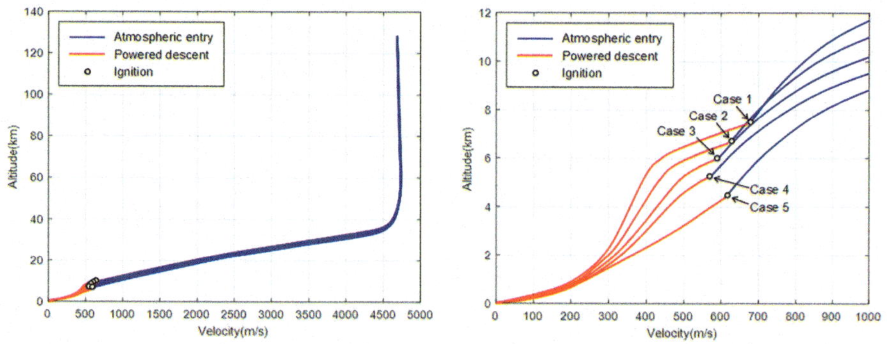

Fig. 7.3 Altitude-velocity profiles in Mars atmospheric entry and powered descent phases for five cases in Table 7.5. (The right side subplot is a local magnified view of the left side subplot.)

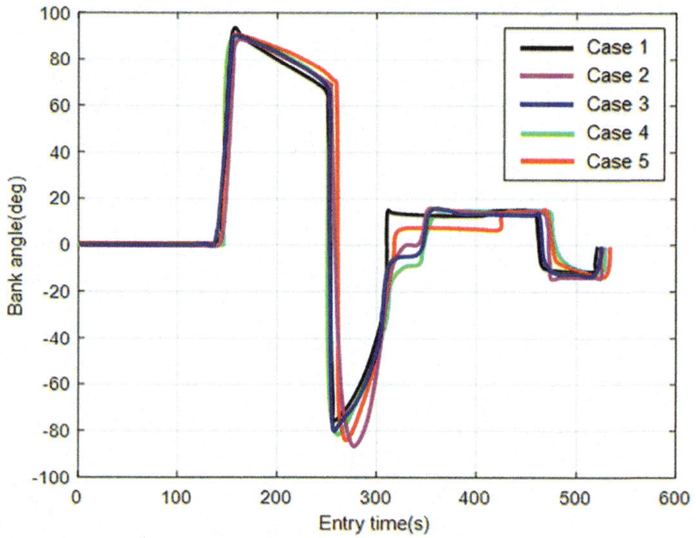

Fig. 7.4 Guidance profiles in Mars atmospheric entry phase for five cases in Table 7.5

The mass comparison of total propellant consumption in five cases is presented in Fig. 7.6. For case-1 the propellant consumption is 11267 kg, while 9812 kg for case-2, 8753 kg for case-3, 8215 kg for case-4, and 11,860 kg for case-5, with the same entry condition and landing condition for the same vehicle. From Fig. 7.4, the case-5 has the smallest height in the five cases and a larger speed than case-3 and case-4, so the thrust throttle for case-5 keeps full and lasts 62 s. Whereas, the full throttle is only 34 s for case-4. The minimal propellant consumption (case-4) saves over 3 metric tons than the maximal propellant consumption (case-5).

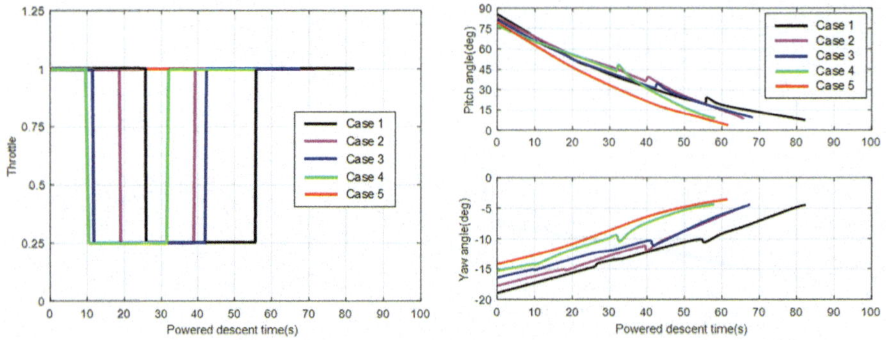

Fig. 7.5 Guidance profiles in Mars powered descent phase for five cases in Table 7.5

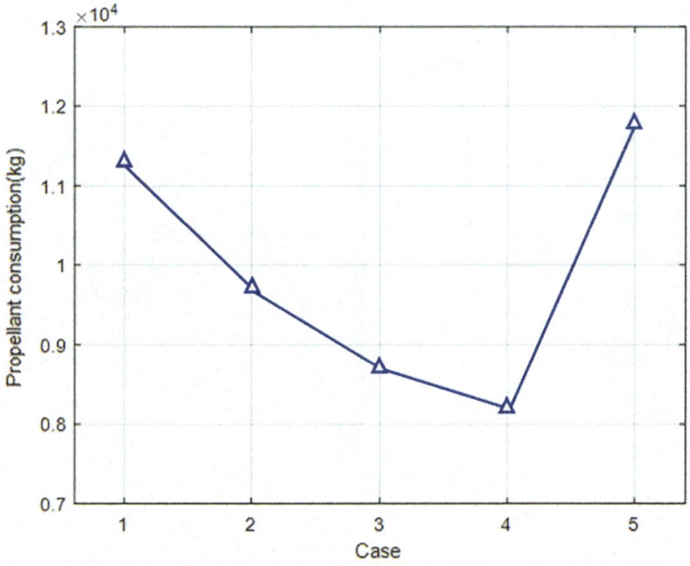

Fig. 7.6 Total propellant consumption in Mars powered descent phase for five cases in Table 7.5

Scenario 2. Optimal Integrated Guidance

In this simulation scenario, the optimal integrated guidance with optimal handover selection developed in Sect. 7.3 is demonstrated. Figure 7.7 shows the combined three-dimensional trajectory. Figure 7.8 presents altitude versus relative velocity trajectory. The bank angle for the atmospheric entry phase and thrust for the powered descent phase are plotted in Fig. 7.9 and Fig. 7.10 respectively. The thrust throttle profile of the powered descent causes the fuel consumption profile illustrated in Fig. 7.11.

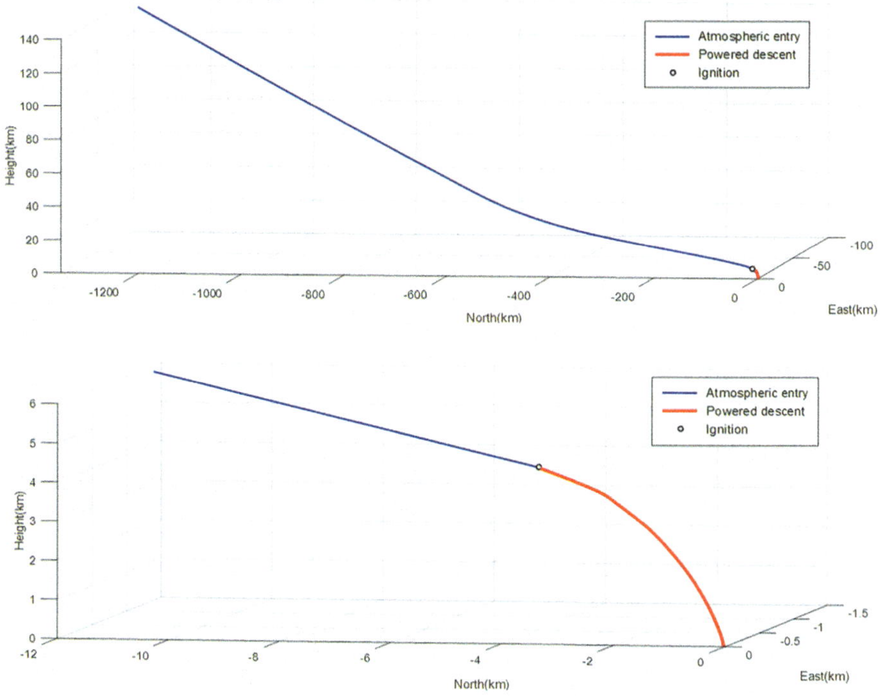

Fig. 7.7 Trajectories in Mars atmospheric entry and powered descent phases. (The lower subplot is a local magnified view of the upper subplot.)

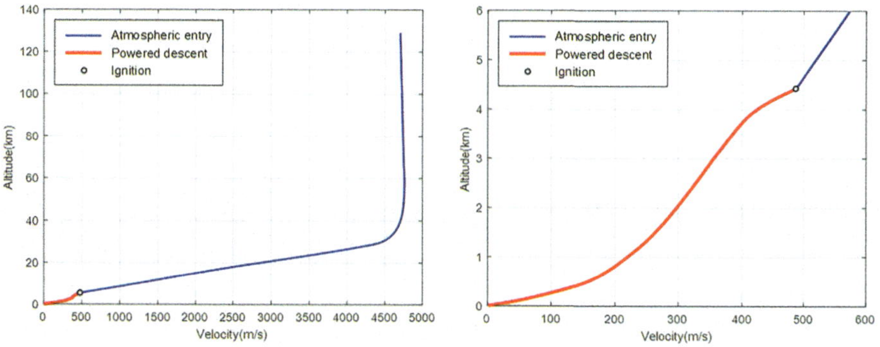

Fig. 7.8 Altitude-velocity profiles in Mars atmospheric entry and powered descent phases. (The right side subplot is a local magnified view of the left side subplot.)

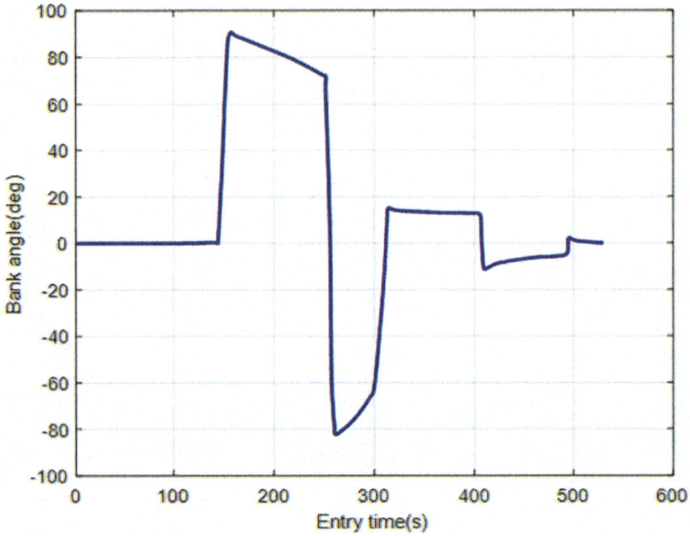

Fig. 7.9 Guidance profile in Mars atmospheric entry phase

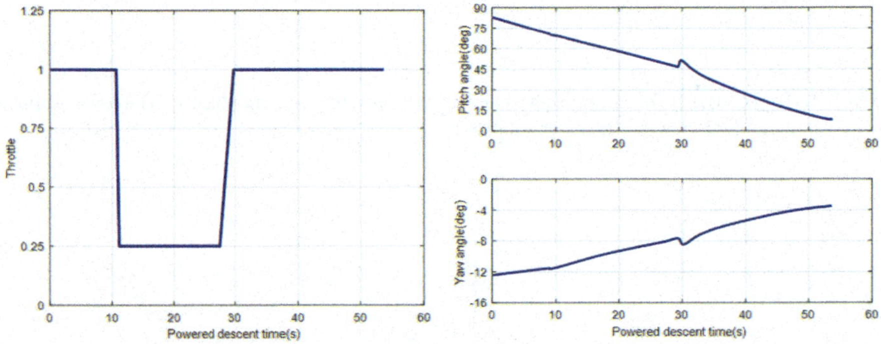

Fig. 7.10 Guidance profile in Mars powered descent phase

Comparing optimal results with Fig. 7.2, 7.3, 7.4, 7.5, and 7.6 in Scenario 1, we can found that the proposed approach is able to effectively find the optimal handover location (ignition condition with low-velocity and just right sufficient range), so as to save the propellant usage in powered descent. Under the same conditions, the optimal handover has lower speed and smaller range to the landing site than those handovers in scenario 1. In other words, the more energy depleted by entry guidance to reach the optimal handover location, the lower propellant consumption is required by powered descent phase. Figure 7.9 shows the total propellant consumption in this scenario is 7923 kg which is less than anyone in scenario 1. Especially, the optimal

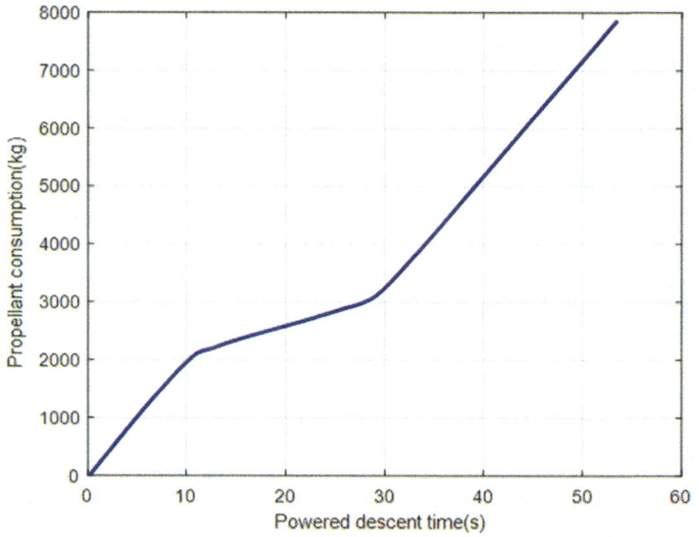

Fig. 7.11 Propellant consumption in Mars powered descent phase

integrated guidance reduces 3937 kg of propellant consumption than the worst case in scenario 1, while it can save additionally 292 kg of propellant consumption than the best case in scenario 1.

7.4.3 Discussions

Although the optimal entry guidance and optimal powered descent guidance can be connected together by the manually given handover positions in scenario 1, and the performance index can reach the optimum on the given conditions for each phase separately, the total fuel consumption cannot guarantee to be a true optimum for the integrated guidance. Therefore, extra propellant usage is inevitable. Scenario 2 illustrates the optimal integrated guidance can save nearly 4 metric tons of propellant consumption than the worst case in scenario 1. Along this entry trajectory, the reinforcement learning algorithm adaptively determines where the powered descent should start, and this start is a best choice for optimal-fuel soft landing. However, with regard to the same vehicle and mission constraints, the propellant consumption will exceed 12 tons if the ignition condition is not well chosen. That is to say, there are 3 to nearly 4 tons of propellant (about 35% to 45% propellant consumption) will be reduced using the proposed technique. Such a contrast highlights the potentially substantial benefit of the integrated guidance approach. The main reasons for the superior performance can be found from the design view point. The optimal handover location provides a specific velocity, motion direction, and remaining range to the powered descent. The remaining range is merely right enough for the powered descent to reduce vehicle's speed to zero while arriving at the target site. Therefore, there is almost no need to take extra propellant for extra maneuvers.

In this study, only handover location is considered to be optimized, while the velocity and direction at the handover location are determined by the flight trajectory steered by the optimal guidance. It means a high/far location would not always match a high speed while a low/near location would not always accompany with a low speed. This is just another indication that a good ignition condition is a function of the overall flight state of the vehicle, not just one single variable. The two scenarios are also demonstrated that the optimal handover in Sect. 7.3 cannot be replaced by a simple trigger on any of the trajectory parameters or time.

The main point in this chapter is how the collaborative optimization framework automatically determines an optimal handover and optimal integrated guidance based on the given initial entry condition and target landing state. And all this effective optimization is done by the reinforcement learning and hp-adaptive pseudospectral algorithms. Since the two algorithms are intelligent and adaptive, there is no need for any manual tweaking. Admittedly, the convergence rate of collaborative optimization can be greatly improved if some rules are added into the optimization procedure design. Such as the bank angle profile is specified as a linear function with respect to entry energy, the thrust profile is prescribed as the form of bang-bang function with regard to descent time. But, it may miss some potential optimal solutions of the integrated guidance problem.

7.5 Conclusions

This chapter employed reinforcement learning and hp-adaptive pseudospectral method to investigate the interplay between entry guidance and powered descent guidance. The study demonstrates how did they integrated working can bring dramatic reduction in propellant consumption of several tons in powered descent phase for a mid-lift high-mass Mars landing mission. The integrated guidance strategy for atmospheric entry and powered descent is a departure from traditional divide-and-conquer strategy in EDL studies. Unlike traditional robotic Mars landing missions with a parachute that requires final altitude maximization in entry phase, integrated guidance designed in this chapter is to minimize the final energy of entry phase for reducing propellant usage in powered descent. On the optimal handover position, the vehicle has a remaining range, velocity, and motion direction which are merely right enough for the powered descent to reduce vehicle's speed to zero while arriving at the target site. There is almost no extra usage of propellant except for conducting this behavior. The minimization of propellant consumption is thus realized through collaborative optimization. Simulation results demonstrate this integrated guidance approach can effectively bring great benefit in reducing propellant consumption. However, the proposed approach requires large-scale computations to converge to the optimal solution. In the future work, we will attempt to add some rules into the optimal integrated guidance framework to reduce unnecessary optimization computations and improve convergence rate.

Part III
Uncertainty Optimization

Chapter 8
Mars Entry Trajectory Optimization with Desensitized Optimal Control

The chapter is to develop the desensitizing optimal Mars atmospheric entry trajectories by use of Desensitivity Optimal Control (DOC) and Direct Collocation Nonlinear Programming (DCNLP). Only the entry trajectory from the atmospheric interface to the parachute deployment is considered here and the state estimated errors at the atmospheric entry point, uncertainties in the atmospheric density and aerodynamic characteristics are considered as entry perturbations. It aims at reducing the sensitivity of the height, velocity and downrange at the parachute deployment with respect to the entry perturbation, while maximizing the height at parachute deployment. This paper is structured as follows: Sect. 8.1 introduces the Mars entry equations and longitudinal dynamic equations used in this paper. Section 8.2 defines the Mars entry problem using traditional optimal control with the maximum height index at parachute deployment. Section 8.3 redefines the Mars entry problem using desensitizing optimal control strategy to reduce the sensitivity of final states with respect to disturbance. Determination of optimal nominal trajectory and control using DCNLP are described in Sect. 8.4. In Sect. 8.5, simulation system compositions are described and results are discussed in detail. Finally, Sect. 8.6 contains conclusion.

8.1 Mars Entry Dynamics

The entry phase of a Mars EDL typically begins when the vehicle entry the Mars atmosphere interface (defined as a radius of 3522 km from the center of Mars) and ends with the deployment of supersonic parachute. As the entry process only lasts a short time, and the decrease of height is much less than the radius of the Mars, a uniform gravitational field is used here and the rotation of the planet is neglected. It is also assumed that the entry vehicle is commanded to fly at a constant angle of attack and the controlled guidance of the vehicle is achieved only through modulating the bank angle of the entry vehicle with a reaction control system (RCS). The dynamic

S. Li et al., *Trajectory Optimization and Guidance Methods for Mars Entry*,
https://doi.org/10.1007/978-981-99-6282-2_8

equations of Mars entry vehicle defined with respect to the Mars fixed coordinate system are given by

$$\begin{aligned}\frac{dr}{dt} &= v\sin\gamma\\ \frac{dv}{dt} &= -(D + g_M \sin\gamma)\\ \frac{d\gamma}{dt} &= \left(\frac{v}{r} - \frac{g_M}{v}\right)\cos\gamma + \frac{1}{v}L\cos\sigma\\ \frac{d\theta}{dt} &= \frac{v\cos\gamma\sin\psi}{r\cos\lambda}\\ \frac{d\lambda}{dt} &= \frac{v}{r}\cos\gamma\cos\psi\\ \frac{d\psi}{dt} &= \frac{v}{r}\sin\psi\cos\gamma\tan\lambda + \frac{L\sin\sigma}{v\cos\gamma}\end{aligned} \tag{8.1}$$

where r is the distance from the center of the Mars to the vehicle center of mass, density, v is the velocity, θ is the longitude and λ is the latitude, γ is the flight path angle. The heading (azimuth) angle is ψ, defined as a clockwise rotation angle starting at due north, and σ is the bank angle defined as the angle about the velocity vector from the local vertical plane to the lift vector. An inverse square gravitational acceleration $g_M = \mu/r^2$, where $\mu = GM_{Mars}$ is assumed. L and D are the aerodynamic lift and drag accelerations, defined by

$$L = \frac{1}{2}\rho v^2 \frac{C_L S}{m}, \quad D = \frac{1}{2}\rho v^2 \frac{C_D S}{m} \tag{8.2}$$

where q is the Mars atmospheric density, C_L and C_D are the aerodynamic lift and drag coefficients, respectively, S represents the vehicle reference surface area, and m is the mass of the entry vehicle. $\rho v^2/2$ is defined as dynamic pressure and $m/(C_D S)$ is called ballistic coefficient. An exponential Mars atmospheric density model is assumed as follows:

$$\rho = \rho_0 \exp\left(-\frac{r - r_M}{h_s}\right) \tag{8.3}$$

where ρ_0 is the density on the surface of the Mars, r_M is the radius of the Mars and h_s is the constant scale height.

$$\frac{ds}{dt} = v\cos\gamma \tag{8.4}$$

Thus, longitude and latitude integrations are not required in this reduced dynamics formulation since they are replaced by the downrange. As the follow-up optimal performance index is a function of height, the state variable r is replaced with the height h in the dynamics equations. The longitudinal model is mainly used for dynamics analysis as follows

$$\frac{dh}{dt} = v\sin\gamma$$

$$\frac{dv}{dt} = -(D + g_M \sin\gamma)$$
$$\frac{d\gamma}{dt} = (\frac{v}{r_M + h} - \frac{g_M}{v})\cos\gamma + \frac{L}{v}u$$
$$\frac{ds}{dt} = v\cos\gamma \tag{8.5}$$

where $u = \cos\gamma$ is the system control input and defined as the following form for simplification.

The Mars entry vehicles have a very limited life force available for control since the atmospheric density is two orders of magnitude lower than on earth. Furthermore, the life coefficient of the entry vehicle is limited by its low-life configuration because the space restriction on the launcher does not allow a winged configuration like for the Space Shuttle. So, the control saturation can be expected for Mars atmospheric entry phase, especially at the higher altitude. As mentioned as above, the control of the entry trajectory is achieved solely through modulating the bank angle σ of the entry vehicle. In order to keep control system with a certain control margin, it is assumed that the control input u is constrained with prescribed bounds

$$u_{\min} \le u \le u_{\max} \tag{8.6}$$

Define state variable $\boldsymbol{x} = [h, v, \gamma, s]^T$, and the entry dynamics can be rewritten as a more concise form:

$$\frac{d\boldsymbol{x}}{dt} = \boldsymbol{f}(\boldsymbol{x}, u, t) \tag{8.7}$$

8.2 Mars Entry Optimal Control Problem

MSL is planned to increase landed mass capability to 0.8 metric tons while improving landed accuracy to 10 km and landing at a site as high as +2 km elevation. Target capabilities for future sample return and manned landing missions include landing 40–80 tons payload with a precision of tens of meters, possible at high-altitude [35, 36]. It is critical that the speed of the entry vehicle with larger payload is slowed down to meet the parachute deploying conditions as soon as possible. So, for future Mars entry with larger payload, the objection function of the optimal control is to quickly guide the entry vehicle from the Mars atmosphere interface to the site of deploying supersonic parachute with the maximum height

$$J_1 = -\min h(t_f) \tag{8.8}$$

where $h(t)$ denotes the height at time t, and t_f is the time of parachute deployment.

In the course of entry, the normal acceleration, dynamic pressure and heat flux should satisfy certain constraints

$$\begin{aligned}\dot{Q} &= \alpha\rho^{0.5}v^k \le \dot{Q}_{\max}\\ \overline{q} &= 0.5\rho v^2 \le \overline{q}_{\max}\\ n_a &= \sqrt{L^2 + D^2} \le n_{\max}\end{aligned} \tag{8.9}$$

where $\dot{Q}_{\max}$, $\overline{q}_{\max}$, $n_{\max}$ are the allowable maximum values of restricted physical quantity mentioned above, and α, k are heat flux-related coefficients.

In order to arrive at the prescribed position and succeed deploying parachute, certain state variables at the end of entry phase should satisfy certain constraints:

$$v(t_f) \le 350,\ s(t_f) \le 935000 \tag{8.10}$$

Thus, the Mars entry optimal control problem with maximum height as index performance is completely formulated. Many indirect methods, such Pontryagin's minimum principle, can be adopted to solve the optimal problem with performance index (8.8) and subject to initial state variables $\boldsymbol{x}(t_0) = [h_0, v_0, \gamma_0, s_0]^T$, system Eq. (8.7), and constraints Eqs. (8.9) and (8.10).

DOC methodology is developed based on the conventional optimal control formulation except that the original objective function is augmented with the sensitivity of a given function with respect to state perturbations encountered along the trajectory. The result is an optimal trajectory with reduced sensitivity of the given quantity with respect to the state perturbations, but without an excessive compromise in terms of the original objective value [164–166].

For all $t \in [t_0, t_f]$, let $\boldsymbol{x}(t) = \boldsymbol{X}(t|t_0, x_0)$ denotes the solution of Eq. (8.7) subject to the initial condition $\boldsymbol{x}(t_0) = \boldsymbol{x}_0 = [h_0, v_0, \gamma_0, s_0]^T$. Then, the sensitivity matrix $\boldsymbol{S}(t|t_0, x_0)$ can be defined as follows

$$\boldsymbol{S}(t|t_0, x_0) = \frac{\partial \boldsymbol{x}(t)}{\partial \boldsymbol{x}(t_0)} = \frac{\partial \boldsymbol{X}(t|t_0, x_0)}{\partial \boldsymbol{x}_0} \tag{8.11}$$

The sensitivity matrix $\boldsymbol{S}(t|t_0, x_0)$ represent the sensitivity of the state variable $\boldsymbol{x}(t)$ with respect to perturbations in the initial state variable $\boldsymbol{x}_0$. In the following context, a simpler notation $\boldsymbol{S}(t|t_0)$ is used to denote the sensitivity matrix $\boldsymbol{S}(t|t_0, x_0)$. The sensitivity matrix $\boldsymbol{S}(t|t_0)$ has two important properties

$$\boldsymbol{S}(t|t_0) = \boldsymbol{I},\ \ \boldsymbol{S}(t_2|t_1) = \boldsymbol{S}(t_2|t_0) \cdot \boldsymbol{S}(t_1|t_0)^{-1} \tag{8.12}$$

where $\boldsymbol{I}$ is identity matrix and $(\cdot)^{-1}$ denotes inverse operation.

According to [164], the sensitivity matrix $\boldsymbol{S}(t|t_0)$ satisfies the following differential equation

$$\frac{\partial \boldsymbol{S}(t|t_0)}{\partial t} = \left(\frac{\partial \boldsymbol{f}(\boldsymbol{x})}{\partial \boldsymbol{x}} + \frac{\partial \boldsymbol{f}(\boldsymbol{x})}{\partial u}\frac{\partial u}{\partial \boldsymbol{x}}\right) \cdot \boldsymbol{S}(t|t_0) \tag{8.13}$$

It is assumed in DOC methodology that the perturbed trajectories are within the bunch of the optimal nominal trajectories and reachable by a linear feedback control law [164], that is,

$$u(t) = u^*(t) + \boldsymbol{K}(t)\big(\boldsymbol{x}(t) - \boldsymbol{x}^*(t)\big) \tag{8.14}$$

where $\boldsymbol{x}^*(t)$ and $u^*(t)$ are the state and control variable of the optimal nominal trajectory at time t, respectively, which will be solved by use of DCNLP algorithm in the next section. $\boldsymbol{K}$ denotes the feedback gain matrix, which can be either prescribed a prior or can be determined optimally in parallel to the calculation of the optimal nominal trajectory.

Substitute Eq. (8.14) into Eq. (8.13), and then the sensitivity matrix differential equation becomes

$$\frac{\partial \boldsymbol{S}(t|t_0)}{\partial t} = \left(\frac{\partial \boldsymbol{f}(\boldsymbol{x})}{\partial \boldsymbol{x}} + \frac{\partial \boldsymbol{f}(\boldsymbol{x})}{\partial u}\boldsymbol{K}\right) \cdot \boldsymbol{S}(t|t_0) \tag{8.15}$$

Define the index performance about the sensitivity of the states at the end of entry with respect to perturbations in the states at any time

$$J_2 = \min_u\Big(\sum f\left(\boldsymbol{S}_{i,j}(t_f, t)\right)\Big) \tag{8.16}$$

where $\boldsymbol{S}_{i,j}(\cdot)$ denotes the (i, j) elements of sensitivity matrix.

The addition of the new performance index changes the maximum height Mars entry problem into a multi-objective optimization problem, that is, to minimize J_1 and J_2 at the same time. This paper uses the penalty factor method to combine the two conflicting objectives, resulting in the following performance index

$$J = J_1 + c_0 J_2 \tag{8.17}$$

where $c_0 \geq 0$ denotes the penalty factor on the sensitivity performance index.

It is easily concluded that desensitized optimal control (DOC) differs from tradition optimal control in the former's inclusion of the sensitivity dynamics, the additional performance index and the initial condition of the sensitivity matrix. Obviously, the solution of desensitized optimal control is the result of compromise between the two objectives of minimizing the J_1 and J_2. Thus, the desensitized optimal trajectory defined in this section is different from the optimal trajectory described in Sect. 8.4, that is, the optimal trajectory is reshaped to one with reduced sensitivity to state perturbations. Note that penalty factor c_0, along with the feedback gain matrix $\boldsymbol{K}$, affects how the optimal trajectory is reshaped. When $c_0 = 0$, all related to the sensitivity minimization become irrelevant, and the desensitized optimal control (DOC) degrades to traditional optimal control. As c_O increases, more and more emphasis is focused on minimizing the sensitivity performance index, consequently, the optimal trajectories become less and less sensitive to disturbances.

8.3 Mars Entry Optimal Control with Sensitivity Penalties

In order to successfully deploy the parachute, the sensitivity of the final height, velocity and downrange with respect to perturbations should be reduced as much as possible. To this end, let $\boldsymbol{\xi}(\boldsymbol{x}(t_f), t_f)$ denotes the state vector of the entry vehicle at time t_f

$$\boldsymbol{\xi}(\boldsymbol{x}(t_f), t_f) = [h(t_f), v(t_f), \gamma(t_f), s(t_f)]^T \tag{8.18}$$

The sensitivity of $\xi(\boldsymbol{x}(t_f), t_f)$ with respect to disturbances at any time t can be written as

$$\frac{\partial \boldsymbol{\xi}(\boldsymbol{x}(t_f), t_f)}{\partial \boldsymbol{x}(t)} = \frac{\partial \boldsymbol{\xi}(\boldsymbol{x}(t_f), t_f)}{\partial \boldsymbol{x}(t_f)} \boldsymbol{S}(t_f|t) \tag{8.19}$$

According to inversion and multiplication rule of sensitivity matrix, we can obtain

$$\boldsymbol{S}(t_f|t) = \boldsymbol{S}(t_f|t_0) \cdot \boldsymbol{S}(t|t_0)^{-1} \tag{8.20}$$

As the length of the state variable contains is 4 in this study, the partial differential in the right hand side of Eq. (8.19) can be easily obtained

$$\frac{\partial \boldsymbol{\xi}(\boldsymbol{x}(t_f), t_f)}{\partial \boldsymbol{x}(t_f)} = \boldsymbol{I}_{4\times 4} \tag{8.21}$$

Therefore, Eq. (8.19) can be further simplified to the following form

$$\frac{\partial \boldsymbol{\xi}(\boldsymbol{x}(t_f), t_f)}{\partial \boldsymbol{x}(t)} = \boldsymbol{S}(t_f|t) \tag{8.22}$$

It is obvious that $\boldsymbol{S}(\cdot|\cdot)$ is 4×4 square matrix and $\partial \boldsymbol{\xi}(\boldsymbol{x}(t_f), t_f) / \partial \boldsymbol{x}(t)$ contains the first four rows of matrix $\boldsymbol{S}(t_f|t)$. So, each element of these rows needs to be minimized at all times, which is reflected by the choice of the following performance index

$$J_2 = \int_{t_0}^{t_f} \sum_{i=1}^{4} \left(c_i \cdot \sqrt{\sum\nolimits_{j=1}^{4} S_{i,j}^2(t_f, t)} \right) dt \tag{8.23}$$

where $c_i \geq 0$, $i = 1, 2, 3, 4$ are weighted coefficients, which leverage the significance of the final height, velocity, flight path angle and downrange dispersions to the success of the entry mission. For the Mars entry problem, the height, velocity and downrange dispersions at the end of entry will play the decisive role to successfully deploy parachute. Therefore, weighted coefficients c_1, c_2 and c_4 are set greater than c_3 to place more weighting on reducing the final height, velocity and downrange dispersion.

With the consolidated performance index, the maximum-height Mars entry problem with the sensitivity penalty can be summarized as follows. Performance index with sensitivity penalty is as follows:

$$J_T = -r(t_f) + c_0 \cdot \int_{t_0}^{t_f} \sum_{i=1}^{4} \left(c_i \cdot \sqrt{\sum\nolimits_{j=1}^{4} S_{i,j}^2(t_f, t)} \right) dt \tag{8.24}$$

It subjects to system Eq. (8.7), sensitivity dynamics Eq. (8.15), and physical constraints Eqs. (8.9) and (8.10); terminal time t_f is free here;

$$\begin{aligned} &\boldsymbol{x}(t_0) = [h_0, v_0, \gamma_0, s_0]^T, \quad v(t_f) \leqslant 350 \\ &s(t_f) = 935000, \quad \boldsymbol{S}(t_0|t_0) = \boldsymbol{I}_{4\times 4} \end{aligned} \tag{8.25}$$

8.4 Optimal Nominal Trajectory and Command Generation Using DCNLP

DOC tries to reduce the sensitivity of a physical quantity with respect to perturbations along the trajectory, in addition to optimizing the original performance index. DOC does this by incorporating the optimal control solution into the nominal trajectory and control optimization. Thus, the nominal trajectory and control and the optimal control solution are obtained in one optimization run. The section determinates the optimal nominal trajectory and control using Direct Collocation and Nonlinear Programming (DCNLP) algorithm.

The optimal control problem can be transformed into nonlinear programming problem using direct collocation method or, in other words, dynamic optimization problem is transformed into static parameter optimization problem [167]. To do this, the state variables and control are represented by piecewise polynomials and collocation is used to satisfy Eq. (8.7).

The time interval $[t_0, t_f]$ is sub-divided into N segments, and the (i + 1)th segment is marked as $[t_i, t_{i+1}], i = 0, 1, \cdots, N-1$. Cubic polynomials are defined by Hermite interpolation for each state on each segment using values of the state and the state time derivatives at the nodes. The values of the states are then selected to force the interpolated derivatives to agree with the differential equation at the center of the segment. The basic procedure can be derived as follows. Let the states $\boldsymbol{x}$ be represented on each segment by cubic polynomial of the form [168]

$$\boldsymbol{x} = a_0 + a_1 s + a_2 s^2 + a_3 s^3 \tag{8.26}$$

where $s \in [0, 1]$ is dimensionless time variable and defied as follows

$$s = (t - t_i)/(t_i - t_{i+1}) \tag{8.27}$$

Boundary conditions are set as follows [168]

$$x_1 = x(0), \boldsymbol{x}_1 = \boldsymbol{x}(0), \dot{\boldsymbol{x}}_1 = \left.\frac{dx}{ds}\right|_{s=0}, \dot{\boldsymbol{x}}_2 = \left.\frac{dx}{ds}\right|_{s=1} \tag{8.28}$$

Differentiating Eq. (8.26) and evaluating at 0 and 1 gives

$$\begin{bmatrix} \boldsymbol{x}_1 \\ \dot{\boldsymbol{x}}_1 \\ \boldsymbol{x}_2 \\ \dot{\boldsymbol{x}}_2 \end{bmatrix} = \begin{bmatrix} 1\,0\,0\,0 \\ 0\,1\,0\,0 \\ 1\,1\,1\,1 \\ 0\,1\,2\,3 \end{bmatrix} \begin{bmatrix} a_0 \\ a_1 \\ a_2 \\ a_3 \end{bmatrix} \tag{8.29}$$

Inverting the 4 × 4 matrix gives

$$\begin{bmatrix} a_0 \\ a_1 \\ a_2 \\ a_3 \end{bmatrix} = \begin{bmatrix} 1 & 0 & 0 & 0 \\ 0 & 1 & 0 & 0 \\ -3 & -2 & 3 & -1 \\ 2 & 1 & -2 & 1 \end{bmatrix} \begin{bmatrix} \boldsymbol{x}_1 \\ \dot{\boldsymbol{x}}_1 \\ \boldsymbol{x}_2 \\ \dot{\boldsymbol{x}}_2 \end{bmatrix} \tag{8.30}$$

Now using Eq. (8.30), evaluating Eq. (8.26) at $s = 1/2$, and transforming to segment length $h_i = t_i - t_{i+1}$, the interpolated value of $\boldsymbol{x}$ at the center of the segment can be obtained

$$\boldsymbol{x}_{ci} = \frac{\boldsymbol{x}_i + \boldsymbol{x}_{i+1}}{2} + \frac{h_i(f_i - f_{i+1})}{8} \tag{8.31}$$

$$\dot{\boldsymbol{x}}_{ci} = -\frac{3(\boldsymbol{x}_i - \boldsymbol{x}_{i+1})}{2h_i} - \frac{(f_i + f_{i+1})}{4} \tag{8.32}$$

where f_i and f_{i+1} denote the function value evaluated at $\boldsymbol{x}_i$ and $\boldsymbol{x}_{i+1}$ respectively, that is

$$f_i = f(\boldsymbol{x}_i, u_i, t_i) \tag{8.33}$$

$$f_{i+1} = f(\boldsymbol{x}_{i+1}, u_{i+1}, t_{i+1}) \tag{8.34}$$

According to Eq. (8.7), $f(\boldsymbol{x}_{ci}, u_{ci}, t_{ci})$ can be computed after the control variable u_{ci} and state variable $\boldsymbol{x}_{ci}$ are provided. Then, Hermite–Simpson defect at the midpoint of the segment can be defined as follows [168]

$$\boldsymbol{\delta}_i = f(\boldsymbol{x}_{ci}, u_{ci}, t_{ci}) - \dot{\boldsymbol{x}}_{ci} \tag{8.35}$$

where $\boldsymbol{x}_{ci}$ is the state variable at midpoint time and given in Eq. (8.31), u_{ci} is the control at the middle time of (i + 1)th segment and defined as

$$u_{ci} = \frac{u_i + u_{i+1}}{2} \tag{8.36}$$

Third-order Simpson method considers the state variable $\boldsymbol{x}_i$ and control variable u_i of each node as the optimal decision variables, and the final time t_f is also looked as decision variable. Here, the control variables at the midpoint of subinterval u_{ci} are no longer taken as the collocation points, but it can be obtained by linear interpolation according to Eq. (8.36). For this problem to be solved in this paper, there are $5 \times (N + 1) + 1$ decision variables. Therefore, nonlinear programming variables are defined as follows

$$\boldsymbol{Z} = [\boldsymbol{x}_0^T, u_0, \cdots, \boldsymbol{x}_N^T, u_N, t_f]_{5\times(N+1)+1} \tag{8.37}$$

Hermite–Simpson defect reflects the approximate level of the fitted polynomial Eq. (8.26) with respect to the entry system dynamics Eq. (8.7). If the cubic polynomial is capable of representing the solution on the given segment, then selecting $\boldsymbol{x}_i$ and $\boldsymbol{x}_{i+1}$ to drive nonlinear equation constraints $\boldsymbol{\delta}_i = 0$ will produce an accurate approximation the solution of Eq. (8.7). Then, the optimal control problem is transformed into nonlinear programming problem. As a large amount of literature has been devoted to the solution of nonlinear programming problem, the specific solution process will be no longer repeated here.

8.5 Simulation and Results

In order to confirm the validity of Mars entry trajectory optimization algorithm developed in this paper, Monte Carlo simulation and analysis in MATLAB environment has been carried out. The Mars entry dynamics modeling and entry error analysis are performed by use of MATLAB/Simulink package, and the desensitized optimal control problem of Mars entry is solved by SNOPT, which is a software package for solving large-scale linear and nonlinear optimization problems by SQP methods and invented by Philip Gill, Walter Murray and Michael Saunders [169]. SNOPT is especially effective for nonlinear problems whose functions and gradients are expensive to evaluate. The functions should be smooth but need not be convex. In this study, MATLAB version SNOPT software package is adopted to solve the Mars entry trajectory optimization problem. In order to use this software, the entry time interval $[t_0, t_f]$ is subdivided into 20 segments, and then there are $5 \times (20 + 1) + 1$ decision variables to optimize. Since the final height, velocity and downrange have significant effect on successful deploying parachute, and flight path angle has little impact on parachute deployment, the weight coefficients c_i, $i = 1, 2, 3, 4$ are set as follows: $c_1 = 0.1$, $c_2 = 0.3$, $c_3 = 0$, $c_4 = 1$. The entry vehicle model is selected similar to MSL lander and the physical parameters are given in Table 8.1. The boundary condition used in this study are summarized in Table 8.2 respectively, and other parameters used in the simulation are set as follows: $r_0 = 3397$ km, $\mu = 42{,}828.29 \times 10^9$ m^3/s^2, $u_{\min} = -0.8$, $u_{\max} = 0.8$, $\boldsymbol{K} = [-0.01, -0.005, -50, -0.0001]$.

The optimal nominal controls u^* and the state variables $[h^*, v^*, s^*]$ along the optimal nominal trajectories are solved firstly using the DCNLP with different penalty factor c_0. Optimal nominal controls u^* with different penalty factor c_0 are shown in Fig. 8.1, and the corresponding optimal nominal state variables h^*, v^*, s^* are depicted in Figs. 8.1, 8.2, 8.3 and 8.4, respectively. The optimal nominal control u^* resembles the Bang–Bang control when penalty factor $c_0 = 0$, because the desensitized optimal control degrades to the traditional optimal control in this case. This kind of optimal maximum–minimum–maximum control profile makes the nominal control always stopping at saturated states and greatly degrades the feedback control effect. When there are disturbances along the entry trajectory, this control mode of maximum–minimum–maximum profile is difficult to timely track the nominal trajectory and will result in relatively large tracking error. It can be seen by comparing the four sub-graph of Fig. 8.1 that the desensitized optimal nominal controls gradually go away from the saturation boundary of the control maximum/minimum with the increase of the penalty factor. Therefore, it can be concluded that the degree of saturation of optimal nominal control gradually decreased with the increase of the penalty factor, and the control effect becomes better and better. The penalty factors have a significant impact on the entry terminal time t_f, which is evidently observed from Figs. 8.2, 8.3 and 8.4. The reason why the entry terminal time at $c_0 = 0$ is a little longer than those at $c_0 \neq 0$ may be lies on the fact that the desensitized control reshapes the entry trajectories and makes the entry vehicle arriving at lower altitude where the Mars atmosphere has a higher density as soon as quickly, then, the aerodynamics

Table 8.1 Entry vehicle physical parameters

Parameters	Mass, m	Diameter, d	Ballistic coefficient, $m/C_D S$	Lift-drag ratio, L/D
Value	2920 kg	4.6 m	115 kg/m^2	0.24

Table 8.2 Initial and final conditions of the entry trajectories

Parameters	Initial value	Final values
Height, h	125 km	Free
Velocity, v	6000 m/s	<350 m/s
Downrange, s	0	935 km

force can be used to reduce the sensitivity and improve the entry precision. Partial enlarged results in Fig. 8.2 show that the penalty factors have a significant effect on the final height of the entry phase trajectory, the more detailed heights at the end of entry corresponding with different penalty factor c_0 are given in Table 8.3. However, it can be seen from Figs. 8.3 and 8.4 that the final velocity and downrange at the end of entry have little relation with penalty factor. According to the above analysis, we can obtain the following conclusions that increasing the penalty factor c_0 will improve the sensitivity performance index and simultaneously will lower down height performance index, that is, the sensitivity will reduce and the height at the end of the entry will decrease with the increase of the penalty factor c_0.

After the optimal nominal controls and trajectories are solved by using DCNLP, Monte Carlo simulations of entry error analysis have been performed in order to estimate the sensitivity of the terminal state variable with respect to disturbances encountered in the course of entry. In the 500 simulations runs, uncertainties and perturbations come from the navigation errors at entry interface and aerodynamics related parameters uncertainties along the entry trajectories, Monte Carlo simulation parameters distribution type and range are set in Table 8.4. The actual control inputs u with penalty factors $c_0 = 0$, $c_0 = 0.002$ and $c_0 = 0.02$ are shown in Figs. 8.5, 8.7,

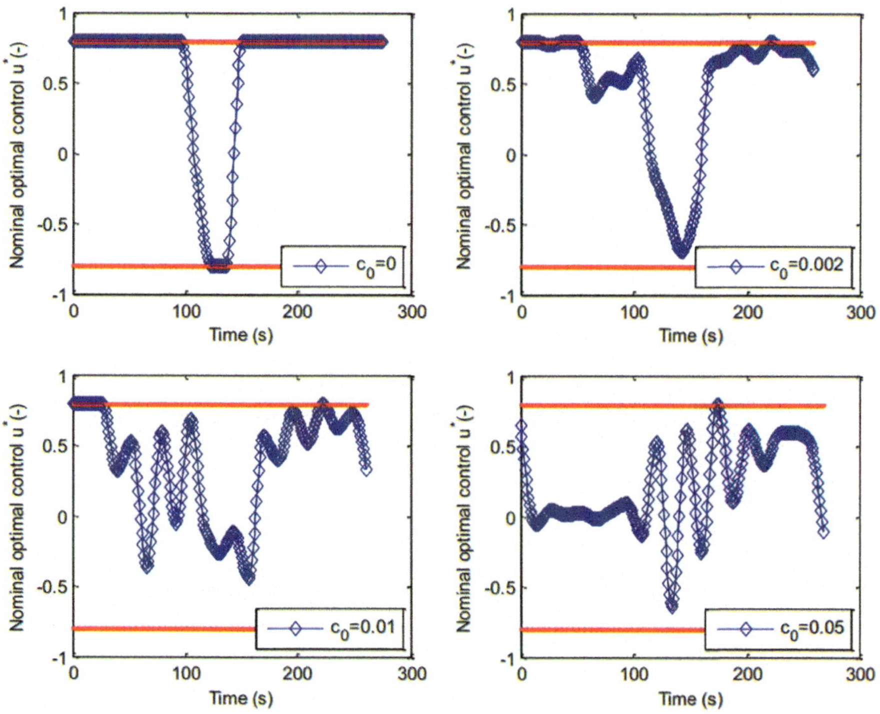

Fig. 8.1 Nominal optimal control u^* corresponding to different penalty factor c_0

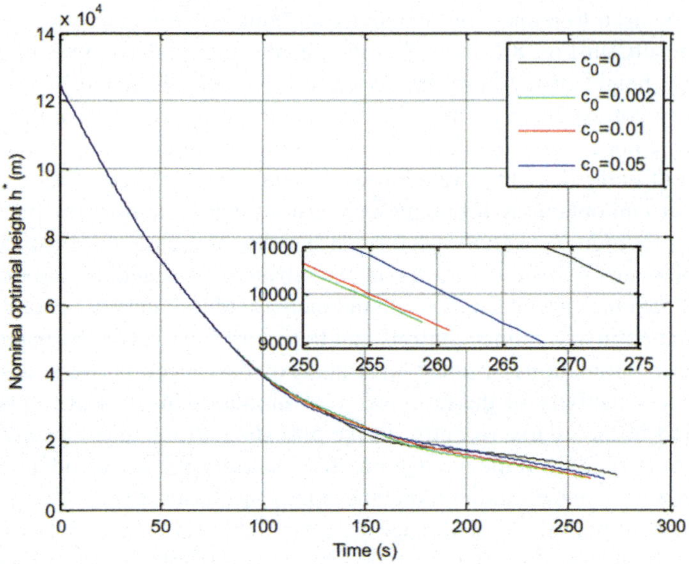

Fig. 8.2 Nominal optimal control h^* corresponding to different penalty factor c_0

Table 8.3 The final height of entry phase corresponding with different penalty factor

	I	II	III	IV	V	VI
Penalty factor, c_0	0	0.002	0.006	0.01	0.02	0.05
Final height, h_f	10,174	9428	9201	9137	9108	8925

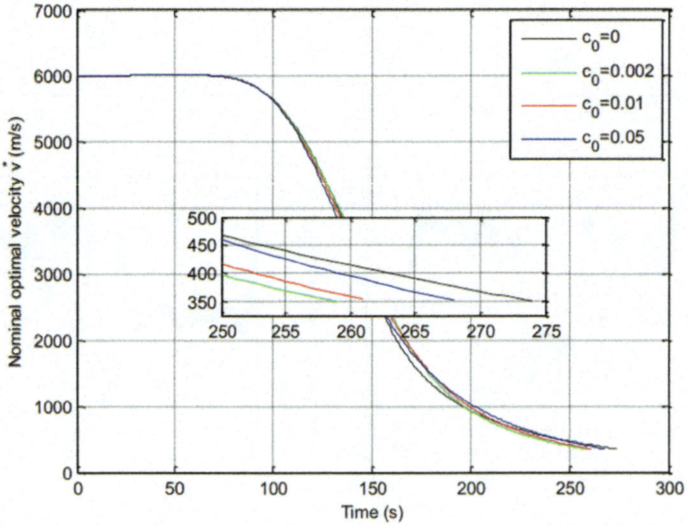

Fig. 8.3 Nominal optimal control v^* corresponding to different penalty factor c_0

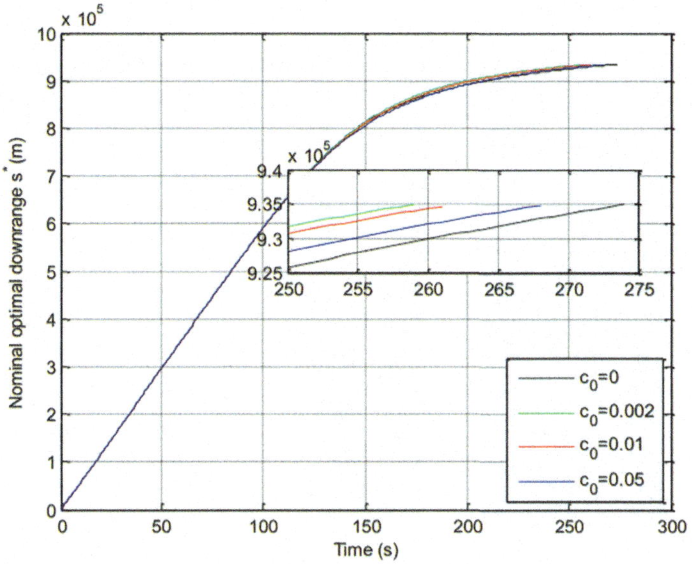

Fig. 8.4 Nominal optimal control s^* corresponding to different penalty factor c_0

and 8.9, respectively, and the corresponding to final state errors (height error, velocity error and downrange error) are depicted in Figs. 8.6, 8.8 and 8.10, respectively.

The Monte Carlo simulations show that the distribution statistics of the actual control input over time resemble those of the nominal optimal control, that is, the degree of saturation of the actual control inputs are gradually decreased with the increase of the penalty factors. Therefore, it can be inferred that the final state errors will be reduced as the penalty factors are increased, the results in Figs. 8.6, 8.8 and 8.10 also proves the aforementioned conclusion. Another interesting result is that the actual control inputs appear disorderly and unsystematic in the first third part of entry trajectory; however, the rest almost follows the unified variation trend. The reason why the actual control inputs display disorderly firstly and orderly later lies in the fact that the Mars atmospheric density becomes bigger and bigger with continuous reduce

Table 8.4 Monte carlo simulation parameters settings

Parameters with uncertainty	Uncertainty distribution	
	Range (3σ)	Type
Height, h	$\pm$ 0.5 km	Normal
Velocity, v	$\pm$ 10 m/s	Normal
Downrange, s	$\pm$ 0.5 km	Normal
Atmospheric density, ρ	$\pm$ 10%	Normal
Ballistic coefficient, $m/C_D S$	$\pm$ 5%	Normal
Lift-drag ratio, L/D	$\pm$ 5%	Normal

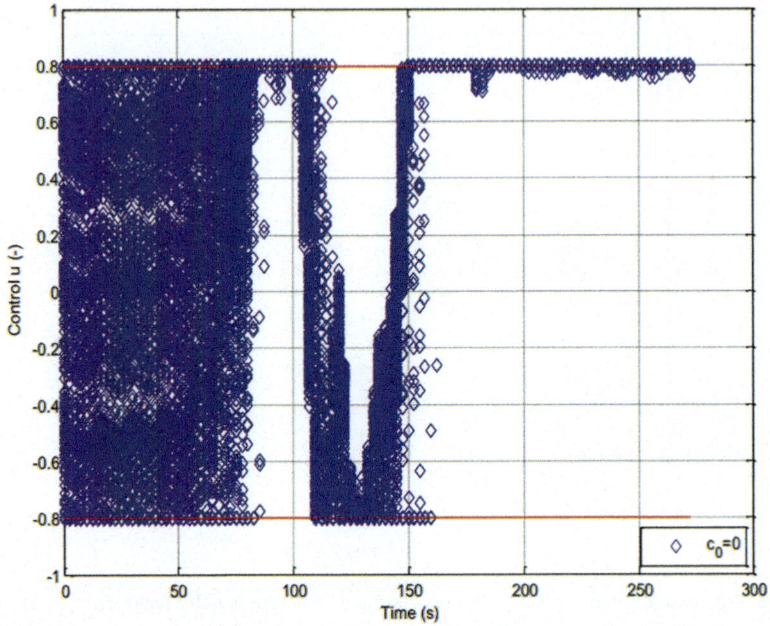

Fig. 8.5 Control u corresponding to penalty factor $c_0 = 0$

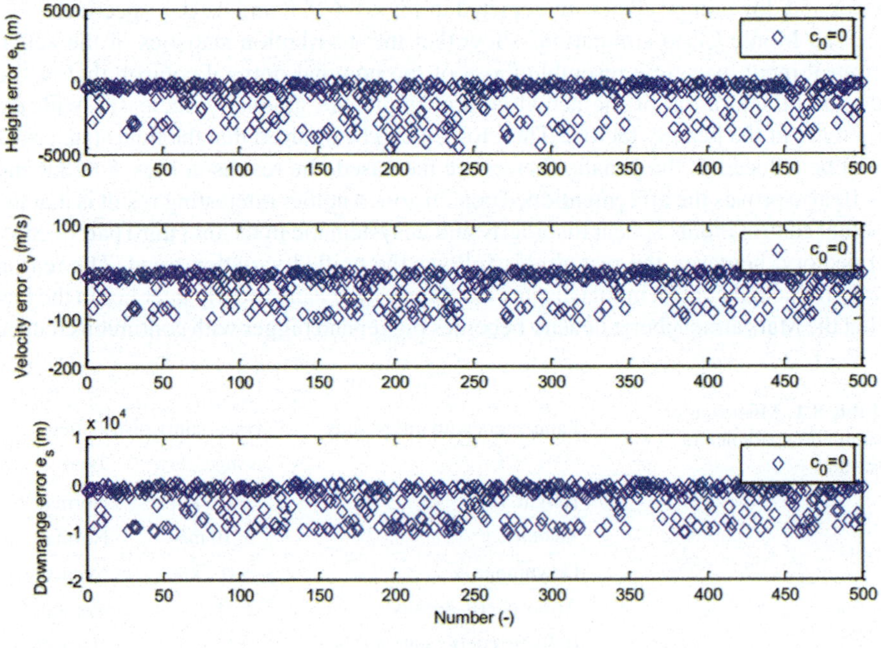

Fig. 8.6 State variable errors $[e_h, e_v, e_s]$ corresponding to penalty factor $c_0 = 0$

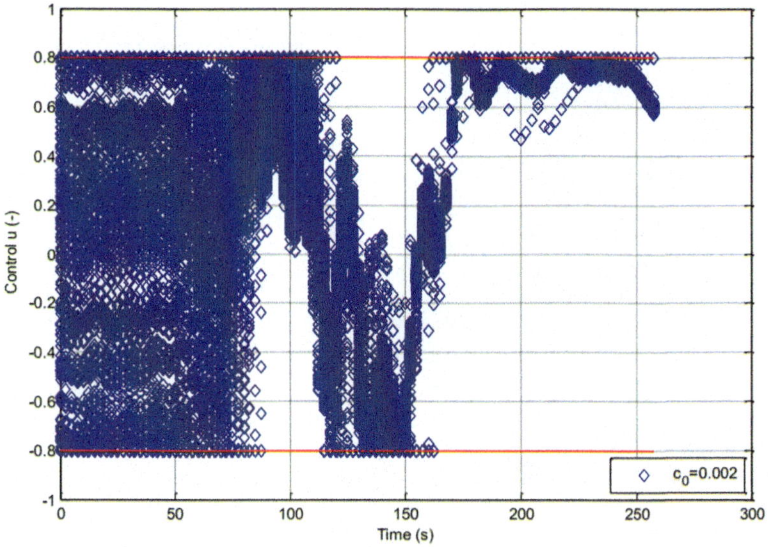

Fig. 8.7 Control u corresponding to penalty factor $c_0 = 0.002$

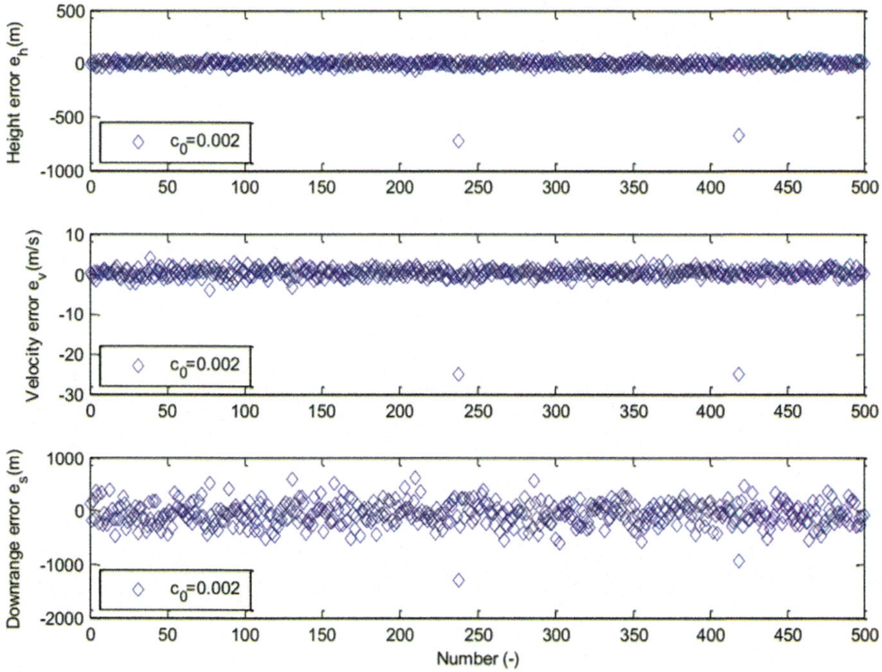

Fig. 8.8 State variable errors $[e_h, e_v, e_s]$ corresponding to penalty factor $c_0 = 0.002$

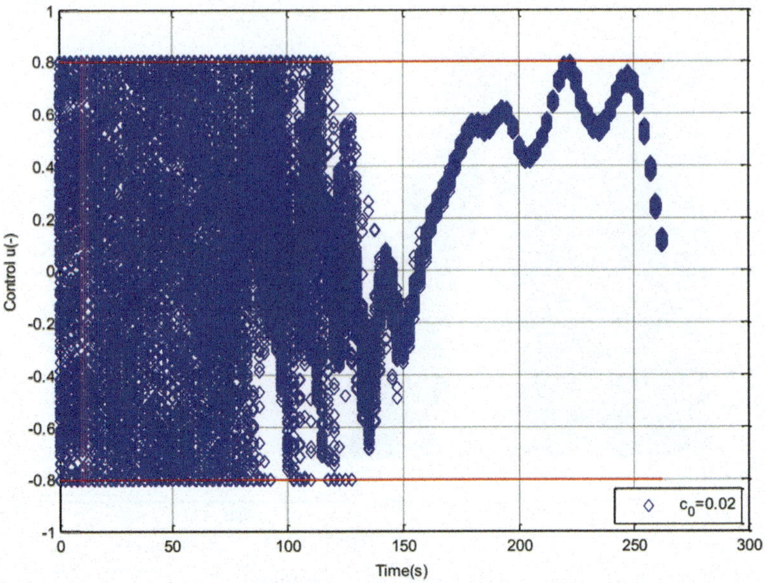

Fig. 8.9 Control u corresponding to penalty factor $c_0 = 0.02$

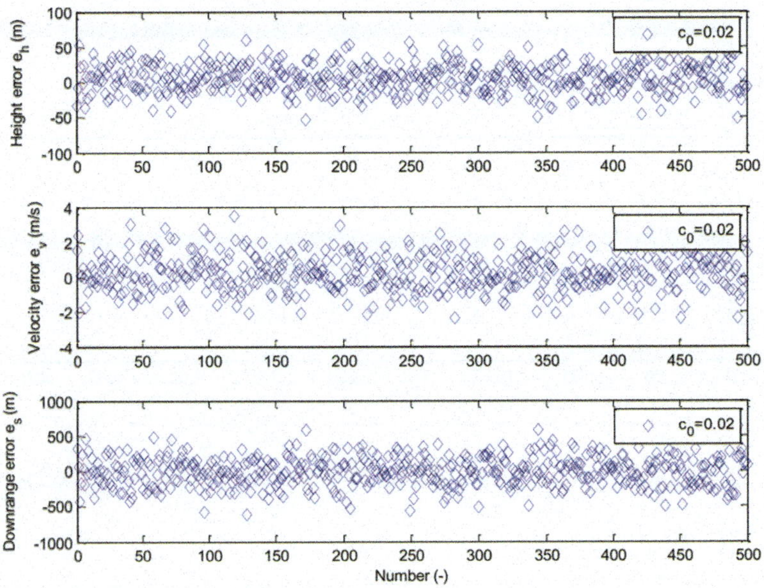

Fig. 8.10 State variable errors $[e_h, e_v, e_s]$ corresponding to penalty factor $c_0 = 0.02$

of vehicle height, and then the control effect of aerodynamic force become better and better. When the penalty factor is set to zero, the maximum value of the final height error, velocity error and downrange error are respectively close to 5000 m, 100 m/s and 10,000 m. When the penalty factor c_0 is increased to 0.002, the corresponding final eight error, velocity error and downrange error will be reduced to 100 m, 5 m/s and 750 m, respectively. Figure 8.8 shows that there are two cases with bad results when $c_0 = 0.002$, which corresponds to the situation of control saturation and entry errors increases. It should be pointed out that augmented penalty factor will lead to more bank reversals and result in higher fuel consumption, so, an appropriate value for c_0 is important in practice. In this paper, $c_0 = 0.002$ is considered as a reasonable tradeoff between the entry errors and the fuel consumption, the corresponding final height error, velocity error and downrange error are less than 50 m, 4 m/s and 500 m, respectively, which meets the parachute deployment condition.

8.6 Discussions

Reducing the sensitivity of the height, velocity and downrange at parachute deployment with respect to uncertainties and perturbations along the entry trajectory is essential to improve the accuracy of the Martian atmosphere entry. This chapter describes the Mars entry trajectory optimization in presence of uncertainties and control saturation by use of desensitized optimal control (DOC) methodology and Direct Collocation and Nonlinear Programming (DCNLP). The final height and the state sensitivities are simultaneously included into the weighted performance index. Monte Carlo simulations show that the final height error, velocity error and downrange error can be effectively decreased from original 5000 m, 100 m/s and 10,000 to 50 m, 4 m/s and 500 m, respectively. Thus, it can be concluded that the optimization algorithm developed in this paper can effectively reduce the sensitivity of the final state at the end of entry phase with respect to the uncertainties on the initial state variables and the disturbances on the Mars atmospheric density and aerodynamic parameters. Meanwhile, the impact of weighted parameter on the height, velocity and downrange is also studied in detail. For the sake of simplicity, the feedback gain matrix $K(t)$ in Eq. (8.14) is assumed to be a prescribed constant matrix. In order to achieve better tracking of the optimal nominal trajectory, the feedback matrix should have some adaptive capacity according to the state and control deviations. Since there are many physical quantities and their magnitudes are different, normalization may be necessary before optimization in order to obtain better solutions. At the same time, only 20 modes are set for collocation to reduce the compute burden in our current simulations, the number of node and control decision variable in the course of collocation should be moderately increased in order to improve the accuracy of solving nonlinear programming. These three aspects will be analyzed in the further study.

Chapter 9
Uncertainty Quantification for Mars Entry

9.1 Stochastic Nonlinear Dynamics for Mars Atmospheric Entry

Mars atmospheric entry is the first phase of the Mars EDL process. It begins when the entry vehicle reaches the Mars atmosphere boundary and ends when parachute deploys. In order to describe the trajectory of the center-of-mass of the Mars entry vehicle, we consider the three degree-of-freedom Mars atmospheric entry dynamics as follows:

$$\dot{h} = v \sin\gamma \tag{9.1}$$

$$\dot{v} = -\frac{\varepsilon\rho v^2}{2B_c} - \frac{\mu_M \sin\gamma}{(r_0 + h)^2} \tag{9.2}$$

$$\dot{\gamma} = \frac{k\varepsilon\rho v \cos\sigma}{2B_c} - \frac{\mu_M \cos\gamma}{v(r_0 + h)^2} + \frac{v\cos\gamma}{r_0 + h} \tag{9.3}$$

where h is the flight altitude, v is the velocity of the entry vehicle, γ is the flight path angle, σ is the bank angle, μ_M denotes the gravitational constant of the Mars (set to be $4.282829 \times 1013\ \text{m}^3/\text{s}^2$), and ε denotes the uncertain coefficient of atmospheric density. k is the lift-to-drag ratio, $k = C_L/C_D$, where C_L and C_D are the aerodynamic lift and drag coefficients respectively. The ballistic coefficient is $B_c = C_D S/m$, where S represents vehicle's reference surface area, and m denotes the mass of the entry vehicle. The Mars atmospheric density is defined as $\rho = \rho_0 \exp(-h/h_s)$, where ρ_0 is the density on the surface of Mars (set to be 0.0158 kg/m^3), and h_s is the constant scale height (set to be 9354.5 m).

S. Li et al., *Trajectory Optimization and Guidance Methods for Mars Entry*,
https://doi.org/10.1007/978-981-99-6282-2_9

The errors from orbit determination and control inevitably result in uncertain dispersion for the initial state at the entry interface. Moreover, the dynamical parameters cannot be exactly determined in advance due to the uncertainty of Martian atmospheric environment. Therefore, the uncertainties in the initial state and Mars entry dynamics are considered and modeled as

$$h(t_0) = \overline{h}_0 + \delta_{h_0}\Delta \tag{9.4}$$

$$v(t_0) = \overline{v}_0 + \delta_{v_0}\Delta \tag{9.5}$$

$$\gamma(t_0) = \overline{\gamma}_0 + \delta_{\gamma_0}\Delta \tag{9.6}$$

$$B_c = (1 + \delta_{B_c}\Delta)\overline{B}_c \tag{9.7}$$

$$k = (1 + \delta_k\Delta)\overline{k} \tag{9.8}$$

$$\varepsilon = (1 + \delta_\varepsilon\Delta)\overline{\varepsilon} \tag{9.9}$$

where the random variable $\Delta \in [-1, 1]$, and δ_{h_0}, δ_{v_0}, δ_{γ_0}, δ_{B_c}, δ_k, δ_ε represent the perturbations with regard to the corresponding nominal values $\overline{h}_0$, $\overline{v}_0$, $\overline{\gamma}_0$, $\overline{B}_c$, $\overline{k}$, $\overline{\varepsilon}$, respectively. Thus, the Eqs. (9.1)–(9.3) are random differential equations (RDEs), and h, v, and γ are random processes.

9.2 Adaptive Generalized Polynomial Chaos Approach

9.2.1 Equivalent Deterministic Differential Equations via GPC

Under the GPC framework, the RDEs should be transcribed into equivalent deterministic differential equations (DDEs). The parametric uncertainty of Mars entry dynamics is mainly related to the pre-flight design of the entry vehicle and the current state of the Martian environment. However, the abovementioned uncertainty is independent from the nonlinear terms exhibited by the dynamics. Hence, in this study, the statistic characteristic of parametric uncertainty is assumed to be unchanged. To capture the PDFs of the uncertain parameters, two basic functions are utilized and the uncertain parameters are expressed as follows:

$$B_c(\zeta) = B_{c,0}H_0(\zeta) + B_{c,1}H_1(\zeta) \tag{9.10}$$

$$k(\zeta) = k_0 H_0(\zeta) + k_1 H_1(\zeta) \tag{9.11}$$

$$\varepsilon(\zeta) = \varepsilon_0 H_0(\zeta) + \varepsilon_1 H_1(\zeta) \tag{9.12}$$

where ζ is a random variable, and $H_0(\zeta)$ and $H_1(\zeta)$ denote the orthogonal polynomials basis, which subject to Table 9.1 [170].

For the states of the system, the initial uncertainty is assumed to be a Gaussian distribution. However, during the hypersonic entry phase, such states are mutually coupled and their statistic characteristics are heavily influenced by the nonlinear terms comprising the dynamical equations. As a result, the statistical characteristics of system states vary along the Mars entry trajectory, and higher order polynomials are required to approximate the propagation of the flight states with initial uncertainty:

$$h(t, \zeta) = \sum_{0}^{P} h_i(t) H_i(\zeta) \tag{9.13}$$

$$v(t, \zeta) = \sum_{0}^{P} v_i(t) H_i(\zeta) \tag{9.14}$$

$$\gamma(t, \zeta) = \sum_{0}^{P} \gamma_i(t) H_i(\zeta) \tag{9.15}$$

In order to reach the optimal solution, the stochastic dynamical system in $\mathbb{R}^N$ is transformed into deterministic dynamical system in $\mathbb{R}^{N(P+1)}$ which then can be solved numerically [7]. N denotes the number of the stochastic state variables. $P+1$ is the number of the polynomial chaos coefficients (h_i, v_i, γ_i), and the series is truncated after $P+1$ term. Here, $P = [(d+r)!]\big/(d!r!) - 1$, d is the dimension of random variable ζ, and r is the order of the orthogonal polynomials $\{\phi_i\}$ [6,7]. In Eqs. (9.10) through (9.15), the random variable ζ is normally distributed, and $H_i(\zeta)$ denotes the orthogonal polynomials basis. According to the correspondence between the choice of polynomials for a given distribution of random variable under the GPC framework (as listed in Table 9.1) [29, 30], Hermite polynomials are employed as the

Table 9.1 Distribution types of random variables and corresponding orthogonal polynomials

Distributions of ξ	PDFs	$\Phi_j(\xi_l)$ of Askey scheme	Expansions	Intervals
Gaussian	$e^{-\xi^2/2}\big/\sqrt{2\pi}$	Hermite	$e^{-\xi^2/2}$	$[-\infty, +\infty]$
Gamma	$\xi^\alpha e^{-\xi}\big/[\Gamma(\alpha+1)]$	Laguerre	$\xi^\alpha e^{-\xi}$	$[0, +\infty]$
Uniform	1/2	Legendre	1	$[-1, +1]$
Beta	$\frac{(1-\xi)^\alpha(1+\xi)^\beta}{2^{\alpha+\beta+1}B(\alpha+1,\beta+1)}$	Jacobi	$(1-\xi)^\alpha(1+\xi)^\beta$	$[-1, +1]$

trail basis family firstly. Substituting Eqs. (9.10) through (9.15) into the stochastic dynamical system in Eqs. (10.1)–(9.3) and taking the Galerkin projection on the basis functions $H_m(\zeta)$ yield the following DDEs:

$$\dot{h}_m = \frac{1}{\langle H_m^2 \rangle} \sum_{i=0}^{P} v_i \left\langle H_m H_i \sin\left(\sum_{i=0}^{P} \gamma_i H_i \right) \right\rangle \tag{9.16}$$

$$\begin{aligned} \dot{v}_m = &-\frac{1}{\langle H_m^2 \rangle} \sum_{i,j,k=0}^{P} \langle x_i H_i v_j H_j v_k H_k H_m \rangle \\ &-\frac{1}{\langle H_m^2 \rangle} \left\langle \sin\left(\sum_{i=0}^{P} \gamma_i H_i \right) \sum_{i,j=0}^{P} z_i H_i z_j H_j H_m \right\rangle \end{aligned} \tag{9.17}$$

$$\begin{aligned} \dot{\gamma}_m = &\frac{\cos\varphi}{\langle H_m^2 \rangle} \sum_{i,j,k=0}^{P} \langle x_i H_i k_j H_j v_k H_k H_m \rangle \\ &-\frac{\mu}{\langle H_m^2 \rangle} \left\langle \cos\left(\sum_{i=0}^{P} \gamma_i H_i \right) \sum_{i,j,k=0}^{P} y_i H_i z_j H_j z_k H_k H_m \right\rangle \\ &+\frac{1}{\langle H_m^2 \rangle} \left\langle \cos\left(\sum_{i=0}^{P} \gamma_i H_i \right) \sum_{i,j=0}^{P} z_i H_i v_j H_j H_m \right\rangle \end{aligned} \tag{9.18}$$

with

$$\sum_{i=0}^{P} x_i H_i = \frac{\rho(h_i) \sum_{i=0}^{P} \varepsilon_i H_i}{2 \sum_{i=0}^{P} B_{c,i} H_i}, \sum_{i=0}^{P} y_i H_i = \left(\sum_{i=0}^{P} v_i H_i \right)^{-1}, \sum_{i=0}^{P} z_i H_i = \left(r_0 + \sum_{i=0}^{P} h_i H_i \right)^{-1} \tag{9.19}$$

where $m = 0, 1, \ldots, P$, and $\langle \cdot \rangle$ is the mean operator. The Eqs. (9.16)–(9.18) are the equivalent DDEs of the stochastic dynamical system (Eqs. (10.1)–(9.3)) which is approximated by GPC expansions. Therefore, the solutions of the RDEs in Eqs. (9.16)–(9.18) can be employed to characterize the Mars entry flight states with uncertainties, i.e. $h(t, \zeta)$, $v(t, \zeta)$, and $\gamma(t, \zeta)$.. Moreover, the uncertainties (i.e. the statistical characteristic of each state) can be quantitatively described accordingly.

9.2.2 *Spectral Decomposition*

The GPC employs a set of variables that conform to the invariant basis which usually does not consider the change of the statistical property of the system state along the nonlinear stochastic dynamics. Then, the truncation error will exponentially increase and lead to divergence, especially when limited order polynomials are selected. As evidenced from Eqs. (9.16)–(9.19), the RDEs are highly nonlinear, and the solution polynomials increase in magnitude with long-term integration. This implies that the stochastic characteristics of the solution are changing [173, 174]. Consequently, the original basis family loses its representativeness, resulting in a diverging algorithm.

The spectral decomposition (SD) technique [173, 174] is introduced to enhance the GPC and enable accounting for the change of the stochastic characteristics of the system state. To sense the significant changes of the stochastic characteristics (i.e. the failure of the original basis family), the divergence of truncation error is employed as an indicator. The basis family used by GPC [175] will be updated adaptively whenever the truncation error oversteps a threshold. By this means, the divergence phenomenon can be effectively suppressed, and the accuracy of the GPC is thus improved.

An adaptive criterion is designed as follows to trigger the SD procedure. Assuming that the threshold is reached at time t, then the adaptive criterion is expressed as follows [177, 178]:

$$\max_{j=2,\ldots,P} \left|h_j(t)\right| \geq \kappa_1 |h_1(t)| \tag{9.20}$$

$$\max_{j=2,\ldots,P} \left|v_j(t)\right| \geq \kappa_2 |v_1(t)| \tag{9.21}$$

$$\max_{j=2,\ldots,P} \left|\gamma_j(t)\right| \geq \kappa_3 |\gamma_1(t)| \tag{9.22}$$

where κ_1, κ_2, κ_3 are the selected tolerance coefficients. If any one of Eqs. (9.20)–(9.22) is met, then the SD procedure is subsequently activated. Three new random variables ξ_1, ξ_2, ξ_3 are introduced as new orthogonal basis:

$$\xi_1 = \sum_{i=0}^{P} h_i(t) H_i(\zeta) := Z_1(\zeta) \tag{9.23}$$

$$\xi_2 = \sum_{i=0}^{P} v_i(t) H_i(\zeta) := Z_2(\zeta) \tag{9.24}$$

$$\xi_3 = \sum_{i=0}^{P} \gamma_i(t) H_i(\zeta) := Z_3(\zeta) \tag{9.25}$$

In Eqs. (9.23)–(9.25), the time t is the special value at which one of the Eqs. (9.20)–(9.22) is met. Here, we take the PDF $f_{\xi_i}(\xi_i)$, $i = 1, 2, 3$, as the weight function, which are calculated by [178]

$$f_{\xi_j}(\xi_j) = \sum_n \frac{f_\zeta(\zeta_n)}{\left| \frac{dZ_j}{d\zeta}\Big|_{\zeta=\zeta_n} \right|} \tag{9.26}$$

where ζ_n ($n = 1,2 ,\ldots$) represents all the roots of $\xi_j - Z_j(\zeta) = 0$. By applying the Schmidt orthogonalization rule, one can get the orthogonal polynomial family $\{\phi_P^{(\xi_i)}\}$ in terms of the random variable ξ_i. Consequently, the state variables of Mars entry vehicle are updated as follows [178]

$$h(t, \xi_1, \xi_2, \xi_3) = \sum_{0\le l+m+n\le P}^{P} h_{lmn}(t)\phi_l^{(\xi_1)}(\xi_1)\phi_m^{(\xi_2)}(\xi_2)\phi_n^{(\xi_3)}(\xi_3) \tag{9.27}$$

$$v(t, \xi_1, \xi_2, \xi_3) = \sum_{0\le l+m+n\le P}^{P} v_{lmn}(t)\phi_l^{(\xi_1)}(\xi_1)\phi_m^{(\xi_2)}(\xi_2)\phi_n^{(\xi_3)}(\xi_3) \tag{9.28}$$

$$\gamma(t, \xi_1, \xi_2, \xi_3) = \sum_{0\le l+m+n\le P}^{P} \gamma_{lmn}(t)\phi_l^{(\xi_1)}(\xi_1)\phi_m^{(\xi_2)}(\xi_2)\phi_n^{(\xi_3)}(\xi_3) \tag{9.29}$$

where the coefficients $h_{lmn}(t)$, $v_{lmn}(t)$, $\gamma_{lmn}(t)$ are obtained by [178]

$$h_{lmn}(t) = \begin{cases} -\phi_0^{(\xi_1)}, & l = m = n = 0; \\ 1, & l = 1 \ \& m = n = 0; \\ 0, & \text{otherwise.} \end{cases} \tag{9.30}$$

$$v_{lmn}(t) = \begin{cases} -\phi_0^{(\xi_2)}, & l = m = n = 0; \\ 1, & m = 1 \ \& \ l = n = 0; \\ 0, & \text{otherwise.} \end{cases} \tag{9.31}$$

$$\gamma_{lmn}(t) = \begin{cases} -\phi_0^{(\xi_3)}, & l = m = n = 0; \\ 1, & n = 1 \ \& \ l = m = 0; \\ 0, & \text{otherwise.} \end{cases} \tag{9.32}$$

respectively, where $\phi_0^{(\xi_i)}$ denotes the zeroth-order term of the orthogonal polynomial of degree one with the weight $f_{\xi_i}(\xi_i)$.

Under the GPC framework, the equivalent DDEs expressed using the updated basis are reconstructed as follows:

$$\sum_{0\le l+m+n\le P} \dot{h}_{lmn}\phi_l^{(\xi_1)}\phi_m^{(\xi_2)}\phi_n^{(\xi_3)} = \left(\sum_{0\le l+m+n\le P} v_{lmn}\phi_l^{(\xi_1)}\phi_m^{(\xi_2)}\phi_n^{(\xi_3)}\right) \cdot \sin\left(\sum_{0\le l+m+n\le P} \gamma_{lmn}\phi_l^{(\xi_1)}\phi_m^{(\xi_2)}\phi_n^{(\xi_3)}\right) \tag{9.33}$$

$$\begin{aligned}\sum_{0\le l+m+n\le P} & \dot{V}_{lmn}\phi_l^{(\xi_1)}\phi_m^{(\xi_2)}\phi_n^{(\xi_3)} \\ &= -\frac{\varepsilon\rho}{2B_c}\sum_{0\le l+m+n\le P}\sum_{0\le u+v+w\le P} v_{lmn}\phi_l^{(\xi_1)}\phi_u^{(\xi_1)}\phi_m^{(\xi_2)}v_{uvw}\phi_v^{(\xi_2)}\phi_n^{(\xi_3)}\phi_w^{(\xi_3)} \\ &+ \frac{\mu}{\left(r_0+\sum_{0\le l+m+n\le P} v_{lmn}\phi_1^{(\xi_1)}\phi_m^{(\xi_2)}\phi_n^{(\xi_3)}\right)^2 \sin\left(\sum_{0\le l+m+n\le P}\gamma_{lmn}\phi_1^{(\xi_1)}\phi_m^{(\xi_2)}\phi_n^{(\xi_3)}\right)}\end{aligned} \tag{9.34}$$

$$\begin{aligned}\sum_{0\le l+m+n\le P} \dot{\gamma}_{lmn}\phi_l^{(\xi_1)}\phi_m^{(\xi_2)}\phi_n^{(\xi_3)} &= \frac{\kappa\varepsilon\rho\cos\varphi}{2B_c}\sum_{0\le l+m+n\le P} v_{lmn}\phi_l^{(\xi_1)}\phi_m^{(\xi_2)}\phi_n^{(\xi_3)} \\ &+ \frac{\sum_{0\le l+m+n\le P} v_{lmn}\phi_l^{(\xi_1)}\phi_m^{(\xi_2)}\phi_n^{(\xi_2)}\cos\left(\sum_{0\le l+m+n\le P}\gamma_{lmn}\phi_l^{(\xi_1)}\phi_m^{(\xi_2)}\phi_n^{(\xi_2)}\right)}{\left(r_0+\sum_{0\le l+m+n\le P} v_{lmn}\phi_l^{(\xi_1)}\phi_m^{(\xi_2)}\phi_n^{(\xi_2)}\right)} \\ &- \frac{\mu\cos\left(\sum_{0\le l+m+n\le P}\gamma_{lmn}\phi_l^{(\xi_1)}\phi_m^{(\xi_2)}\phi_n^{(\xi_2)}\right)}{\sum_{0\le l+m+n\le P} v_{lmn}\phi_l^{(\xi_1)}\phi_m^{(\xi_2)}\phi_n^{(\xi_2)}\left(r_0+\sum_{0\le l+m+n\le P} v_{lmn}\phi_l^{(\xi_1)}\phi_m^{(\xi_2)}\phi_n^{(\xi_2)}\right)}\end{aligned} \tag{9.35}$$

Multiplying both sides of Eqs. (9.33)–(9.35) by $\phi_l^{(\xi_1)}f_{\xi_1}\phi_m^{(\xi_2)}f_{\xi_2}\phi_n^{(\xi_3)}f_{\xi_3}$ and performing the triple integration on ξ_1, ξ_2, ξ_3 invoke the orthogonality of the basis functions. Therefore, the updated equivalent DDEs are obtained as follows:

$$\dot{h}_{lmn} = \frac{\int\int\int F(\xi_1,\xi_2,\xi_3)\phi_l^{(\xi_1)}\phi_m^{(\xi_2)}\phi_n^{(\xi_3)} f_{\xi_1\xi_2\xi_3}d\xi_1 d\xi_2 d\xi_3}{\left\langle \phi_l^{(\xi_1)2}\right\rangle\left\langle \phi_m^{(\xi_2)2}\right\rangle\left\langle \phi_n^{(\xi_3)2}\right\rangle} \tag{9.36}$$

$$\dot{v}_{lmn} = \frac{\int\int\int G(\xi_1,\xi_2,\xi_3)\phi_l^{(\xi_1)}\phi_m^{(\xi_2)}\phi_n^{(\xi_3)} f_{\xi_1\xi_2\xi_3}d\xi_1 d\xi_2 d\xi_3}{\left\langle \phi_l^{(\xi_1)2}\right\rangle\left\langle \phi_m^{(\xi_2)2}\right\rangle\left\langle \phi_n^{(\xi_3)2}\right\rangle} \tag{9.37}$$

$$\dot{\gamma}_{lmn} = \frac{\int\int\int T(\xi_1,\xi_2,\xi_3)\phi_l^{(\xi_1)}\phi_m^{(\xi_2)}\phi_n^{(\xi_3)} f_{\xi_1\xi_2\xi_3}d\xi_1 d\xi_2 d\xi_3}{\left\langle \phi_l^{(\xi_1)2}\right\rangle\left\langle \phi_m^{(\xi_2)2}\right\rangle\left\langle \phi_n^{(\xi_3)2}\right\rangle} \tag{9.38}$$

where $F(\xi_1, \xi_2, \xi_3)$, $G(\xi_1, \xi_2, \xi_3)$, $T(\xi_1, \xi_2, \xi_3)$ denote the right-hand sides of Eqs. (9.33)–(9.35), respectively, $f_{\xi_1\xi_2\xi_3}$ represents the joint PDF of ξ_1, ξ_2, ξ_3, and $f_{\xi_1\xi_2\xi_3} = f_{\xi_1}f_{\xi_2}f_{\xi_3}$, because ξ_1, ξ_2, ξ_3 are statistically independent.

In practice, an integral conversion operation is necessary to reduce the computational load of the multiple integrals and the joint PDF $f_{\xi_1\xi_2\xi_3}$ in Eqs. (9.36)–(9.38). Since the relationship between $f_{\xi_1\xi_2\xi_3}$ and f_ζ is [113]

$$f_{\xi_1\xi_2\xi_3}d\xi_1 d\xi_2 d\xi_3 = \int f_\zeta d\zeta \tag{9.39}$$

the three triple integrals and joint PDFs in Eqs. (9.36)–(9.38) can be transformed into

$$\int\int\int F(\xi_1, \xi_2, \xi_3)d\xi_1 d\xi_2 d\xi_3 = \int F(Z_1(\zeta), Z_2(\zeta), Z_3(\zeta)) f_\zeta d\zeta \tag{9.40}$$

$$\int\int\int G(\xi_1, \xi_2, \xi_3)d\xi_1 d\xi_2 d\xi_3 = \int G(Z_1(\zeta), Z_2(\zeta), Z_3(\zeta)) f_\zeta d\zeta \tag{9.41}$$

$$\int\int\int H(\xi_1, \xi_2, \xi_3)d\xi_1 d\xi_2 d\xi_3 = \int H(Z_1(\zeta), Z_2(\zeta), Z_3(\zeta)) f_\zeta d\zeta \tag{9.42}$$

and the $\phi_l^{(\xi_1)}(\xi_1)$, $\phi_m^{(\xi_2)}(\xi_2)$, $\phi_n^{(\xi_3)}(\xi_3)$ are obtained by

$$\phi_l^{(\xi_1)}(\xi_1) = \phi_l^{(\xi_1)}(Z_1(\xi)), \phi_m^{(\xi_2)}(\xi_2) = \phi_m^{(\xi_2)}(Z_2(\xi)), \phi_n^{(\xi_3)}(\xi_3) = \phi_n^{(\xi_3)}(Z_3(\xi)) \tag{9.43}$$

Therefore, Eqs. (9.36)–(9.38) can be transformed into the following updated equivalent DDEs:

$$\dot{h}_{lmn} = \frac{\int F(Z_1(\zeta), Z_2(\zeta), Z_3(\zeta))\phi_l^{(\xi_1)}(Z_1(\zeta))\phi_m^{(\xi_2)}(Z_2(\zeta))\phi_n^{(\xi_3)}(Z_3(\zeta)) f_\zeta(\zeta)\mathrm{d}\zeta}{\left\langle\phi_l^{(\xi_1)}(Z_1(\zeta))^2\right\rangle\left\langle\phi_m^{(\xi_2)}(Z_2(\zeta))^2\right\rangle\left\langle\phi_n^{(\xi_3)}(Z_3(\zeta))^2\right\rangle} \tag{9.44}$$

$$\dot{v}_{lmn} = \frac{\int G(Z_1(\zeta), Z_2(\zeta), Z_3(\zeta))\phi_l^{(\xi_1)}(Z_1(\zeta))\phi_m^{(\xi_2)}(Z_2(\zeta))\phi_n^{(\xi_3)}(Z_3(\zeta)) f_\zeta(\zeta)\mathrm{d}\zeta}{\left\langle\phi_l^{(\xi_1)}(Z_1(\zeta))^2\right\rangle\left\langle\phi_m^{(\xi_2)}(Z_2(\zeta))^2\right\rangle\left\langle\phi_n^{(\xi_3)}(Z_3(\zeta))^2\right\rangle} \tag{9.45}$$

$$\dot{\gamma}_{lmn} = \frac{\int T(Z_1(\zeta), Z_2(\zeta), Z_3(\zeta))\phi_l^{(\xi_1)}(Z_1(\zeta))\phi_m^{(\xi_2)}(Z_2(\zeta))\phi_n^{(\xi_3)}(Z_3(\zeta)) f_\zeta(\zeta)\mathrm{d}\zeta}{\left\langle\phi_l^{(\xi_1)}(Z_1(\zeta))^2\right\rangle\left\langle\phi_m^{(\xi_2)}(Z_2(\zeta))^2\right\rangle\left\langle\phi_n^{(\xi_3)}(Z_3(\zeta))^2\right\rangle}. \tag{9.46}$$

9.2.3 *Random Space Decomposition*

In addition to the improvement in accuracy by the SD, the increase of equation dimension in long-term integration is suppressed by the random space decomposition technique. Random space decomposition presented here is a specially modified version of the multi-element generalized polynomial chaos [179, 180] for enhancing the performance of the uncertainty quantification of Mars entry dynamics. The main idea behind the adaptive decomposition of random space is basically to decompose the random space of inputs whenever the relative error in variance exceeds a threshold [179, 180]. An adaptive criterion is designed to trigger such decomposition, while all other random dimensions remain unchanged. Thus, the total element amount and computational cost can be significantly reduced.

We denote the system state vector as $\boldsymbol{x} = [h, v, \gamma]^{\mathrm{T}}$. The superscript (ι) represents ιth system state with $\iota = 1,2,3$. The random space is decomposed into two equal parts once the following conditions are satisfied:

$$\left(\frac{1}{\sigma_{k,p}^2}\sum_{i=1}^{3}\hat{x}_i^{(\iota)2}\langle\phi_i^2\rangle\right)^{\alpha}\Pr(I_k = 1) \geq \vartheta_1 \tag{9.47}$$

$$s_i \geq \vartheta_2 \cdot \max_{j=1,2,3} s_j \tag{9.48}$$

where

$$I_k = \begin{cases} 1, & \text{if } \xi \in B_k; \\ 0, & \text{otherwise.} \end{cases}, \; s_i = \hat{x}_{i,p}^{(\iota)2}\langle\phi_{i,p}^2\rangle \Big/ \sum_{j=1}^{3}\hat{x}_j^{(\iota)2}\langle\phi_j^2\rangle, \; \Pr(I_k = 1) := \prod_{i=1}^{3}\frac{b_i^k - a_i^k}{2} \tag{9.49}$$

The threshold parameters $\alpha \in (0, 1)$, $\vartheta_1 \in (0, 0.1]$, $\vartheta_2 \in (0, 0.1]$, ϑ_2 is selected to determine the most sensitive random dimension, and s_i is the sensitivity of i-th random dimension. Here, the subscript i, p represents the mode consists of random dimension ξ_i with polynomial order p, and $p = 2, 3, \ldots 10$. Subscripts $k = 1, 2, \ldots, N_k$. N_k denotes non-overlapping elements. $\boldsymbol{B}_k = [(a_\iota, b_\iota)]_{3\times 1}$, $i = 1, 2, 3$ and a_ι, b_ι denotes the ιth upper and lower bounds of the initial state variables, respectively. Whenever the adaptive criteria are satisfied, the random field is mapped from the original mesh of elements to a new mesh of elements. The GPC expansion of the current random field is expressed as

$$\hat{x}_k^{(\iota)}\left(\hat{\boldsymbol{\xi}}\right) = \sum_{i=0}^{N_p}\hat{x}_{k,i}^{(\iota)}\phi_i\left(\hat{\boldsymbol{\xi}}\right) \tag{9.50}$$

The SD-updated-GPC expression of the next random field can be written as follows:

$$\tilde{x}_k^{(\iota)}\left(\tilde{\boldsymbol{\xi}}\right) = \sum_{i=0}^{N_p}\tilde{x}_{k,i}^{(\iota)}\phi_i\left(\tilde{\boldsymbol{\xi}}\right) \tag{9.51}$$

where $\tilde{\boldsymbol{\xi}} \in [-1, 1]^d$. To determine the coefficients $\tilde{x}_k^{(\iota)}$, we choose $(N_p + 1)$ grid points $\hat{\boldsymbol{\xi}}_i$ in $[-1, 1]^d$ and solve the following statespace equations:

$$\begin{bmatrix} \phi_{00} & \phi_{10} & \cdots & \phi_{N_p0} \\ \phi_{01} & \phi_{11} & \cdots & \phi_{N_p1} \\ \vdots & \vdots & \vdots & \vdots \\ \phi_{0N_p} & \phi_{1N_p} & \cdots & \phi_{N_pN_p} \end{bmatrix} \begin{bmatrix} \tilde{x}_{k,0}^{(\iota)} \\ \tilde{x}_{k,1}^{(\iota)} \\ \vdots \\ \tilde{x}_{k,N_p}^{(\iota)} \end{bmatrix} = \begin{bmatrix} \sum_{i=0}^{N_p} \hat{x}_{k,i}^{(\iota)} \phi_i\left(\chi^{-1}(\tilde{\xi}_0)\right) \\ \sum_{i=0}^{N_p} \hat{x}_{k,i}^{(\iota)} \phi_i\left(\chi^{-1}(\tilde{\xi}_1)\right) \\ \vdots \\ \sum_{i=0}^{N_p} \hat{x}_{k,i}^{(\iota)} \phi_i\left(\chi^{-1}(\tilde{\xi}_{N_p})\right) \end{bmatrix} \tag{9.52}$$

where $\phi_{ij} = \phi_i(\tilde{\xi}_j)$. The total amount of basis modes is calculated by [7, 17]

$$N_p = \frac{(p+d)!}{p!d!} - 1 \tag{9.53}$$

where p denotes the highest order of the polynomial chaos, and d is the dimension of the random variable.

9.2.4 *Obtaining Statistics of State Trajectory for Mars Entry*

According to the computational procedure described above, the statistics of the state trajectory can be determined from the GPC expansions. Here, the approximate global mean is calculated by

$$\overline{x}^{(\iota)} = \sum_{k=1}^{N_p} \hat{x}_{k,0}^{(\iota)} \Pr(I_k = 1) \tag{9.54}$$

where N denotes non-overlapping elements. The local variance is approximated by p-order polynomial chaos using the orthogonality of the GPC as follows:

$$\sigma_{k,p}^{(\iota)2} = \sum_{i=1}^{N_p} \hat{x}_{k,i}^{(\iota)2} \left\langle \phi_i^2 \right\rangle \tag{9.55}$$

Consequently, the approximate global variance can be calculated as

$$\left(\sigma^{(\iota)}\right)^2 = \sum_{k=1}^{N_p} \left[\sigma_{k,p}^{(\iota)2} + \left(\hat{x}_{k,0}^{(\iota)} - \overline{x}^{(\iota)} \right)^2 \right] \Pr(I_k = 1) \tag{9.56}$$

Let $e_k^{(\iota)}$ denote the error of the term $\sigma_{k,p}^{(\iota)2} + \left(\hat{x}_{k,0}^{(\iota)} - \overline{x}^{(\iota)}\right)^2$. Then, the exact global variance can be expressed as

$$\left(\sigma^{(\iota)}\right)^2 = \left(\overline{\sigma}^{(\iota)}\right)^2 + \sum_{k=1}^{N} e_k^{(\iota)} \Pr(I_k = 1) \tag{9.57}$$

As such, the uncertainty propagation of the state trajectory for Mars atmospheric entry can be quantitatively described, and the PDF of the flight state can be determined from time-varying GPC coefficient.

9.3 Numerical Simulations

9.3.1 Simulations Setup

A set of Mars entry simulations are implemented to test the performance of the proposed adaptive GPC procedure. To verify the effectiveness of the method, 10,000-run Monte-Carlo simulations are employed as the baseline. Additionally, results obtained by the original GPC approach [175] are compared with those from the proposed methodology. A set of parametric investigations are performed to analyze and unveil the behavior of the uncertainty along the state trajectory for the Mars atmospheric entry. We consider the sensitivity of the state trajectories to two categories of uncertainties: input uncertainty and inherent uncertainty. Importantly, the propagation of uncertainty due to initial conditions (h_0, v_0, γ_0) and systematic parameters (B_c, k, ρ) is analyzed separately. Coupled uncertainty, which considers all the uncertain parameters at the same time, is evaluated. Results for uncertainty quantification are presented for two kinds of probabilistic distributions, including $\pm 5\%$ uniform dispersion and Gaussian dispersion about the nominal with 10% variance for each dimension.

All necessary conditions and parameters' values of the problem and algorithm have been mentioned in the paper. Most of those conditions and parameters' values are presented in the context and the rest of them is set herein. To match the uncertainties expressed by Eqs. (9.4)–(9.9), the uncertain parameters are set and shown in Table 9.2. The system dynamics are propagated until $v > 450$ m/s and $h > 6.5$ km, i.e. until the conditions for parachute deployment are met. The parameters necessary for the criteria are set to be $\kappa_1 = \kappa_2 = \kappa_3 = 2/3$, $\vartheta_1 = 10^{-2}$, and $\vartheta_2 = 10^{-1}$, respectively. All simulations are performed on a laptop with Intel Core (TM) i7-7820HQ CPU at 2.9 GHz and 32 GB of memory.

The Gaussian integral algorithm [170, 176] is the selected numerical integrator for the random variable propagation. Fourth-order Runge–Kutta integrator is used to propagate the dynamics and for solving the DDEs. The latter is achieved via the MATLAB's ode45 integrator with a variable step-size. MATLAB's statistics

Table 9.2 Simulation setup

Parameters	Mean	Case 1: Uniform distribution Interval	Case 2: Gaussian distribution 3σ	Unit
Initial height h_0	125	[118.75, 131.25]	12.5	km
Initial velocity V_0	5500	[5225, 5775]	550	m/s
Initial flight path angle γ_0	−15.2	[−15.96, −14.44]	1.52	°
Ballistic coefficient B_c	121.6	[115.52, 127.68]	12.16	kg/m^2
Lift-to-drag ratio k	0.24	[0.228, 0.252]	0.024	–
Density uncertainty ε	1	[0.95, 1.05]	0.1	–

toolbox is utilized to compute the mean and variance (or deviation) of the Monte-Carlo results. The uncertainties are specified by the evolution of the PDFs based on the setup reported in Table 9.2. To avoid the curse of dimensionality, pseudorandom number generator using low-discrepancy sequences is the preferred choice versus the more standard grid-based discretization. Here, samples governed by the uniform PDFs are generated using the Halton sequence technique [181] which is implemented via MATLAB's haltonset function. Additionally, samples from the Gaussian PDF are generated via the Box-Muller probability integral transform [182]. The PDFs are visualized by scatter and pseudo-color plots according to their numerical results in each profile. Limited by the space of the paper, the simulation results of the evolutions of only dynamical parameter uncertainty are not presented herein, but this work has done.

9.3.2 Case 1: Uniform Uncertainty

For this case, we assume $\pm 5\%$ uniformly parametric variation in the initial conditions (h_0, v_0, γ_0) and in the dynamical parameters (B_c, k, ρ) with regard to their nominal values, as reported in Table 9.2. To observe the uncertainty evolution behavior and its basic laws, uniform uncertainty in initial conditions and dynamical parameters are quantified along Mars entry profile. To evaluate the performance of the proposed methodology, we compare the results with those obtained from MC simulations, GPC, and Adaptive Generalized Polynomial Chaos (AGPC) approaches.

1. **Uniform uncertainty in initial conditions**

Figure 9.1 shows the spatio-temporal evolution of the initial uncertainty along the simulated Mars entry trajectories. We can found that the states along the trajectories are greatly affected by the size of the uniform uncertainty in the initial conditions (h_0, v_0, γ_0). Restricted by the conditions of parachute deployment, the magnitudes of final height and velocity converge well and are generally smaller than their initial dispersion. This phenomenon agrees with our physical intuition as Mars entry

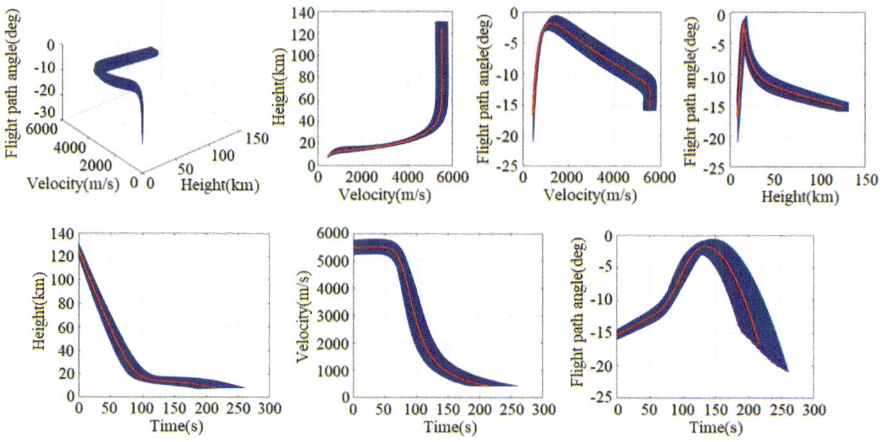

Fig. 9.1 Evolution of uncertainty along Mars entry profile: sample (blue), mean (red)

vehicle, with high probability, slows down through the denser atmosphere at lower altitude.

Figure 9.2 shows how the probability distributions (i.e. PDFs) change along the Mars entry trajectory. Here, by visualizing the appropriate cross-sections, one can see how the PDFs vary along the Mars entry profile due to initial uncertainties. The joint PDFs are visualized using both scatter plots obtained by MC and pseudo color plots obtained by AGPC. The color intensity represents the value of the instantaneous joint PDF. Here, we can see that the results from the AGPC method agree well with those obtained by MC simulations, which is a testament of the goodness of the proposed approach. Figure 9.2 (b) shows that probability density for each section is significantly different. In the AGPC procedure, because uniform PDFs are generated using the Halton sequence technique, a series of domains exhibiting small Gaussian peaks are uniformly distributed in the defined random space. As the system evolves, the joint PDFs for subsequent sections are no longer characterized by uniform distributions. The PDFs' peaks rise with time, accompanied with shrinkage of the PDFs' supports. More precisely, the probability mass for each bivariate joint PDF gradually accumulates to a domain with the peak around the mean.

Since the probability distribution plots obtained by GPC are similar to the results by the AGPC approach, a quantitative comparison on the mean and deviation trajectories obtained by MC, AGPC, and GPC under the condition of uniform initial uncertainty is performed and reported in Fig. 9.3. Here, the plots show that the results obtained from AGPC more closely represent the distribution described by MC. In terms of the sensitivity of the trajectories to these parameter variations, it is reported that the magnitude levels of the errors in the mean profiles are a few orders smaller than those in the deviation profiles. When comparing the results obtained by AGPC and GPC, it can be seen that the accuracy of the AGPC approach is almost one-order

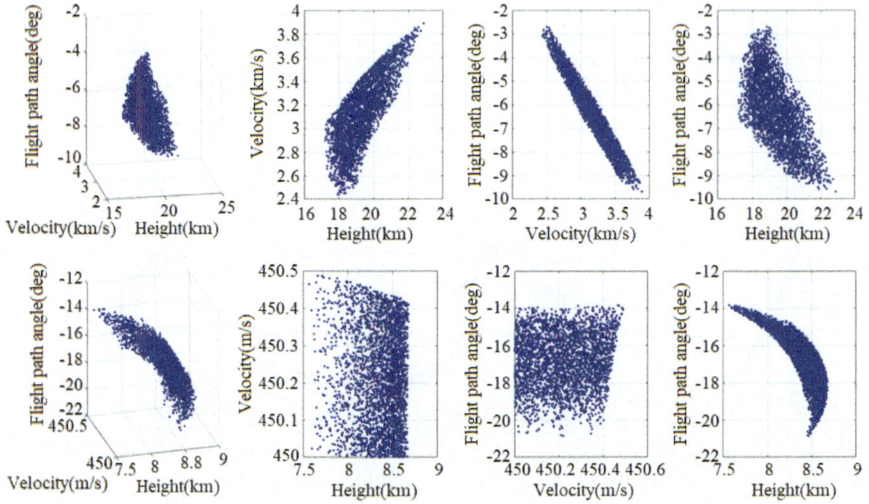

(a) Monte-Carlo scatter plots of joint PDFs: rows show middle time (upper) and final time (lower).

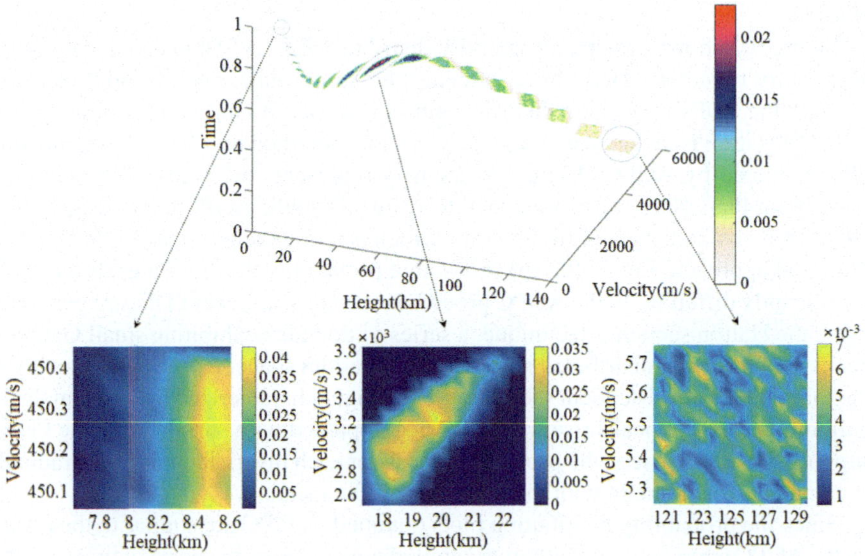

(b-1) AGPC pseudo-color plots of bivariate joint PDFs (height-velocity profile).

Fig. 9.2 Probabilistic distribution plots along Mars entry profile

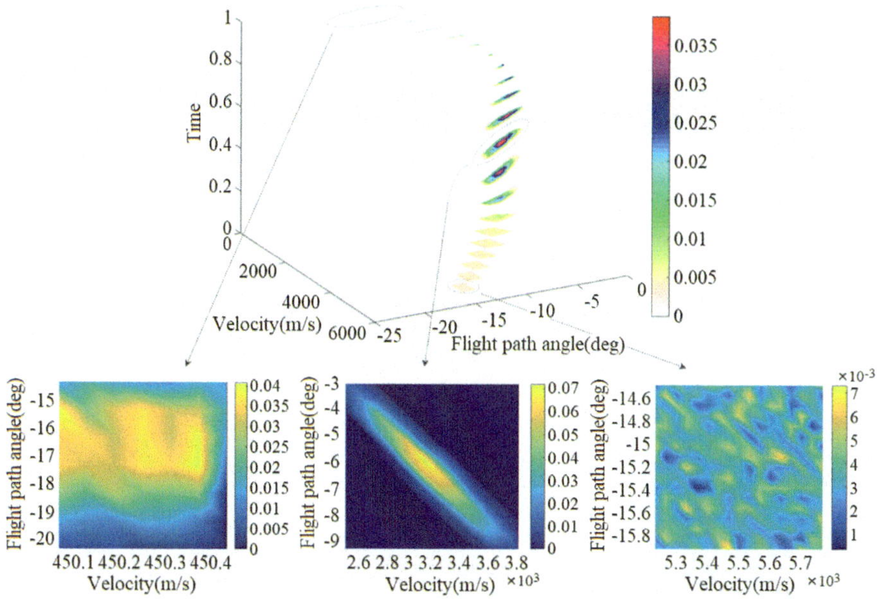

(b-2) AGPC pseudo-color plots of bivariate joint PDFs (velocity-flight path angle profile).

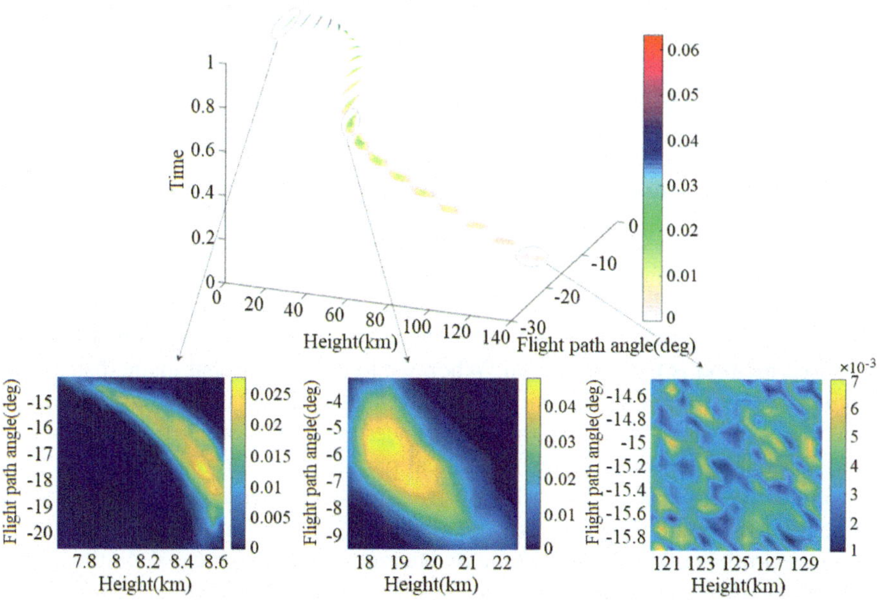

(b-3) AGPC pseudo-color plots of bivariate joint PDFs (height-flight path angle profile).

Fig. 9.2 (continued)

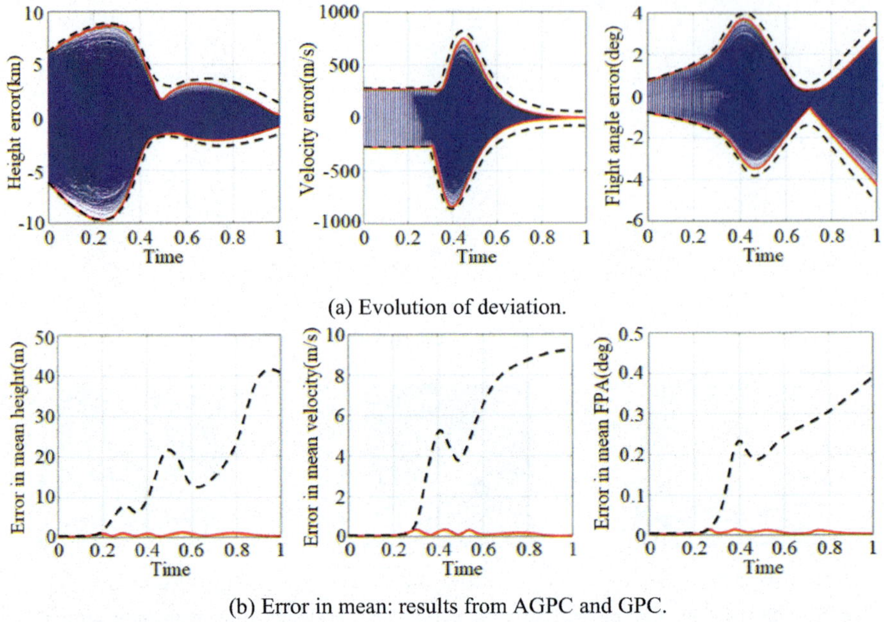

Fig. 9.3 Comparison of the results obtained from MC (blue dots), AGPC (red solid), and GPC (black dashed)

higher than that of the GPC method. Divergence is observed in the implementation of the GPC technique.

2. **Coupled uniform uncertainty in system parameters and initial conditions**

Figure 9.4 shows the spatio-temporal evolution of the uniform uncertainties in both initial conditions (h_0, v_0, γ_0) and dynamical system parameters (B_c, k, ρ) along the Mars entry profile. The joint PDFs visualized using scatter plots obtained by MC simulations match well with those obtained by AGPC. The performance of the proposed method is thus confirmed. When we look at the probabilistic distribution profiles as reported in Fig. 9.4, one can observe that the effect of the coupled uniform uncertainty is similar to that of the initial uncertainty. Therefore, the initial uncertainty is dominant instead of the dynamical parameter uncertainty when the same uniform distribution is considered. However, one can observe that the evolution of the coupled uncertainty is not the simple addition of the separate evolutions of uncertainties in initial conditions and in the dynamical system parameters. As the system dynamics evolves, the distributions for coupled uncertainty change from uniform to non-uniform. As a result, the visualized joint PDFs are similar to those represented in the evolution profile of the initial uncertainty. Consequently, the probabilistic distribution of the initial uncertainty is also dominant for the case of the coupled uncertainty propagation.

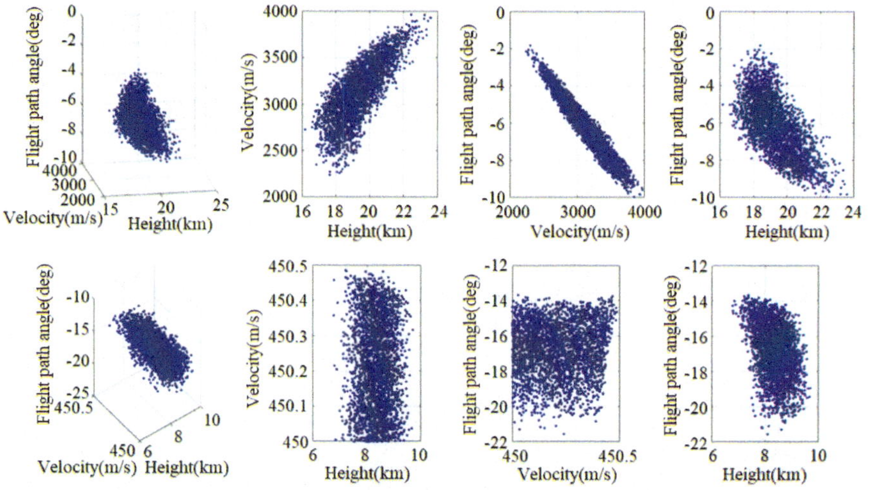

(a) Monte-Carlo scatter plots of joint PDFs: rows show middle time (upper) and final time (lower).

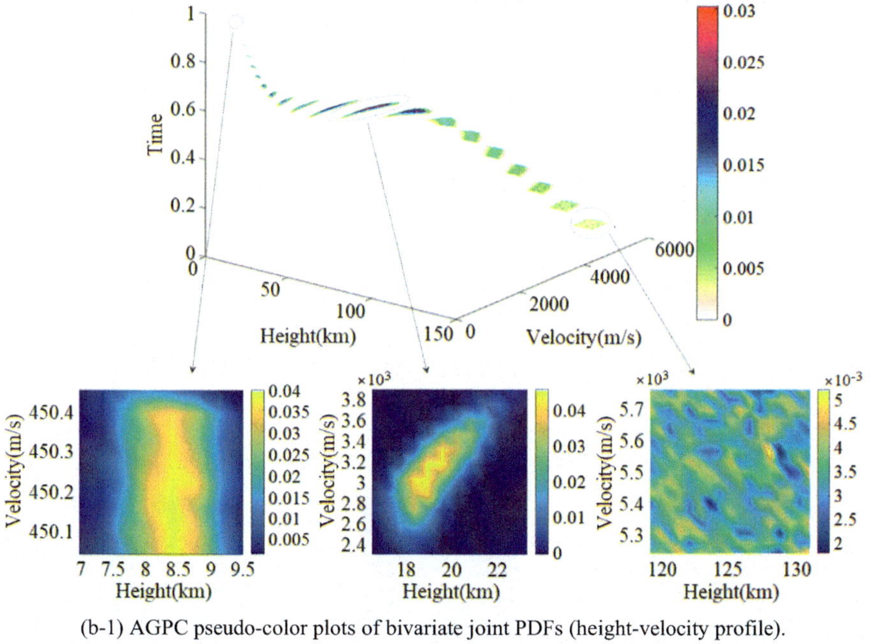

(b-1) AGPC pseudo-color plots of bivariate joint PDFs (height-velocity profile).

Fig. 9.4 Probabilistic distribution plots along Mars entry profile

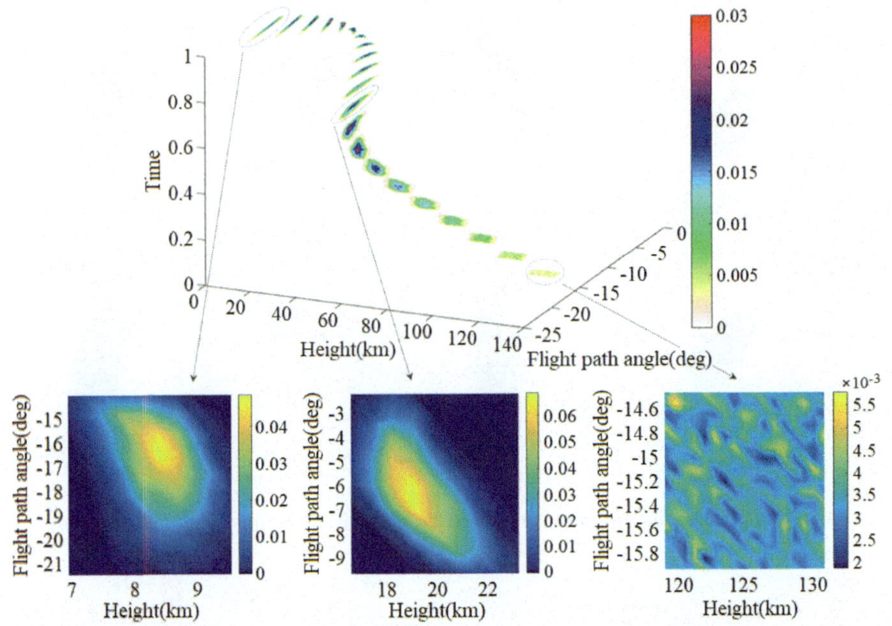

(b-2) AGPC pseudo-color plots of bivariate joint PDFs (height-flight path angle profile).

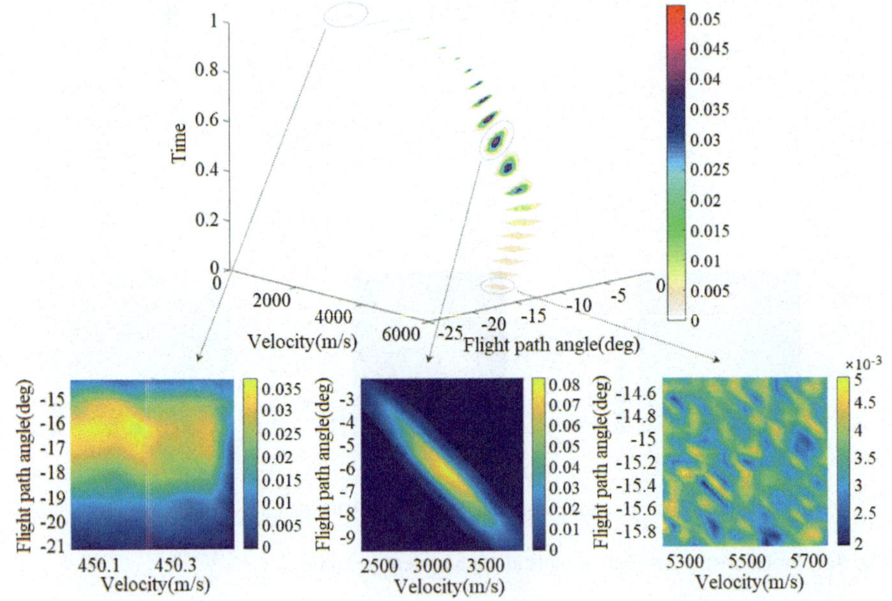

(b-3) AGPC pseudo-color plots of bivariate joint PDFs (velocity-flight path angle profile).

Fig. 9.4 (continued)

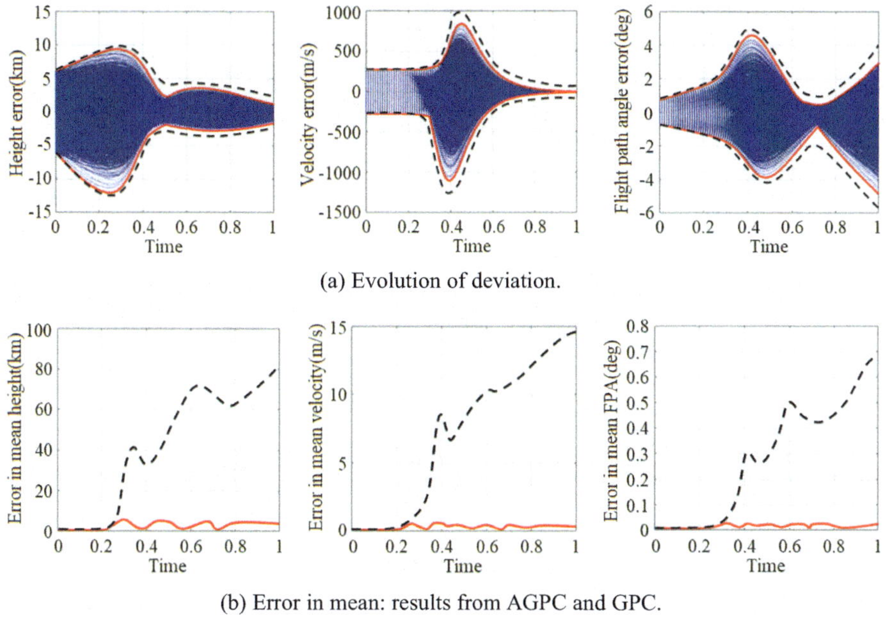

Fig. 9.5 Comparison of the results obtained from MC (blue dots), AGPC (red solid), and GPC (black dashed)

Figure 9.5 quantitatively shows the abovementioned dominance which can be observed by comparing the mean and variance profiles obtained by MC, AGPC, and GPC methods under the condition of uniformly coupled uncertainty. In terms of the sensitivity of the statistical properties of the trajectories to these uncertainties, it is found that the deviations existing in the variance profiles are a few orders larger than the relative errors occurring in the mean profiles. Importantly, the error of the AGPC is always smaller than that of the GPC.

9.3.3 Case 2: Gaussian Uncertainty

In this section, we assume that the initial conditions (h_0, v_0, γ_0) and dynamical system parameters (B_c, k, ρ) are governed by Gaussian distributions, with their mean and variance reported in Sect. 9.2. To evaluate the effect on the state trajectories of these two types of distributions, quantifications of Gaussian uncertainties in both initial conditions and dynamical system parameters evolving along Mars entry profile are performed first. Subsequently, the evolution of the coupled Gaussian uncertainties in both initial conditions and dynamical system parameters as they propagate along the Mars entry profile are computed. We compare the results obtained from MC simulations, GPC, and AGPC approaches.

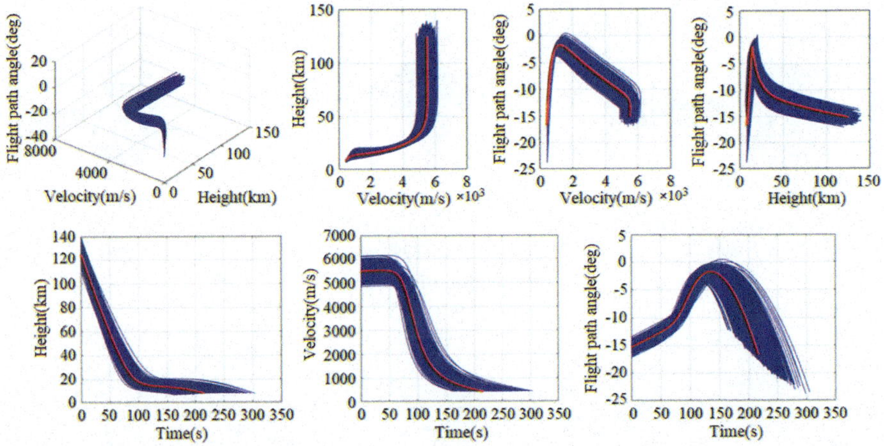

Fig. 9.6 Evolution of uncertainty along Mars entry profile: sample (blue), mean (red)

1. **Gaussian uncertainty in initial conditions**

Figure 9.6 shows the spatio-temporal evolution of the Gaussian initial uncertainty along the Mars entry profile. It is observed that the state trajectories of the Mars entry vehicle are significantly affected by the Gaussian uncertainty in initial conditions (h_0, v_0, γ_0). The dispersion of final time achieves values between 50 through 70 s relative to the nominal value. However, the variance of flight-path-angle diverges greatly which would inevitably result in large final dispersion. Restricted by the stopping conditions imposed by the parachute aperture, the variances of final altitude and final speed converge well and are generally smaller than their initial uncertainty.

Figure 9.7 reports the evolution of the PDFs. The results from the AGPC method agree well with those obtained by MC simulations. Figure 9.7a shows that the PDFs of Gaussian initial uncertainties also vary along Mars entry profile. As the system evolves, the joint PDFs gradually become non-standard Gaussian distributions. Therefore, to adapt to the evolution of Gaussian initial uncertainty, it is necessary to update the basis function family. When comparing Fig. 9.7 with Fig. 9.2, one observes that their probabilistic distributions are similar, although the Gaussian initial uncertainty is completely different from the uniform one. Therefore, we can infer that the nonlinear dynamics of the Mars entry change the joint PDFs of the uncertainty and drive them to converge to analogous terminal probabilistic distributions as shown in Fig. 9.7b.

Figure 9.8 provides a quantitative comparison about the mean and variance profiles obtained from MC simulations, AGPC, and GPC. It is found that the errors on mean profiles are always several orders greater than the deviations occurring in the variance profiles. Finally, it is shown that the AGPC always has smaller error than the GPC when compared with MC simulations.

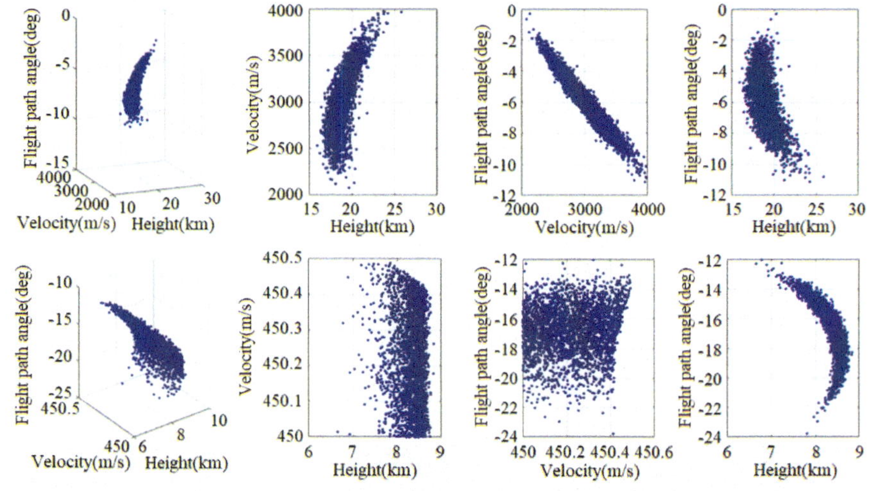

(a) Monte-Carlo scatter plots of joint PDFs: rows show middle time (upper) and final time (lower).

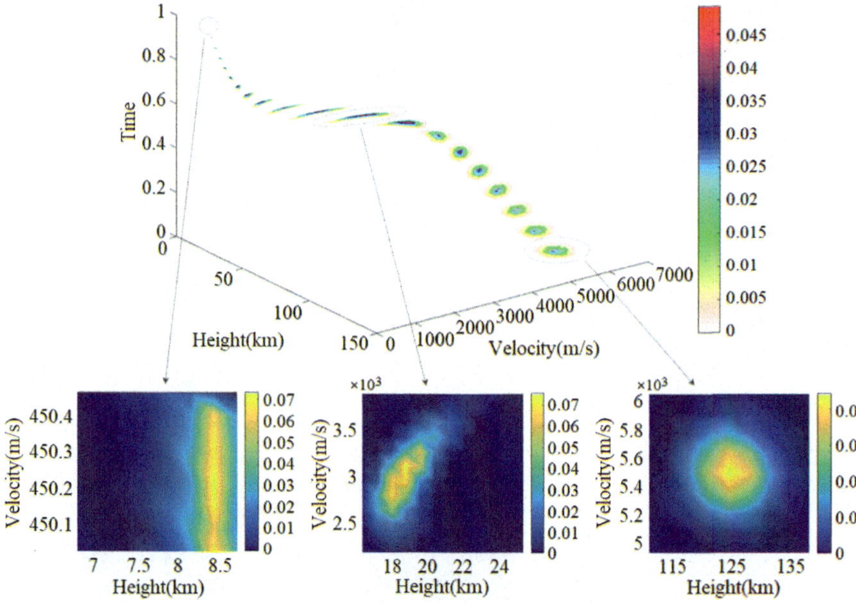

(b-1) AGPC pseudo-color plots of bivariate joint PDFs (height-velocity profile).

Fig. 9.7 Probabilistic distribution plots along Mars entry profile

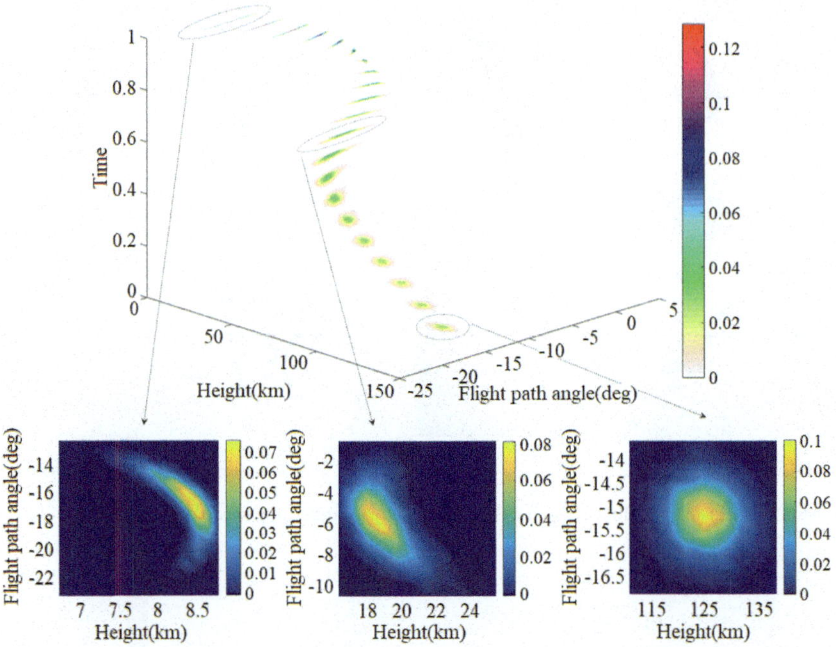

(b-2) AGPC pseudo-color plots of bivariate joint PDFs (height-flight path angle profile).

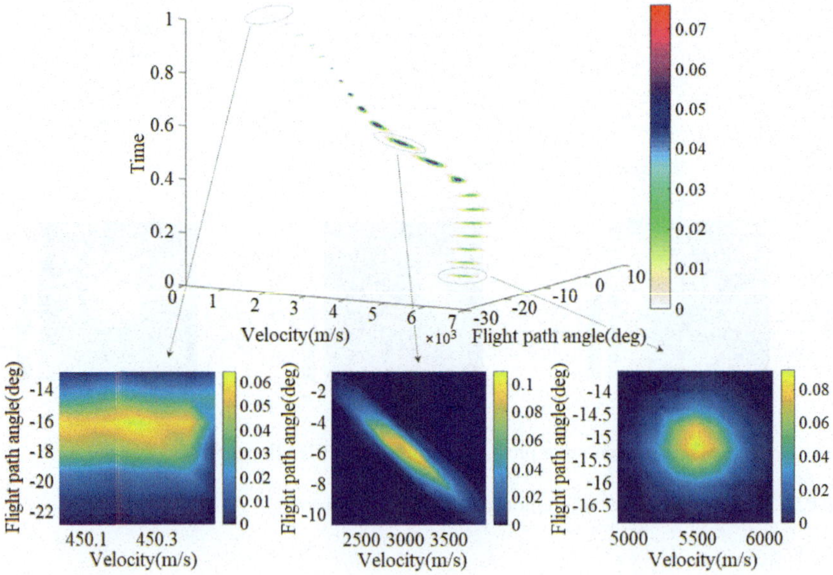

(b-3) AGPC pseudo-color plots of bivariate joint PDFs (velocity-velocity profile).

Fig. 9.7 (continued)

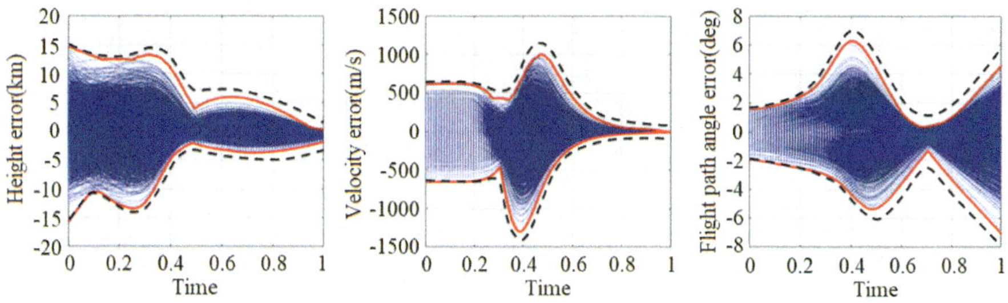

(a) Evolution of deviation.

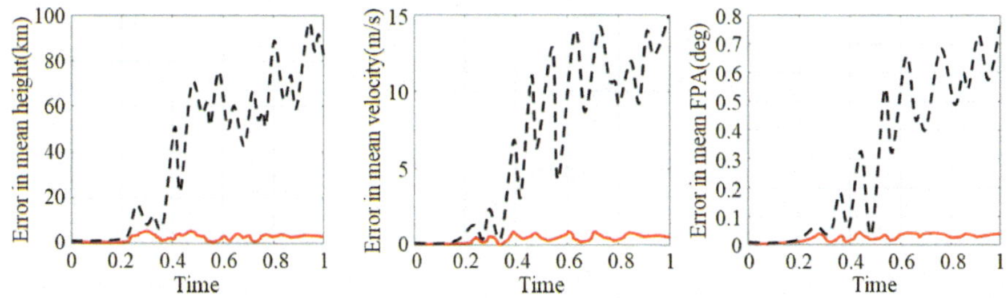

(b) Error in mean: results from AGPC and GPC.

Fig. 9.8 Comparison of the results obtained from MC (blue dots), AGPC (red solid), and GPC (black dashed)

2. **Coupled Gaussian uncertainties**

Figure 9.9 shows the spatio-temporal evolution of Gaussian uncertainties in both initial conditions (h_0, v_0, γ_0) and dynamical system parameters (B_c, k, ρ) along the Mars entry profile. It can also be seen from Fig. 9.9 that the joint PDFs evolution obtained by AGPC matches well with that by MC simulations (scatter plots). As the system dynamics evolve, the Gaussian distributions for coupled uncertainties gradually change to non-Gaussian ones. When we compare Fig. 9.9 with Fig. 9.4, it can be found that the final probabilistic distributions for coupled uniform uncertainty scenario are similar to those observed for coupled Gaussian uncertainty scenario. Therefore, we can infer that the nonlinear dynamics of Mars entry drive them to converge to similar probability distributions.

We have also done the simulation analysis of parametric uncertainty, but the space of the paper is limited, and the results of the parametric uncertainty case are not shown here. We found that the effect of the coupled Gaussian uncertainty on the state trajectories is similar to that of the Gaussian initial uncertainty (Fig. 9.7) versus that of the Gaussian dynamical system parameter uncertainty. Therefore, the initial uncertainty is dominant than the influence of the dynamical system parameters uncertainty. The evolutions of the variances for the coupled uncertainties are larger than that caused by initial and parametric uncertainties separately. Nevertheless, it cannot be represented by the simple addition of the evolutions of uncertainties in initial conditions and dynamical parameters.

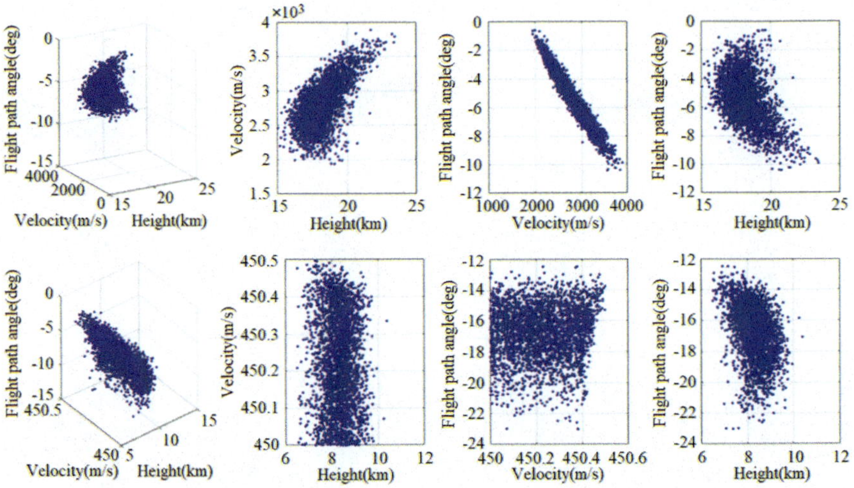

(a) Monte-Carlo scatter plots of joint PDFs: rows show middle time (upper) and final time (lower).

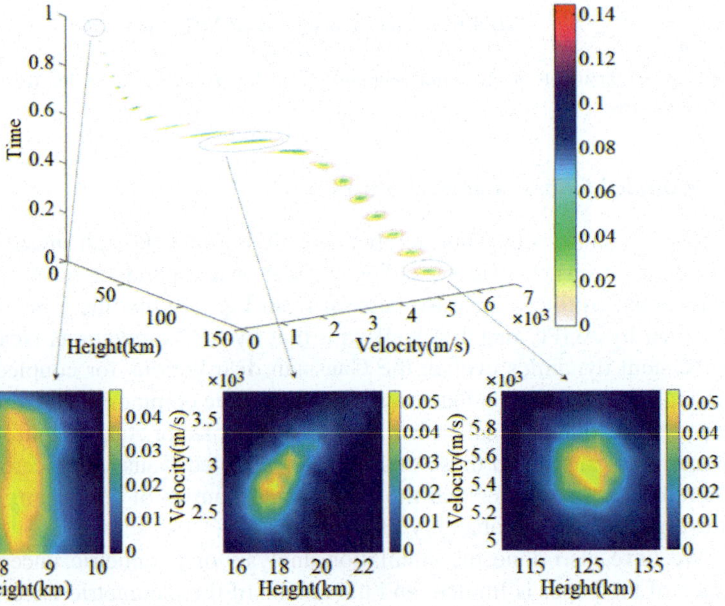

(b-1) AGPC pseudo-color plots of bivariate joint PDFs (height-velocity profile).

Fig. 9.9 Probabilistic distribution plots along Mars entry profile

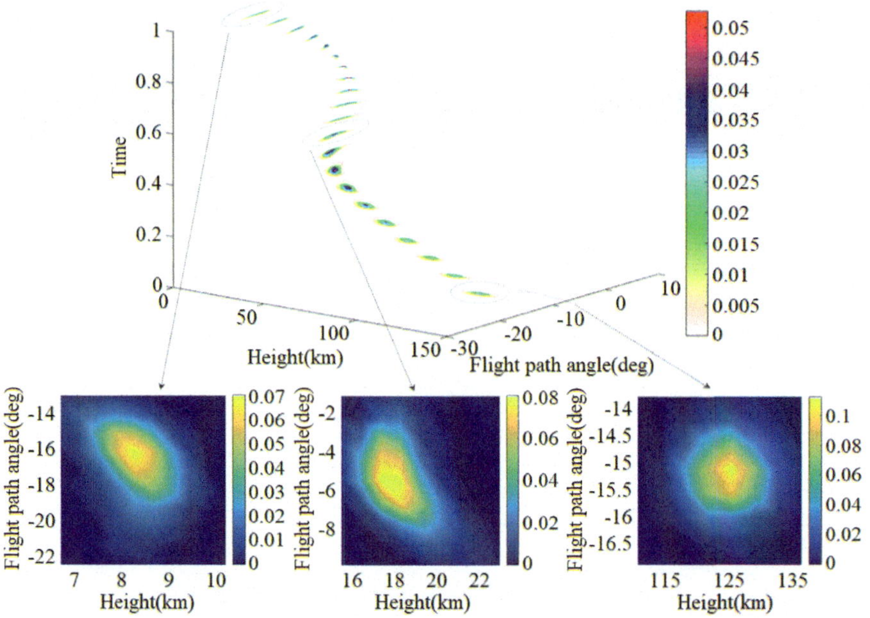

(b-2) AGPC pseudo-color plots of bivariate joint PDFs (height-flight path angle profile).

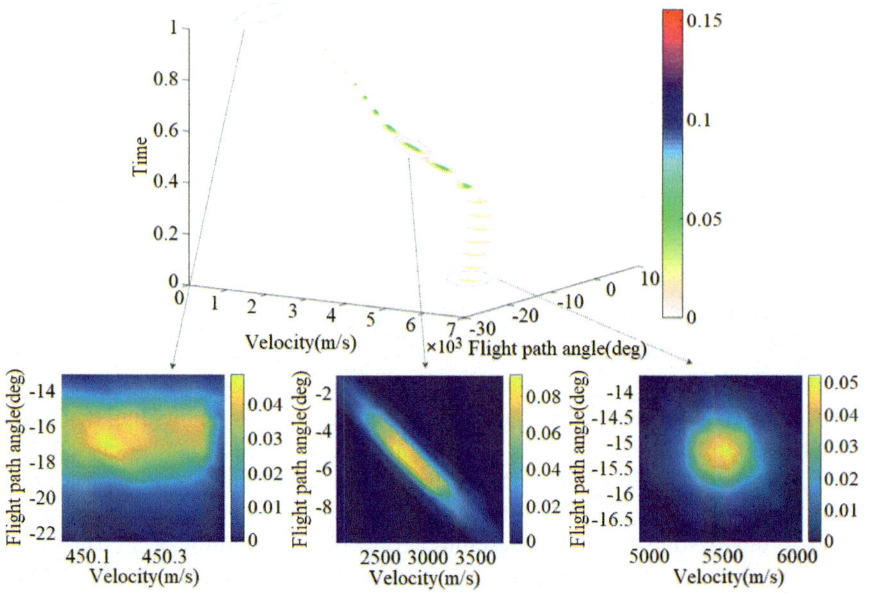

(b-3) AGPC pseudo-color plots of bivariate joint PDFs (velocity-flight path angle profile).

Fig. 9.9 (continued)

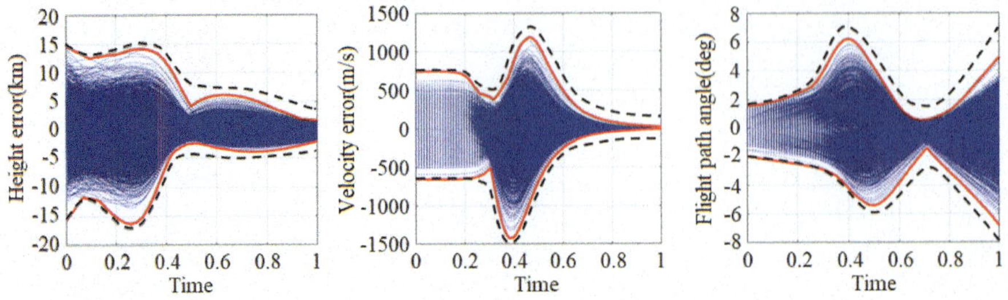

(a) Evolution of deviation.

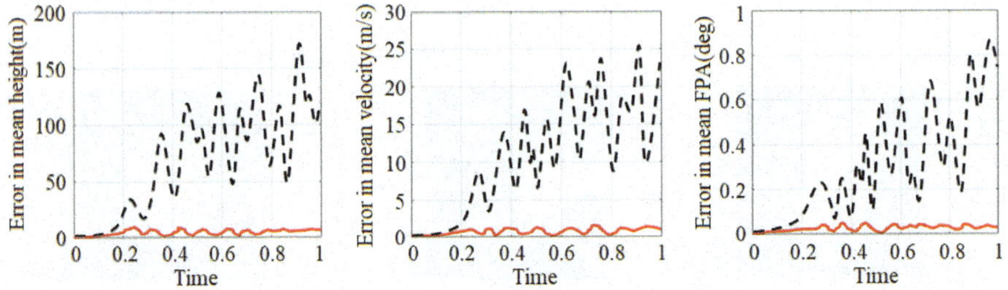

(b) Error in mean: results from AGPC and GPC.

Fig. 9.10 Comparison of the results obtained from MC (blue dots), AGPC (red solid), and GPC (black dashed)

Figure 9.10 shows a quantitative analysis by comparing the mean and variance profiles obtained by MC, AGPC, and GPC methods under the condition of coupled Gaussian uncertainty. It is observed that the errors on mean profiles are a few orders lower than the deviations occurring in the variance profiles. Meanwhile, the error for AGPC is always one-order smaller than that of the GPC.

9.3.4 Analysis and Discussions

From Poincare section point of view [183], the flight trajectory for Mars atmospheric entry is in a chaotic state. This due to the fact that the points on the map that scatter in each section has a hierarchical structure instead of fixed points or closed curve (see Figs. 9.2, 9.4, 9.7, and 9.9). Nevertheless, the uncertainty propagation can be characterized by the presented computational approach in a probabilistic fashion. To unveil the change of the sought uncertainty probability distribution, a large number of brute-force MC simulations is generally needed. Conversely, only one-run of AGPC or GPC is sufficient to characterize the uncertainty evolution. AGPC is able to characterize the uncertainty profile precisely, whereas GPC always gradually diverges and cannot approximate the mutations. The uncertainty quantification for Mars atmospheric entry is achieved with significantly reduced computational cost. Table 9.3

shows the CPU times for the cases under investigation. For lower dimensional uncertainty quantification, the computing time of AGPC is slightly increased than that of the GPC. Conversely, AGPC takes less time than GPC when quantifying higher dimensional uncertainty.

For higher-dimensional uncertainty parameter space (e.g. coupled uncertainty), the advantage of the AGPC is more significant. To obtain the uncertainty quantification results with the same accuracy level as the AGPC, the computational time of MC simulations is always one order of magnitude higher than the AGPC counterpart. In the same uncertainty scenario, GPC achieved nearly ten times lower accuracy than that of the AGPC, while exhibiting similar computational time. As the PDFs of the considered uncertainty are propagated according to the system dynamics, the joint PDF at any given time is represented by the instantaneous distribution of the evolved states. Such a distribution is generally a data set scattered over the extended state space, and the dimension of random space is generally enlarged. Consequently, MC needs more computational time, whereas the accuracy of the GPC is exponentially reduced.

The superior performance exhibited by AGPC is due to the adaptive update of basis function family and to the adaptive decomposition of random space, which preserves the statistical characteristics of the initial uncertainty (i.e. input uncertainty of dynamical system) while suppresses the dimensional increase of RDEs accordingly. Indeed, the adaptive update of basis function family is executed to suppress both the divergence exhibited by GPC and the increase in the order of polynomial chaos during long-term integration through sensing the change of the stochastic characteristics of the inherent state of the system. Such features are the basis for the better performance of AGPC.

In regard to uncertainty quantification for Mars entry, the following four interesting findings are reported: (1) The statistical features of uncertainty of flight state (including mean, variance, probability distribution) always change along Mars entry profile; (2) The joint PDFs of the uncertainty always converge to analogous probability distributions at the final section, no matter if uniform or Gaussian scenario is deployed, which are completely different from the input probability distributions; (3) The final state uncertainty is generally neither standard uniform distribution nor standard Gaussian distribution no matter uniform or Gaussian scenario is deployed; (4) Initial state uncertainty is dominant in the uncertainty propagation for Mars entry in both uniform and Gaussian scenarios.

Table 9.3 Comparison of CPU time

Time usage of each scenario (seconds)		MC	GPC	AGPC
Case 1: Uniform	Initial uncertainty	1786.532	61.962	83.058
	Coupled uncertainty	3052.476	185.031	146.632
Case 2: Gaussian	Initial uncertainty	1967.384	74.618	90.751
	Coupled uncertainty	3595.849	203.695	171.925

9.4 Conclusions

Although it is an important prerequisite for the design of guidance and trajectory of Mars EDL system, research on high-dimensional uncertainty quantification for Mars entry is still lacking. To further contribute to bridge this gap, in this chapter, the propagation of high-dimensional uncertainty in Mars atmospheric entry dynamics is investigated. Based on spectral decomposition and random space decomposition, generalized polynomial chaos is modified to efficiently address the uncertainty quantification task. Comparison simulations demonstrate its effectiveness in quantifying the high-dimensional uncertainty propagated along Mars atmospheric entry trajectories, exhibiting a higher level of accuracy than generalized polynomial chaos as well as better computational efficiency than Monte-Carlo simulations. The laws of the uncertainty propagation due to the initial conditions and dynamical system parameters such as ballistic coefficient, lift-over-drag ratio, and atmospheric density are investigated via case studies. Numerical experiments demonstrate that the statistical characteristics of uncertainties always change as the system dynamics evolve, i.e., the uncertainties change from standard uniform and Gaussian distributions to non-uniform and non-Gaussian ones. No matter if the uncertainties are governed by uniform or Gaussian distribution, the initial uncertainty is always the major contributor to the dispersion of the state trajectory, while each bivariate joint PDF always converges to its own but analogous probability distribution. The effects of the uncertainty in the dynamics system parameters are pronounced at relatively low altitude, where the atmosphere is dense and the aerodynamic effects are dominant. The uncertainty propagation profiles can be used to support uncertainty-based design optimization work for Mars atmospheric entry guidance in the near future.

Chapter 10
Robust Trajectory Optimization for Mars Entry

10.1 Trajectory Optimization Problem Formulation

In order to implement robust optimization of Mars entry trajectory under both initial uncertainty and dynamical parameter uncertainty, a corresponding robust optimization problem should be formulated at first, i.e., Mars entry dynamics, constraints, and objective function.

10.1.1 Dynamics of Mars Entry

The entry phase of low-lift Mars EDL typically begins when the vehicle reaches the Mars atmospheric interface (defined as a radius of 3522 km from the center of Mars) and ends when the parachute deployment. During this process, the MSL-type entry vehicle is usually commanded to fly at a trim angle-of-attack and the guidance is achieved only through modulating the bank angle with a reaction control system. The three degree-of-freedom (DOF) dynamical equations defined with respect to the Mars fixed coordinate frame are given by:

$$\begin{cases} \dot{r} = v\sin\gamma \\ \dot{\theta} = \frac{v\cos\gamma\sin\psi}{r\cos\lambda} \\ \dot{\lambda} = \frac{v}{r}\cos\gamma\cos\psi \\ \dot{v} = -D - g_M\sin\gamma + \omega_M^2 r\cos\lambda(\sin\gamma\cos\lambda - \cos\gamma\sin\lambda\cos\psi) \\ \dot{\gamma} = \frac{L\cos u}{v} + \left(\frac{v}{r} - \frac{g}{v}\right)\cos\gamma + 2\omega_M\cos\lambda\sin\psi \\ \quad + \frac{\omega_M^2 r}{v}\cos\lambda(\cos\gamma\cos\lambda + \sin\gamma\sin\lambda\cos\psi) \\ \dot{\psi} = \frac{L\sin u}{v\cos\gamma} + \frac{v}{r}\sin\psi\cos\gamma\tan\lambda \\ \quad -2\omega_M(\tan\gamma\cos\psi\cos\lambda - \sin\lambda) + \frac{\omega_M^2 r}{v\cos\gamma}\cos\lambda\sin\lambda\sin\psi \end{cases} \tag{10.1}$$

S. Li et al., *Trajectory Optimization and Guidance Methods for Mars Entry*,
https://doi.org/10.1007/978-981-99-6282-2_10

where r is the radial distance between the center of Mars and the vehicle center of mass, θ and λ are the longitude and latitude respectively, v is vehicle's velocity magnitude, γ denotes the flight path angle, ψ represents the heading (azimuth) angle, u is the bank angle, and ω_M is the self-rotation angular rate of Mars. $g = GM/r^2$ denotes Mars gravitational acceleration, and GM is the gravitational constant of Mars. The lift and drag accelerations are respectively defined by

$$\begin{cases} L = \frac{1}{2}\rho v^2 \frac{C_L S_{ref}}{m} \\ D = \frac{1}{2}\rho v^2 \frac{C_D S_{ref}}{m} \end{cases} \tag{10.2}$$

where ρ denotes the Mars atmospheric density, C_L and C_D are the lift and drag coefficients respectively, S_{ref} and m are the reference area and mass of entry vehicle respectively. Lift-over-drag ratio is defined as C_L/C_D. Ballistic coefficient is defined as $B_c = m/(C_D S_{ref})$. The exponential Mars atmospheric density model with respect to the geographic altitude is defined as:

$$\rho = \rho_0 \exp\left(-\frac{r - R_M}{h_s}\right) \tag{10.3}$$

where ρ_0 is the density on the surface of the Mars, R_M is the radius of the Mars and h_s is the constant scale height.

In practice, uncertainties generally contain external/input uncertainty and inner/inherent uncertainty [112, 184], i.e. initial state uncertainty and dynamical parameter uncertainty regarding Mars entry [172, 185]. These uncertainties are usually modeled as stochastic deviations with respect to their nominal models, which are governed by certain probabilistic distributions. Thus, the initial state uncertainty and dynamical parameter uncertainty for Mars entry are formulated as below.

$$\begin{aligned} \boldsymbol{x}_0 &= \overline{\boldsymbol{x}}(t_0) + \boldsymbol{\delta}_{x0}, \quad \boldsymbol{x}_0 = \left[r_0, \theta_0, \lambda_0, v_0, \gamma_0, \psi_0\right]^{\mathrm{T}}, \\ \boldsymbol{\delta}_{x0} &= \left[\delta_{r0}, \delta_{\theta 0}, \delta_{\lambda 0}, \delta_{v0}, \delta_{\gamma 0}, \delta_{\psi 0}\right]^{\mathrm{T}} \end{aligned} \tag{10.4}$$

$$C(t) = \overline{\mathbf{C}}(t) \cdot (\mathbf{1}_{3\times 1} + \boldsymbol{\delta}_C), \quad \mathbf{C} = [\rho, B_c, LD]^{\mathrm{T}}, \quad \boldsymbol{\delta}_C = \left[\delta_\rho, \delta_{BC}, \delta_{LD}\right]^{\mathrm{T}} \tag{10.5}$$

where t denotes any time during Mars entry, t_0 is the initial time, represents stochastic deviation which is governed by corresponding probabilistic distribution, and the cap '-' means nominal value.

10.1.2 Constraints

Limited by the design of entry vehicle, normal overload n_a, dynamic pressure q, aerodynamic heating rate $\dot{Q}$ and total heat Q should satisfy certain constraints as follows.

$$n_a = \sqrt{L^2 + D^2} \le n_{\max} \tag{10.6}$$

$$q = \frac{1}{2}\rho v^2 \le q_{\max} \tag{10.7}$$

$$\dot{Q} = c_q\sqrt{\frac{\rho}{r_n}}v^3 \le \dot{Q}_{\max} \tag{10.8}$$

$$Q = \int_{t_0}^{t_f} \dot{Q}\,dt \le Q_{\max} \tag{10.9}$$

where $n_{\max}$, $q_{\max}$, $\dot{Q}_{\max}$ and $Q_{\max}$ are the tolerable maximal values correspondingly, r_n is the vehicle's nose radius, c_q is the heat flux-related coefficient, and t_f denotes the final time.

According to the typical safety requirements for parachute deployment, the terminal constraints of entry trajectory can be defined as

$$r(t_f) - R_M \ge h_{f\min} \tag{10.10}$$

$$v(t_f) \le v_{f\max} \tag{10.11}$$

where $h_{f,\min}$ is the tolerable minimal final altitude, and $v_{f,\max}$ is the tolerable maximal final velocity. Besides, the initial state constraints are given as the form of Eq. (10.4).

In order to satisfy the requirements of control system, the bank angle u and its maneuver rate u are constrained with prescribed bounds as follows.

$$|u| \le u_{\max}, |\dot{u}| \le \dot{u}_{\max}, \tag{10.12}$$

where $u_{\max}$ and $\dot{u}_{\max}$ are the tolerable maximum magnitude correspondingly.

10.1.3 Objective Functions

In this study, two typical cases are considered. The first case is to maximize the final altitude such that definitely sufficient margin can be reserved for parachute deceleration and powered descent to guarantee safe landing. The second case is to minimize the final horizontal location error such that enough accurate initial state can be provided to subsequent descent phase for achieving pin-point landing under the premise of security. Therefore, the objective functions for case 1 and 2 are respectively defined as

$$J_1 = -\left[r(t_f) - R_M\right] \tag{10.13}$$

$$J_2 = \left[\theta(t_f) - \theta_f\right]^2 + \left[\lambda(t_f) - \lambda_f\right]^2 \tag{10.14}$$

where θ_f and λ_f are the longitude and latitude of the target position of parachute deployment respectively. The predefined target position is in the reachable domain of the entry vehicle.

10.1.4 Optimization Problem Formulations

For the convenience of subsequent comparison simulation studies, we give out the descriptions of DO, RBO, and RO for the two objectives [112, 186, 187].

(1) Deterministic Optimization Problem

$$\begin{cases} \text{find } u(t) \\ \min \ J_1 \text{ or } J_2 \\ \text{s.t. } \ \dot{\boldsymbol{x}} = \boldsymbol{f}(\boldsymbol{x}, u, t) \\ \qquad \boldsymbol{g}(\boldsymbol{x}, u, t) \le 0 \\ \qquad \boldsymbol{x}(t_0) = \overline{\boldsymbol{x}}_0 \end{cases} \tag{10.15}$$

where $\boldsymbol{f}(\boldsymbol{x}, u, t)$ denotes the dynamics (Eq. 10.1), $\boldsymbol{g}(\boldsymbol{x}, u, t)$ represents the inequality constraints (Eqs. 10.6~10.12), and $\overline{\boldsymbol{x}}_0$ is the nominal initial state.

(2) Reliability-Based Optimization Problem

$$\begin{cases} \text{find } u(t) \\ \min \ \mu(J_1) \text{ or } \mu(J_2) \\ \text{s.t. } \ \dot{\boldsymbol{x}} = \boldsymbol{f}(\boldsymbol{x}, \boldsymbol{C}, u, t) \\ \qquad \Pr\{\boldsymbol{g}(\boldsymbol{x}, u, t) \le 0\} \ge \mathbf{P} \\ \qquad \boldsymbol{x}(t_0) = \boldsymbol{x}_0 \end{cases} \tag{10.16}$$

where both $\boldsymbol{C}$ and $\boldsymbol{x}_0$ contain uncertainty (see Eqs. 10.4 and 10.5); $\mu(\cdot)$ denotes mean value, $\Pr(\cdot)$ represents the probability of the statement within the braces to be true; $\mathbf{P}$ is the reliability vector which is consist of the requested probabilities for satisfaction of constraints accordingly. Note that $\mu(\cdot)$ and $\Pr(\cdot)$ can be captured generally using uncertainty quantification techniques. Due to the influence of uncertainties, the possibility that the actual Mars entry trajectory completely flies along the nominal design is very small, so there is often **1-P > 0**.

(3) Robust Optimization Problem:

$$\begin{cases} \text{find } u(t) \\ \min \mu(J_1) + \nu \cdot \sigma(J_1) \text{ or } \mu(J_2) + \nu \cdot \sigma(J_2) \\ \text{s.t. } \dot{\boldsymbol{x}} = \boldsymbol{f}(\boldsymbol{x}, \boldsymbol{C}, u, t) \\ \quad\quad \Pr\{\boldsymbol{g}(\boldsymbol{x}, u, t) \leq 0\} \geq \mathbf{P} \\ \quad\quad \boldsymbol{x}(t_0) = \boldsymbol{x}_0 \end{cases} \tag{10.17}$$

where both $\boldsymbol{C}$ and $\boldsymbol{x}_0$ contain uncertainty (see Eqs. 10.4 and 10.5), $\sigma(\cdot)$ denotes standard deviation, robustness level is generally set to be 1, 2, …, 6. In this study, we consider 3-sigma case, so ν the is set to be 3, and the elements of reliability vector **P** are set to be 0.997. By integrating $\sigma(\cdot)$ into the objective function with an appropriate weight, minimization of trajectory sensitivity to uncertainties can be achieved. In this case, to prevent the constraints from violation, reliability constraints should be taken into account. Note that $\mu(\cdot)$ and $\sigma(\cdot)$ can be obtained generally using uncertainty quantification techniques.

Remark 1 In this study, the main reasons behind that we choose RO (instead of RBO) to generate Mars entry trajectory under uncertainty is twofold. (a) The RBO is generally used to obtain the reliable design that can work normally (i.e. no violation of constraints occur within predefined reliability probability requirement) under extreme situations [112]. In other words, the RBO does not care about the performance loss of the optimized result under uncertainty. (b) For the trajectory optimization under uncertainty studied in this paper, its major task is to do a best balance between the optimal objective function value and its loss under reliable satisfaction of all constraints. That is to say, the optimized trajectory has the best possible performance and the lowest possible performance loss, while no violation of constraints occur within predefined reliability probability requirement. Such that the generated optimal trajectory is robust to the uncertainty while tracking control can be achieved readily with little extra correction. Therefore, RO is more beneficial to enhancing the robustness and accuracy of entry guidance under uncertainty, instead of using RBO.

Remark 2 It can be observed from Eqs. (10.15–10.17) that when is set to be zero, robust optimization problem (Eq. 10.17) will degrade to be reliability-based optimization problem (Eq. 10.16). When P is set to be zero vector, and initial state and dynamics parameters are set to be their nominal values, reliability-based optimization problem (Eq. 10.16) will degrade to be deterministic optimization problem (Eq. 10.15).

10.2 Uncertainty Quantification and Propagation

The procedure of PCE for uncertainty propagation in Mars entry dynamics system, considering both initial state uncertainty in $\boldsymbol{x}(t_0)$ and parametric uncertainty in $\boldsymbol{C}$, is designed as follows.

Step 1. State variables $\boldsymbol{x}$ and dynamical parameter $\boldsymbol{C}$ are expressed as P order PCE

$$x_i(t) = \sum_{j=0}^{P} x_{ij}(t)\Phi_j(\boldsymbol{\xi}), \quad i = 1, 2, \cdots, 6 \tag{10.18}$$

$$C_k = \sum_{j=0}^{P} C_{kj}(t)\Phi_j(\boldsymbol{\xi}), \quad k = 1, 2, 3 \tag{10.19}$$

where $\boldsymbol{\xi}$ is a random vector and $\boldsymbol{\xi} = [\xi_1, \cdots, \xi_d]$, the dimension d denotes the number of independent uncertainty sources in the dynamics; $\Phi_j(\boldsymbol{\xi})$ is an element of multi-dimensional orthogonal polynomials obtained by conducting tensor product on univariate orthogonal polynomial basis functions $\Phi_j(\boldsymbol{\xi}_l), l = 1, \cdots, d$ for each dimension; $x_{ij}(t)$ and $x_{kj}(t)$ are the PCE coefficients. The series is truncated after $P + 1$ terms, which is determined by the dimension d of $\boldsymbol{\xi}$ and the order s of the orthogonal polynomials, i.e. satisfying $P + 1 = [(d + s)!]/(d!s!)$ [172, 185]. The orthogonal basis function family $\left\{\Phi_j(\boldsymbol{\xi}_l)\right\}$ corresponding to the probabilistic distribution of random variable is selected according to Table 9.1.

Step 2. Substituting the PCEs (i.e. Eqs. 10.18 and 10.19) into the dynamics $\dot{\boldsymbol{x}} = \boldsymbol{f}(\boldsymbol{x}, u, \boldsymbol{C}, t)$ results in

$$\frac{d\sum_{j=0}^{P} x_{ij}(t)\Phi_j(\xi)}{dt} = f\left(\sum_{j=0}^{P} x_{ij}(t)\Phi_j(\xi), u, \sum_{j=0}^{P} C_{kj}\Phi_j(\xi), t\right) \tag{10.20}$$

Step 3. Conducting the Galerkin projection [188] on each orthogonal polynomial $\Phi_j(\boldsymbol{\xi})$ in Eq. (10.20) results in equivalent deterministic ODEs:

$$\dot{x}_{il}(t) = \frac{\left\langle f\left(\sum_{j=0}^{P} x_{ij}(t)\Phi_j(\xi), u, \sum_{j=0}^{P} C_{kj}\Phi_j(\xi), t\right), \Phi_1(\xi)\right\rangle}{\langle \Phi_l(\xi)\rangle}, \quad l = 0, 1, \cdots, P \tag{10.21}$$

where $\langle\cdot\rangle$ denotes the inner product, and $f(\boldsymbol{\xi})$ is the probability distribution of $\boldsymbol{\xi}$

$$\langle\Phi_i(\xi)\Phi_j(\xi)\rangle = \int f(\xi)\Phi_i(\xi)\Phi_j(\xi)d\xi \tag{10.22}$$

Step 4. Clearly, Eq. (10.21) can be solved by Runge–Kutta algorithm to obtain $x_{il}(t)$, and the integral in Eq. (10.22) is conducted based on Gauss quadrature rules. The initial values are the PCE coefficients of the random variables in the original ODEs at the initial time. The PCE coefficients are calculated based on the given uncertainties. Thus, uncertainties in initial states and dynamical parameters are propagated numerically.

Step 5. The uncertainty size existed at the state $x_i(t)$ is characterized analytically by use of mean $\mu[x_i(t)]$ and variance $\sigma^2[x_i(t)]$ as follows.

$$\mu[x_i(t)] = E\left[\sum_{j=0}^{P} x_{ij}(t)\Phi_j(\xi)\right] = \sum_{j=0}^{P} x_{ij}(t)E\big[\Phi_j(\xi)\big] \tag{10.23}$$

$$\begin{aligned}\sigma^2[x_i(t)] &= E[x_i^2(t)] - E[x_i(t)]^2 \\ &= \sum_{j=0}^{P} x_{ij}^2(t)E[\Phi_j(\xi)\Phi_j(\xi)] - E[x_i(t)]^2\end{aligned} \tag{10.24}$$

where $E[\cdot]$ is mean value operator. The stochastic objective and constraints also can be quantified in this manner.

10.3 Robust Optimization Procedure

In this work, the robust trajectory optimization problem is solved using PCE and the hp-adaptive pseudospectral method [156] in conjunction with the robust design theory [189]. The main steps for solving the robust trajectory optimization are provided in Table 10.1. It is a sequential procedure, so no iteration is required for the outer loop, but polynomial chaos and hp-adaptive pseudospectral steps contain iterations.

Table 10.1 Main steps of robust optimization process

Step	Procedure
1	Formulate robust optimization problem using PCE
2	Transcribe the original stochastic ODEs into high-dimension deterministic ODEs
3	Quantify the objective function and constraints under uncertainties using non-intrusive PCE
4	Solve the transcribed equivalent deterministic optimization problem via hp-adaptive pseudospectral method

10.3.1 Reformulation of Robust Trajectory Optimization Problem

Once the original stochastic dynamics is transcribed into a deterministic one, the objective and constraints are re-written based on the robust design optimization theory [189]. Thus, the optimization problem (i.e., Eq. 10.17) considering initial state uncertainty and dynamical parameters uncertainty can be re-formulated as

$$\begin{cases} \text{find } u(t) \\ \min \mu(J_1) + \nu \cdot \sigma(J_1) \text{ or } \mu(J_2) + \nu \cdot \sigma(J_2) \\ \text{s.t. } \dot{x}_{il}(t) = \frac{\left\langle f\left(\sum_{j=0}^{P} x_{ij}(t)\Phi_j(\xi), u, \sum_{j=0}^{P} C_{kj}\Phi_j(\xi), t\right), \Phi_1(\xi)\right\rangle}{\langle \Phi_l(\xi)\rangle} \\ \quad \dot{\mu}\left[\boldsymbol{g}(\boldsymbol{x}, u, u, \boldsymbol{C}, t)\right] + v \cdot \sigma\left[\boldsymbol{g}(\boldsymbol{x}, u, \boldsymbol{C}, t)\right] \leq 0 \\ \quad x_{i0}(t_0) = x_{i00}, x_{i1}(t_0) = x_{i10}, \cdots, x_{iP}(t_0) = x_{iP0} \end{cases} \tag{10.25}$$

where $x_{i0}(t_0) = x_{i00}, x_{i1}(t_0) = x_{i10}, \cdots, x_{iP}(t_0) = x_{iP0}$ are the PCE coefficients for the stochastic initial state $x_i(t_0)$. Clearly, the uncertainty in the initial state is represented as PCE coefficients, which will be propagated numerically.

Remark 3 Due to the fact that the dynamical parameters uncertainty is an inherent property of the Mars entry dynamics, the dynamical parameters uncertainty is considered to be mutually independent and unchanged during the propagation. Thus, uncertainties from the initial state $\boldsymbol{x}(t_0)$ and system parameters $\boldsymbol{C}$ (see Eqs. 10.4 and 10.5) are propagated in terms of their PCE coefficients (see Eqs. 10.18 and 10.19) in each iteration of the trajectory optimization procedure.

10.4 Quantification of Objective Function and Constraints Under Uncertainties

10.4.1 Quantification of Objective Function with Uncertainties

The quantification of objective function and constraints is conducted using PCE in each iteration. For objective function J_1, we use the general form given below to show how to obtain the mean and standard deviation. Let $J(\boldsymbol{x}_0, u, \boldsymbol{C}, \boldsymbol{\xi})$ be the value of objective function, which has a finite variance. Then, $J(\boldsymbol{x}_0, u, \boldsymbol{C}, \boldsymbol{\xi})$ can be expressed by PCE as

$$J(\boldsymbol{x}_0, u, \boldsymbol{C}, \xi) = \sum_{j=0} \widetilde{J}_j(\boldsymbol{x}_0, u, \boldsymbol{C}, \xi)\Phi_j(\xi) \tag{10.26}$$

where $\widetilde{J}_j(\boldsymbol{x}_0, u, \boldsymbol{C}, \xi)$ is the PCE coefficients of $J_j(\boldsymbol{x}_0, u, \boldsymbol{C}, \xi)$. To obtain the coefficients efficiently, non-intrusive PCE method is adopted here, which treats the stochastic ODEs as a black-box and computes the coefficients by use of appropriate samples [190]. By projecting $J_j(\boldsymbol{x}_0, u, \boldsymbol{C}, \xi)$ on orthogonal polynomial $\Phi_j(\xi)$, according to the rule as Eq. (10.22), the PCE coefficients $\widetilde{J}_j(\boldsymbol{x}_0, u, \boldsymbol{C}, \xi)$ can be computed by

$$\widetilde{J}_i(\boldsymbol{x}_0, u, \boldsymbol{C}, \xi) = \frac{\int J\Phi_j(\boldsymbol{\xi}) f(\boldsymbol{\xi}) d\xi}{\left\langle \Phi_j(\boldsymbol{\xi}) \Phi_j(\boldsymbol{\xi}) \right\rangle} \tag{10.27}$$

Under the non-intrusive PCE framework, the discrete projection of Eq. (10.27) can be expressed as

$$\widetilde{J}_i(\boldsymbol{x}_0, u, \boldsymbol{C}, \xi) \approx \frac{\sum_{l=1}^{N} J\Phi_j(\boldsymbol{\xi}) w_l}{\left\langle \Phi_j(\boldsymbol{\xi}) \Phi_j(\boldsymbol{\xi}) \right\rangle} \tag{10.28}$$

where N represents the total number of samples, and the total amount of grid nodes is also N; w_l denotes the weight coefficient of the integral node ξ_l, and w_l is obtained by the Smolyak sparse grid technique [30].

According to the orthogonality of polynomial basis functions, the mean of $J_j(\boldsymbol{x}_0, u, \boldsymbol{C}, \xi)$ is calculated approximately by

$$\mu(J) \approx \iint \sum_{j=0}^{P} \widetilde{J}_j \Phi_{j1}(\xi_1) f(\xi_1) d\xi_1 \cdots \phi_{jd}(\xi_d) f(\xi_d) d\xi_d \tag{10.29}$$

Since the random variables are independent mutually, according to the rule as Eq. (10.24), the standard deviation of $J_j(\boldsymbol{x}_0, u, \boldsymbol{C}, \xi)$ is calculated approximately by

$$\sigma(J) = \mu[(J - \mu(J))^2] \approx \sum_{j=0}^{P} \widetilde{J}_j^2 \prod_{l=1}^{d} \xi_i \tag{10.30}$$

10.4.2 Quantification of Constraints with Uncertainties

For the constraints $\boldsymbol{g}(\boldsymbol{x}_0, u, \boldsymbol{C}, t)$, the quantification of $\mu[\boldsymbol{g}(\boldsymbol{x}_0, u, \boldsymbol{C}, t)]$ and $\sigma[\boldsymbol{g}(\boldsymbol{x}_0, u, \boldsymbol{C}, t)]$ can be done according to the form of Eqs. (10.23) and (10.24) using PCE. The PCE for the constraints $\boldsymbol{g}(\boldsymbol{x}_0, u, \boldsymbol{C}, t)$ is constructed as

$$\boldsymbol{g}(\boldsymbol{x}, u, \boldsymbol{C}, t) = \sum_{j=0}^{P} \boldsymbol{g}_j(\mathbf{x}, u, \mathbf{C}, t) \Phi_j(\xi) \tag{10.31}$$

where $\boldsymbol{g}_j(\boldsymbol{x}, u, \boldsymbol{C}, t)$ denotes the j-th PCE coefficients of $\boldsymbol{g}(\boldsymbol{x}, u, \boldsymbol{C}, t)$. Substituting Eqs. (10.18) and (10.19) into Eq. (10.31) results in

$$\boldsymbol{g}\left[\sum_{j=0}^{P} x_{ij}(t)\Phi_j(\xi), u, \sum_{j=0}^{P} C_{kj}(t)\Phi_j(\xi), t\right] = \sum_{j=0}^{P} \boldsymbol{g}_j(t)\Phi_j(\xi),$$
$$i = 1, 2, \cdots, 6 \tag{10.32}$$

Then, the PCE coefficients of $\boldsymbol{g}(\boldsymbol{x}, u, \boldsymbol{C}, t)$ can be obtained by conducting Galerkin projection [26] on each side of Eq. (10.32).

$$\boldsymbol{g}_l(t) = \frac{\left\langle \boldsymbol{g}\left[\sum_{j=0}^{P} x_{ij}(t)\Phi_j(\xi), u, \sum_{j=0}^{P} C_{kj}(t)\Phi_j(\xi), t\right], \Phi_l(\xi)\right\rangle}{\langle \Phi_l^2 \rangle},$$
$$l = 0, 1, \ldots, P \tag{10.33}$$

Finally, substituting $\boldsymbol{g}_l(t)$ into Eqs. (10.23) and (10.24), its mean and standard deviation can be respectively calculated by

$$\mu[\boldsymbol{g}_i(t)] = E\left[\sum_{j=0}^{P} \boldsymbol{g}_j(t)\Phi_j(\xi)\right] = \sum_{j=0}^{P} \boldsymbol{g}_j(t)E[\Phi_j(\xi)] \tag{10.34}$$

$$\sigma^2[\boldsymbol{g}_i(t)] = E[\boldsymbol{g}_i^2(t)] - E[\boldsymbol{g}_i(t)]^2 = \sum_{j=0} \boldsymbol{g}_i^2(t)E[\Phi_j(\xi)\Phi_j(\xi)] - E[\boldsymbol{g}_i(t)]^2 \tag{10.35}$$

Remark 4 It is noticed that to address the problems described in Eqs. (10.17) and (10.25), numerous inner products should be calculated. Repeated substitution and approximation are conducted to calculate these inner products by constructing numerous PCEs for intermediate results, which is fairly complicated and the errors would be accumulated due to approximation. Since the system equations and constraints are high-dimension and high-nonlinear, full factorial numerical integration (FFNI) method [29] is employed herein. With FFNI, the inner product is represented as a weighted summation of function values at a couple of nodes by conducting tensor product [191].

10.5 Hp-Adaptive Pseudospectral Method

As a new direct method, hp-adaptive pseudospectral method exhibits high-efficiency for solving continuous-time nonlinear OCP, in which segment width denoted by h and polynomial order denoted by p are updated adaptively [156]. Thus, the accuracy of optimal solution can be improved through increasing adaptively the degree of the approximate polynomial in the mesh interval and/or refining the mesh. It can effectively work out spacecraft trajectory optimization problem with high-nonlinearity and non-convexity [192]. Here, to address the equivalent deterministic OCP with high dimension and high non-linearity, hp-adaptive pseudospectral method is used. Its main steps are reviewed as follows [156, 192]. Moreover, the adaptive criteria of the hp-adaptive pseudospectral method can be found in Sect. 7.3.2.

10.6 Assessment of Reliability and Robustness

In order to assess the performance of the obtained optimal trajectory, the following formulas based on MCS are employed. The reliability is evaluated in the form of probability [193]

$$\Pr\{\boldsymbol{g}(\boldsymbol{\delta}) \leq \boldsymbol{0}\} = \int_{\mathbf{g}(\boldsymbol{\delta})\leq\mathbf{0}} f_X(\boldsymbol{\delta})\mathrm{d}\boldsymbol{\delta} \tag{10.36}$$

where $\boldsymbol{\delta}$ denotes a vector of independent standard normal variables which is given in Eqs. (10.4) and (10.5), and $f_X(\boldsymbol{\delta})$ is the probability density function of $\boldsymbol{\delta}$.

The robustness is evaluated quantitatively by

$$\sigma(J) = \sqrt{\frac{\sum_{i=1}^{N_{MC}} (J_i - \mu(J))^2}{N_{MC}}} \tag{10.37}$$

where $\sigma(J)$ is standard deviation, $\mu(J)$ is mean of objective function, J_i is the value of objective function in i-th MCS iteration, and NMC denotes total number of MCS iterations.

Remark 5 It should be pointed out that the MCS-based robustness assessment is used for verifying the performance of the generated optimal trajectories, instead of embedding into the optimization iterations. In this sense, the accuracy of the assessment is the focus instead of computational overhead. In fact, the frequently-used methodologies for evaluating robustness and reliability include nonintrusive polynomial chaos [172, 177], most probable point (MPP) [193], and MCS [194], etc. The accuracy of non-intrusive PCE heavily depends on the amount of samples, since the non-intrusive PCE is an approximate means [172, 177]. To achieve more accurate assessment, larger amount of samples need to be considered into the nonintrusive

PCE. The first-order reliability method (FORM) and the second-order reliability method (SORM) are based on expanding the limit-state surface (first and second order) at the MPP [194]. If the limitstate surface is highly nonlinear or it has multiple MPPs, instead of linear or quadric limit-state surface, the FORM and SORM usually lead a large error [193]. In this study, MCS is preferred.

10.7 Numerical Simulations

The performance of the proposed robust trajectory optimization method for Mars entry under uncertainty is verified in this section. Two cases with different objective functions are simulated respectively. To better illustrate the effectiveness and superior of the proposed technique, the traditional DO and RBO are also implemented simultaneously for comparison.

10.7.1 Simulation Setup

The configuration and physical parameters of MSL-type entry vehicle are utilized as an example to perform simulation and analysis in this paper. According to Refs. [196][197][198], the detailed parameters of the entry vehicle and Mars environment are set as Table 10.2, and the parameters of constraints are set as presented in Table 10.3. The nominal initial state and the desired final state are listed in Table 10.4. Besides, Table 10.5 provides the setup of initial state uncertainty and dynamics parametric uncertainty.

Table 10.2 Parameters of entry vehicle and Mars environment

Parameters	Values	Units
m	2920	kg
S_{ref}	15.9	m^2
C_D	1.45	–
C_L	0.36	–
r_n	0.66	m
c_q	1.9027×10^{-4}	$\sqrt{kg}/m$
G_M	4.282829×10^{13}	m^3/s^2
R_M	3397	km
ω	7.0882×10^{-5}	rad/s
ρ_0	0.0158	kg/m^3
h_s	9354.5	m
g_M	3.7132	m/s^2

Table 10.3 Parameters of constraints

Parameters	Values	Units
$n_{\max}$	18	g_M
$q_{\max}$	16	kPa
$\dot{Q}_{\max}$	70	W/cm^2
$Q_{\max}$	3.4	kJ/cm^2
$h_{f\,\min}$	6	km
$v_{f\,\max}$	490	m/s
$u_{\max}$	90	°
$\dot{u}_{\max}$	20	°/s

Table 10.4 Nominal initial state and desired final state

Parameters	Values	Units
h_0	125	km
θ_0	−65	°
λ_0	−50	°
v_0	5.5	km/s
γ_0	−15.2	°
ψ_0	15.5	°
θ_f	−61.05	°
λ_f	−39.21	°

Table 10.5 Uncertainties in initial states and dynamical parameters

Parameters	Distributions	Intervals or variances	Units
δ_{r0}	Uniform	[−3, 3]	km
$\delta_{\theta 0}$	Uniform	[−0.1, 0.1]	°
$\delta_{\lambda 0}$	Uniform	[−0.1, 0.1]	°
δ_{v0}	Uniform	[−0.1, 0.1]	km/s
$\delta_{\gamma 0}$	Uniform	[−0.1, 0.1]	°
$\delta_{\psi 0}$	Uniform	[−0.1, 0.1]	°
δ_{ρ}	Gaussian	15%	kg/m^3
δ_{Bc}	Gaussian	10%	kg/m^2
δ_{LD}	Gaussian	5%	–

All simulation computations are performed on a mobile workstation with Intel Core® i7-7820HQ CPU at 2.9 GHz and 32 GB of memory and the MATLAB® 2016b environment. All equivalent deterministic optimization problems and traditional DO are solved by a modified program based on the open-source MATLAB software GPOPS (4.x version) with the adaptive pseudospectral method and the NLP solver SNOPT [119]. The number of discrete points for state and control variables is 20. The open-source MATLAB software OpenCossan is adopted for uncertainty propagation with the polynomial chaos steps [199]. The default configuration of the OpenCossan is used, and MATLAB statistics toolbox is called. To assess the performance of the generated optimal trajectories quantitatively, 1000-run MCS results are utilized to obtain the approximate probability distributions by MATLAB statistics toolbox. Note that less MCS is not enough to obtain the approximate probability distributions and statistics data that are able to support the subsequent analysis and conclusion, while more MCS makes the values of the approximate probability distributions and statistics data (shown in Figs. 10.3, 10.6, Table 10.6, Table 10.7, and Table 10.8) more accurate but does not hinder to pick up the same conclusion.

Table 10.6 Statistics of constraints reliability in Case I

Items		n_a	Q	$\dot{Q}$	Q
Limit values		18 g_M	16 kPa	70 W/cm^2	3.4 kJ/cm^2
Peak values	DO	17.216 g_M	15.904 kPa	69.375 W/cm^2	3.356 kJ/cm^2
	RBO	16.525 g_M	14.958 kPa	65.996 W/cm^2	3.202 kJ/cm^2
	RO	16.819 g_M	15.173 kPa	66.047 W/cm^2	3.281 kJ/cm^2
Reliability $\Pr(\boldsymbol{g} \le \boldsymbol{0})$	DO	0.921	0.905	0.981	0.960
	RBO	0.998	0.999	0.997	0.998
	RO	0.997	0.998	0.998	0.997

Table 10.7 Statistics of constraints reliability in Case II

Items		n_a	q	$\dot{Q}$	Q
Limit values		18 g_M	16 kPa	70 W/cm^2	3.4 kJ/cm^2
Peak values	DO	16.805 g_M	15.104 kPa	67.365 W/cm^2	3.257 kJ/cm^2
	RBO	15.963 g_M	14.657 kPa	64.983 W/cm^2	3.170 kJ/cm^2
	RO	16.021 g_M	14.769 kPa	64.988 W/cm^2	3.179 kJ/cm^2
Reliability $\Pr(\boldsymbol{g} \le \boldsymbol{0})$	DO	0.990	0.985	0.989	0.976
	RBO	0.998	0.999	0.998	0.999
	RO	0.998	0.999	0.997	0.999

Table 10.8 Standard deviations of objective functions

Standard deviations	DO	RBO	RO	Units
$\sigma(J_1)$	3.894	2.675	1.173	km
$\sigma(J_2)$	0.547	0.302	0.196	°

10.7.2 Case I: Final Altitude Maximization

Maximizing final altitude of the entry phase is generally required by Mars soft landing missions [187, 196, 200]. In this section, based on the dynamics with parametric uncertainty, the entry trajectory from the given initial state with stochastic deviation to the parachute deployment state is optimized. And the final longitude and latitude are free. The main tendencies of the results agree well with that reported in the existing studies [187, 196, 200].

Figure 10.1 shows the profiles of six states of the optimal trajectories in nominal scenario generated by DO, RBO, and RO, respectively. Figure 10.2 presents the optimal control profiles (i.e. bank angle) and the path constraints variables' profiles in nominal scenario for the results of DO, RBO, and RO, respectively. Figure 10.3 illustrates the performance assessment in the form of approximate probability distributions for the optimal trajectories generated by DO, RBO, and RO; which includes the robustness of the objective function value to uncertainties, and reliability of satisfying constraints under uncertainties. Table 10.6 summarizes the parametric values with respect to constraints reliability.

From the results shown in Fig. 10.1 through Fig. 10.3 and Table 10.6, some noteworthy observations can be made. Firstly, all the three optimization procedures can produce smooth profiles of control and state variables, and the constraints of RBO and RO are all satisfied with desired reliability margin when considering both initial state uncertainty and dynamics parametric uncertainties (see Table 10.6). However, all the

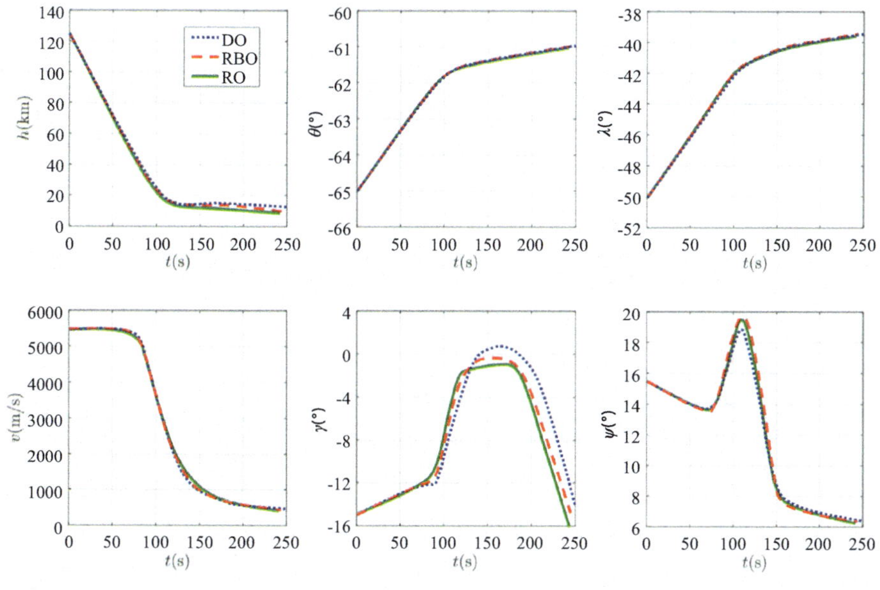

Fig. 10.1 Mars entry trajectories generated by DO, RBO, and RO in Case I

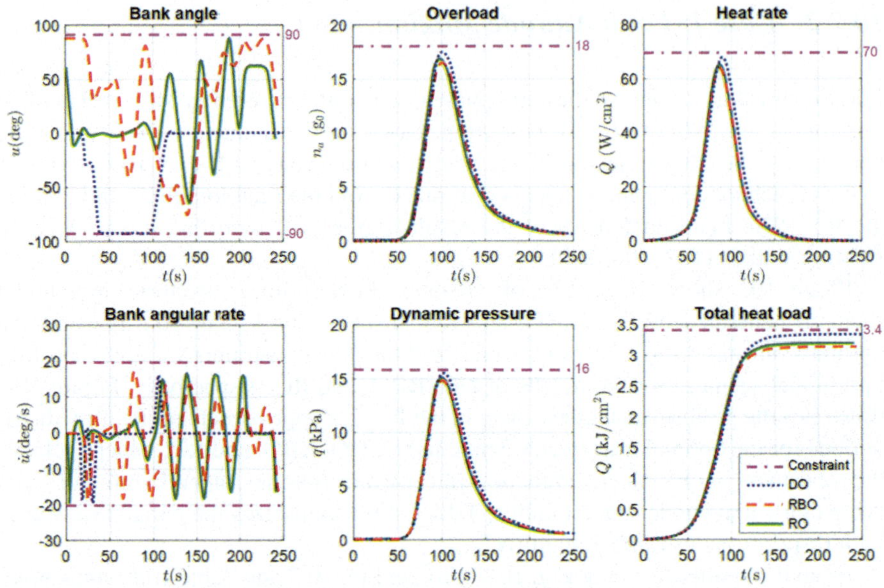

Fig. 10.2 Control and path constraints of the trajectories obtained from DO, RBO, and RO in Case I

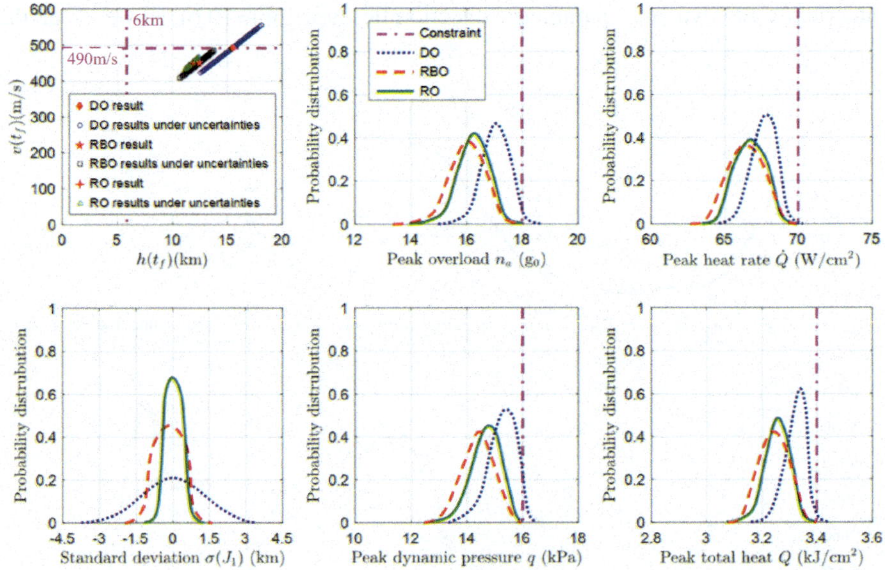

Fig. 10.3 Objectives robustness and constraints reliability of the trajectories from DO, RBO, and RO in Case I

constraints' profiles for DO results are much closer to the corresponding boundaries than that of the RBO and RO results (see Fig. 10.2). In particular, under initial state uncertainty and dynamics parametric uncertainties, almost all the constraints are violated in different extent by the DO results (see Fig. 10.3). For Mars entry, the violation of these constraints would induce failure of the parachute deployment and structural damage to the entry vehicle, which may cause a large risk or even failure to the soft landing mission. Secondly, it is found from Fig. 10.3 that the value of objective function for RO has the smallest variation, i.e. least sensitive to uncertainties, while the value of objective function for DO has the largest variation. Although the DO result reaches the maximal terminal altitude (i.e. the best objective) in nominal scenario than that of RBO and RO, almost half of the DO results under uncertainties exceed the terminal velocity upper bound. The RO result has the minimal value of objective function in nominal scenario, which is close to that of the RBO result. Therefore, the RBO results are the actual optimal trajectories for Mars entry under uncertainties instead of the DO results; while the RO results are close to the RBO results but it's more robust than that of the RBO results. But some values of objective function of the RO results under uncertainties are still larger than that of RBO results. Thirdly, the absolute value of amplitude of the control variable for RO and RBO is smaller than that of DO, but the adjustment times for RO and RBO are more than that of DO. It is reasonable that the robustness and reliability of the trajectory is achieved at the cost of scheduling more control operations in nominal profile. The approximate probability distributions of final states are not standard normal or standard uniform distributions but some types of single peak or single gathering zone distributions. This kind of phenomenon matches well with the findings of the research on uncertainty quantification for Mars entry [71, 185].

10.7.3 Case II: Final Horizontal Position Deviation Minimization

Minimizing final latitude and longitude deviations of the entry phase is generally required by Mars pin-point landing missions [197, 201, 202]. In this section, based on the dynamics with parametric uncertainty, the entry trajectory from the given initial state with stochastic deviation to a predefined site of parachute deployment is optimized. The main tendencies of the results agree well with that reported in the existing studies [185, 201, 202].

Figure 10.4 shows the profiles of six states of the optimal trajectories in nominal scenario generated by DO, RBO, and RO, respectively. Figure 10.5 presents the optimal control profiles (i.e. bank angle) and the path constraints variables' profiles in nominal scenario for the results of DO, RBO, and RO, respectively. Figure 10.6 illustrates the performance assessment in the form of approximate probability distributions for the optimal trajectories generated by DO, RBO, and RO; which includes

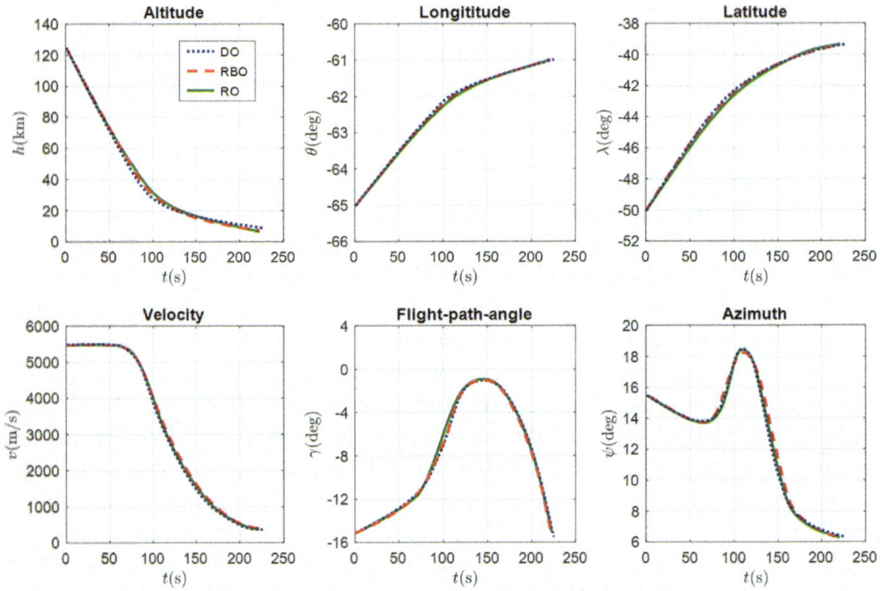

Fig. 10.4 Mars entry trajectories generated by DO, RBO, and RO in Case II

the robustness of the objective function value to uncertainties, and reliability of satisfying constraints under uncertainties. Table 10.7 summarizes the parametric values with respect to constraints reliability.

From the results shown in Fig. 10.4 through Fig. 10.6 and Table 10.7, some significant findings can be made. Firstly, all the three optimization procedures can obtain corresponding optimal trajectories with satisfying all predefined constraints in nominal scenario (see Fig. 10.5). However, once initial state uncertainty and dynamics parametric uncertainties are taken into account, some constraints are violated in different extent by the DO results, while the constraints of RBO and RO are all satisfied with desired reliability margin (see Fig. 10.6 and Table 10.7). Hence, the RBO and RO results are reliable. For the DO results, the violation of these constraints would increase the risk to the safe entry and parachute deployment. Secondly, it can be seen from Fig. 10.6 that the value of objective function for RO has the smallest variation, i.e. most robust to uncertainties, while the value of objective function for DO has the largest variation. Therefore, the RO may cause minimal parachute deployment dispersion than the RBO and DO. Thirdly, the absolute values of amplitude of the control variables for RO and RBO are larger than that of DO. It is reasonable that the robustness and reliability of the trajectory is achieved at the expenses of more nominal control margin at appropriate time. The approximate probability distributions of final states are not standard normal or standard uniform distributions but some types of single peak or single gathering zone distributions, which agree well with the findings of the research on uncertainty quantification for Mars entry [185].

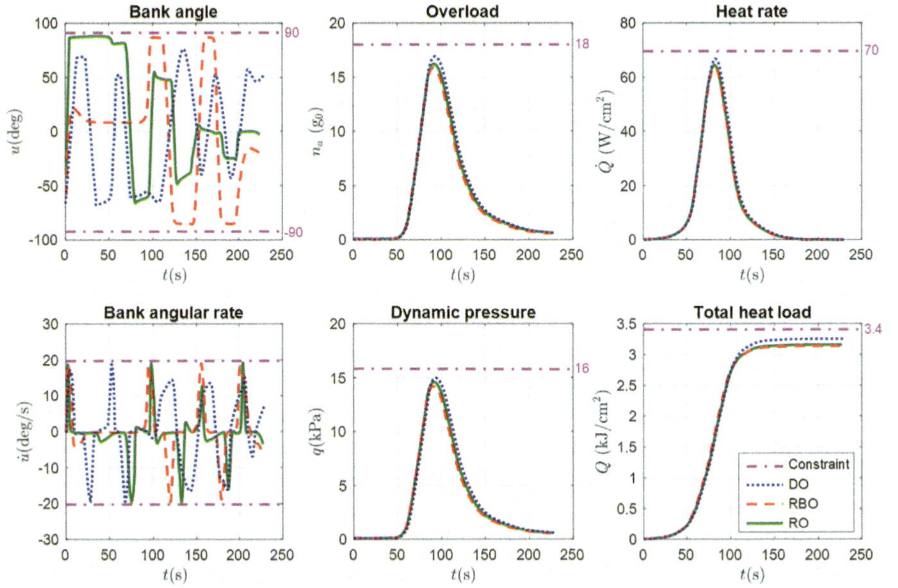

Fig. 10.5 Control and path constraints of the trajectories obtained from DO, RBO, and RO in Case II

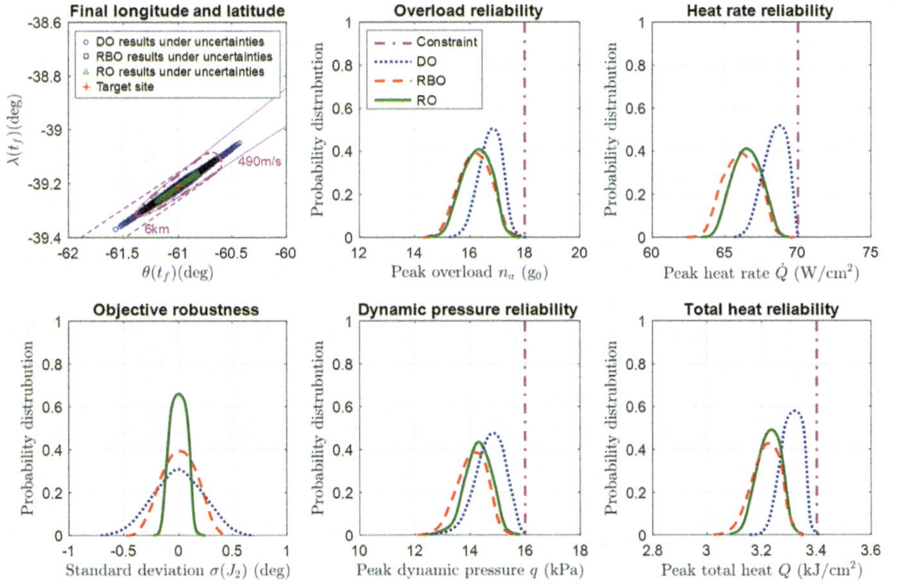

Fig. 10.6 Objectives robustness and constraints reliability of the trajectories from DO, RBO, and RO in Case II

10.7.4 Analysis and Discussion

The essential reasons for the good performance of the proposed RO method are discussed in this section. And also, the main findings are summarized according to the simulation results illustrated above.

From those results of the two examples, the proposed RO method can effectively solve the trajectory optimization problem for Mars entry considering both initial state uncertainty and dynamics parametric uncertainty. It is concluded from Tables 10.6, 10.7, 10.8 that the optimal trajectory generated by the proposed RO method is more reliable and less sensitive to uncertainties compared to the DO and RBO. It can be found that larger nominal control margin is required to increase the robustness of Mars entry trajectory. Therefore, a trade-off should be made between trajectory robustness and practical affordable operations. In addition, equivalent high-dimensional ODEs are produced by the polynomial chaos steps. As a common issue of the direct methodology in solving OCP, it is noticed that if large number of state variables are involved in the equivalent high-dimensional ODEs, the transcribed NLP will be high-dimensional, which would definitely increase the computational cost, see Table 10.9. Fortunately, the increase of the computational time can be effectively suppressed by use of parallel computing technique.

According to the classic optimal control theory, the roots causing the performances of those mentioned optimization procedures are analyzed and summarized as follows. (1) In fact, the reliability constraints reduced the feasible domain of the original OCP, while the objective robustness item reduced the variance or standard deviation of the value of objective function. That is the root behind the phenomenon that the results obtained by the DO, RBO, and RO are always different. (2) Based on the nominal dynamics and nominal constraints, DO is conducted in the original feasible domain. Considering the reliability constraints, RBO and RO are implemented in the reduced feasible domain. That's the reason behind the fact that the results from RBO and RO are reliable than DO results when considering uncertainties. (3) In the same reduced feasible domain, the objective functions for RO and RBO are different. The objective function of RO covers both the mean value and the variance of the original objective function, while the objective function of RBO is just the mean value of the original objective function. For RO, the sensitivity of performance index to uncertainties decreases remarkably whereas the constraints reliability of the optimization problem deteriorates little compared to RBO. That's why the RO results are less sensitive to uncertainties than the RBO results and more robust than the DO results.

Table 10.9 Computational efforts

CPU time (s)	Case 1			Case 2		
	DO	RBO	RO	DO	RBO	RO
Polynomial chaos	–	179.69 s	185.02 s	–	180.01 s	187.11 s
hp-adaptive pseudospectral	56.13 s	324.27 s	399.45 s	61.34 s	357.63 s	412.05 s
Total	56.13 s	503.96 s	584.47 s	61.34 s	537.64 s	599.16 s

10.8 Conclusions

The contribution of this chapter is an innovative robust optimization approach for Mars entry trajectory generation under uncertainties. The proposed technique is verified by designing the terminal altitude maximization trajectory and terminal site deviation minimization trajectory for Mars entry mission. Unlike the existing reliability-based optimization works that only focus on a part of the uncertainties, the proposed robust optimization procedure considers both initial state uncertainty and dynamics parameter uncertainty. The impact of the uncertainties on the trajectory is formulated by polynomial chaos expansion quantitatively. The robust trajectory optimization problem is transcribed into an equivalently optimal control problem in an expanded state space based on the polynomial chaos and robust design theory, which is then solved by the hp-adaptive pseudospectral method. The existing deterministic optimization and reliability-based optimization are employed for comparison studies. The simulation results demonstrate that the proposed technique is capable of planning an optimal trajectory with more robustness to initial state uncertainty and parametric uncertainty than traditional deterministic optimization and reliability-based optimization. The feasible domain for robust optimization is the same as that of reliability-based optimization. The more robust of the designed trajectory, the larger nominal control margin is required. For robust optimization, the sensitivity of performance index to uncertainties decreases remarkably whereas the constraints reliability deteriorates little compared to reliability-based optimization. The computational time for robust optimization has little increase compared to that of the reliability-based optimization. In the future work, the proposed method can be used to deal with other trajectory optimization problems with uncertainty.

Part IV
Robust Optimal Guidance Method

Chapter 11
Direct Model Reference Adaptive Tracking Guidance for Mars Entry

11.1 Mars Atmospheric Entry Dynamics

The Mars entry vehicle is modeled as an unpowered point mass flying in a stationary atmosphere of a nonrotating planet. The dynamic equations of entry vehicle defined with respect to the Mars fixed coordinate system are given by:

$$
\begin{aligned}
\frac{dr}{dt} &= v \sin\gamma \\
\frac{dv}{dt} &= -(D + g_M \sin\gamma) \\
\frac{d\gamma}{dt} &= \left(\frac{v}{r} - \frac{g_M}{v}\right)\cos\gamma + \frac{1}{v} L \cos\sigma \\
\frac{d\theta}{dt} &= \frac{v \cos\gamma \sin\psi}{r \cos\lambda} \\
\frac{d\lambda}{dt} &= \frac{v}{r}\cos\gamma \cos\psi \\
\frac{d\psi}{dt} &= \frac{v}{r}\sin\psi \cos\gamma \tan\lambda + \frac{L \sin\sigma}{v \cos\gamma}
\end{aligned}
\tag{11.1}
$$

where r is the distance from the center of the Mars to the vehicle center of mass, density, v is the velocity, θ is the longitude and λ is the latitude, γ is the flight path angle. The heading (azimuth) angle is ψ, defined as a clockwise rotation angle starting at due north, and σ is the bank angle defined as the angle about the velocity vector from the local vertical plane to the lift vector. An inverse square gravitational acceleration $g_M = \mu/r^2$, where $\mu = GM_{Mars}$ is assumed. L and D are the aerodynamic lift and drag accelerations, defined by

S. Li et al., *Trajectory Optimization and Guidance Methods for Mars Entry*,
https://doi.org/10.1007/978-981-99-6282-2_11

$$L = \frac{1}{2}\rho v^2 \frac{C_L S}{m}, \quad D = \frac{1}{2}\rho v^2 \frac{C_D S}{m} \tag{11.2}$$

where q is the Mars atmospheric density, C_L and C_D are the aerodynamic lift and drag coefficients, respectively, S represents the vehicle reference surface area, and m is the mass of the entry vehicle. $\rho v^2/2$ is defined as dynamic pressure and $m/(C_D S)$ is called ballistic coefficient. An exponential Mars atmospheric density model is assumed as follows:

$$\rho = \rho_0 \exp\left(-\frac{r - r_M}{h_s}\right) \tag{11.3}$$

where ρ_0 is the density on the surface of the Mars, r_M is the radius of the Mars and h_s is the constant scale height.

$$\frac{ds}{dt} = v \cos\gamma \tag{11.4}$$

Thus, longitude and latitude integrations are not required in this reduced dynamics formulation since they are replaced by the downrange. As the follow-up optimal performance index is a function of height, the state variable r is replaced with the height h in the dynamics equations. The longitudinal model is mainly used for dynamics analysis as follows

$$\begin{aligned}
\frac{dh}{dt} &= v \sin\gamma \\
\frac{dv}{dt} &= -(D + g_M \sin\gamma) \\
\frac{d\gamma}{dt} &= \left(\frac{v}{r_M + h} - \frac{g_M}{v}\right)\cos\gamma + \frac{L}{v}u \\
\frac{ds}{dt} &= v \cos\gamma
\end{aligned} \tag{11.5}$$

where $u = \cos\sigma$ is the system control input and defined as the following form for simplification.

In addition, the lateral dynamics are defined through the azimuth equation in Eq. (11.1). The lateral dynamics and guidance law design are not the focus of this paper, which are also included in Sect. 11.6 for integrity of the guidance law design and simulations.

11.2 Reference Drag Acceleration Profile

Hypersonic entry guidance laws are typically formulated in terms of atmospheric drag rather than altitude, because the drag at which the vehicle should be flying does not have to change when the Martian atmosphere is more and less dense than anticipated, whereas the altitude would have to be changed. A detailed discussion of reference drag acceleration planning can be found in [105, 110], a brief overview is given here for completeness.

11.2.1 Drag Dynamics with Energy as Variable

Since aerodynamic drag acceleration is required as a state in the following nominal entry trajectory planning, the longitudinal model using a drag-dependent formulation is brought out firstly. The drag dynamics with energy as a variable are expressed in the following form [105]

$$\frac{d^2D}{dE^2} = a + bu \tag{11.6}$$

where a and b are nonlinear functions of the state variables and $u = \cos\sigma$ is redefined as the control variable. The specific terms a and b are as follows:

$$\begin{aligned} a = {} & D\begin{pmatrix} \frac{1}{h_s}(D + g\sin\gamma)\sin\gamma - \frac{2(D+g\sin\gamma)^2}{v^2} \\ +\frac{4g\sin^2\gamma}{r} + \frac{2}{v^2}g\cos^2\gamma\left(g - \frac{v^2}{r}\right) \\ +\cos^2\gamma\frac{1}{h_s}\left(g - \frac{v^2}{r}\right) \end{pmatrix} \\ & - \frac{dD}{dE}\left(v\sin\gamma\frac{1}{h_s} + \frac{2(D + g\sin\gamma)}{v^2} + \frac{2D}{v}\right) \\ = {} & f_1(D)D + f_2(D)\frac{dD}{dE} \end{aligned} \tag{11.7}$$

$$b = -DL\cos\gamma\left(\frac{1}{h_s} + \frac{2g}{v^2}\right) \tag{11.8}$$

and the first derivative of drag with respect to energy is

$$\frac{dD}{dE} = \frac{\sin\gamma}{h_s} + \frac{2}{v^2}(D + g\sin\gamma) \tag{11.9}$$

11.2.2 Reference Drag Acceleration Profile

Traditionally, the reference trajectory was selected in the drag acceleration–velocity space, since the longitudinal range can be easily expressed as the integral of the velocity over the drag according to Eq. (11.4). Currently, the common approach for reference trajectory determination uses the drag acceleration–energy space, since the constraints can better be represented in this plane and there is no need to assume negligible gravity or small flight path angle [203]. Furthermore, unlike the velocity, the drag deceleration can be directly measured from accelerometers, thus eliminating model approximation errors in determination of the range.

The specific energy can be defined as follows:

$$E = \frac{v^2}{2} + \left(\frac{\mu}{r_0} - \frac{\mu}{r}\right) \tag{11.10}$$

where r_0 is the reference surface radius for Mars and μ is the Mars gravitational constant. Using Eqs. (11.1) and (11.10), it can be shown that

$$\dot{E} = -vD \tag{11.11}$$

Defining downrange as the distance from the initial position of entry vehicle to the prescribed parachute deployment site, and then it is along the flight path in the longitudinal plane. The predicted downrange to go is

$$s = \int_{t_0}^{t_f} v dt = -\int_{E_0}^{E_f} \frac{dE}{D} \tag{11.12}$$

where t_0 is the current time, t_f is the terminal time, E_0 is the current energy, and E_f is the terminal energy. In the drag tracking guidance scheme, the nominal drag profile is adjusted at regular interval using Eq. (11.12), so that the predicted downrange equals the desired range. It is seen that the downrange, as a function of energy, depends only on the drag profile. The drag, in turn, can be controlled through the use of bank angle according to the second order dynamics described in Eq. (11.6). The reference trajectory profile in the drag-energy space is generated inside bounds defined in terms of equilibrium glide and aerodynamic constraints such as maximum dynamic pressure $\overline{q}_{\max}$, maximum overload acceleration $n_{\max}$ and maximum accumulated heat rate $\dot{Q}_{\max}$ [204]. These constraints on the drag acceleration profile are usually defined as

$$D \leq \overline{q}_{\max} \frac{C_D S}{m} \tag{11.13}$$

$$D \leq \frac{n_{\max}}{\sqrt{1 + (L/D)^2}} \tag{11.14}$$

$$D \leq \frac{C_D S v^2}{2m}\left(\frac{\dot{Q}_{\max}}{c v^k}\right)^2 \tag{11.15}$$

where m is the mass of the entry vehicle, c and k are constant parameters. It has been shown that all relevant state constraints for Mars entry can be represented as drag constraints and systematically incorporated into the choice of reference trajectory. When the expected downrange is given in advance, the reference bank angle can be planned with the constraints in Eqs. (11.12)–(11.15), and then the corresponding reference drag acceleration can be obtained by integrating Eq. (11.6).

11.3 CGT-Based Direct Model Reference Adaptive Control

The structure of a command generator tracker (CGT)- based direct adaptive controller is simple and neither an adaptive observer nor full-state feedback is needed. In this section, some background material will be presented on the set-up of a model reference adaptive tracking guidance system, taken from the CGT-based direct adaptive control algorithms [205]. A feasible approach is simple adaptive control based on output feedback. The algorithm aims at matching the response of the system that is to be controlled to that of a reference model which is pre-defined according to various constraints. It is assumed that the system is almost strictly passive (ASP). The controlled system has the representation

$$\begin{aligned}\dot{\boldsymbol{x}}_p(t) &= \boldsymbol{A}_p(\boldsymbol{x}_p)\boldsymbol{x}_p + \boldsymbol{B}_p(\boldsymbol{x}_p)\boldsymbol{u}_p(t)\\ \boldsymbol{y}_p(t) &= \boldsymbol{C}_p(\boldsymbol{x}_p)\boldsymbol{x}_p\end{aligned} \tag{11.16}$$

The output of the system must follow the output of the (possibly) low-order reference model

$$\begin{aligned}\dot{\boldsymbol{x}}_m(t) &= \boldsymbol{A}_m(\boldsymbol{x}_m)\boldsymbol{x}_m(t) + \boldsymbol{B}_m(\boldsymbol{x}_m)\boldsymbol{u}_m(t)\\ \boldsymbol{y}_m(t) &= \boldsymbol{C}_m(\boldsymbol{x}_m)\boldsymbol{x}_m(t)\end{aligned} \tag{11.17}$$

This reference model incorporates the desired input–output behavior of the dynamics system.

Let the output tracking error be defined as

$$\boldsymbol{e}_y = \boldsymbol{y}_m - \boldsymbol{y}_p \tag{11.18}$$

According to the CGT-based direct adaptive control theory, the system input is computed as follows

$$\boldsymbol{u}_p(t) = \boldsymbol{K}_{e_y}(t)\boldsymbol{e}_y(t) + \boldsymbol{K}_{x_m}(t)\boldsymbol{x}_m(t) + \boldsymbol{K}_{u_m}(t)\boldsymbol{u}_m(t) = \boldsymbol{K}(t)\boldsymbol{r}(t) \tag{11.19}$$

where

$$\boldsymbol{K}(t) = \left[\, \boldsymbol{K}_{e_y}(t)\ \boldsymbol{K}_{x_m}(t)\ \boldsymbol{K}_{u_m}(t) \,\right] \tag{11.20}$$

$$\boldsymbol{r}^{\mathrm{T}}(t) = \left[\, \boldsymbol{e}_m^{\mathrm{T}}(t)\ \boldsymbol{x}_m^{\mathrm{T}}(t)\ \boldsymbol{u}_m^{\mathrm{T}}(t) \,\right] \tag{11.21}$$

where $\boldsymbol{K}_{e_y}(t)$, $\boldsymbol{K}_{x_m}(t)$, $\boldsymbol{K}_{u_m}(t)$ are three adaptive gains.

It can be seen the reference model input $\boldsymbol{u}_m$ and state $\boldsymbol{x}_m$ are required to form part of the input signal $\boldsymbol{u}_p$ to the controlled system. Moreover, the output error $\boldsymbol{e}_y$ serves as a feedback quantity to form the first element that composes the system input $\boldsymbol{u}_p$.

To compute the adaptive gains, $\boldsymbol{K}$ is defined to be the sum of an integral and a proportional component as follows:

$$\boldsymbol{K}(t) = \boldsymbol{K}_I(t) + \boldsymbol{K}_P(t) \tag{11.22}$$

where

$$\boldsymbol{K}_P(t) = \boldsymbol{e}_y(t)\boldsymbol{r}^{\mathrm{T}}(t)\boldsymbol{T}_p \tag{11.23}$$

$$\boldsymbol{K}_I(t) = \boldsymbol{K}_{I,0} + \int_0^t \dot{\boldsymbol{K}}_I(t)dt \tag{11.24}$$

$$\dot{\boldsymbol{K}}_I(t) = \boldsymbol{e}_y(t)\boldsymbol{r}^{\mathrm{T}}(t)\boldsymbol{T}_I - \zeta \boldsymbol{K}_I \tag{11.25}$$

where $\boldsymbol{T}_p$ and $\boldsymbol{T}_I$ are the time-invariant positive semi-definite and positive definite weighting matrices respectively, $\boldsymbol{K}_{I,0}$ is the initial integral gain. In order to improve the transient response of the controlled system by integral gain, a constant gain value $\boldsymbol{K}_{I,0}$ was added to $\boldsymbol{K}_I$. At the same time, a forgetting factor ξ is included to adjust the integral gain $\boldsymbol{K}_I$ and then cope with environment disturbance that leads to a persistent non-zero error [205].

11.4 Longitudinal Entry Guidance Using Adaptive Controller

The general layout of CGT-based direct model adaptive tracking guidance is shown in Fig. 11.1. Four distinct elements can be identified: reference model, controlled system model, adaptive gains and feed-forward compensator.

Set $x_1 = D$, $x_2 = \dot{D}$, then the drag dynamics with energy E in Eq. (11.6) as the independent variable can be transformed as follows:

$$\dot{x}_1 = x_2$$

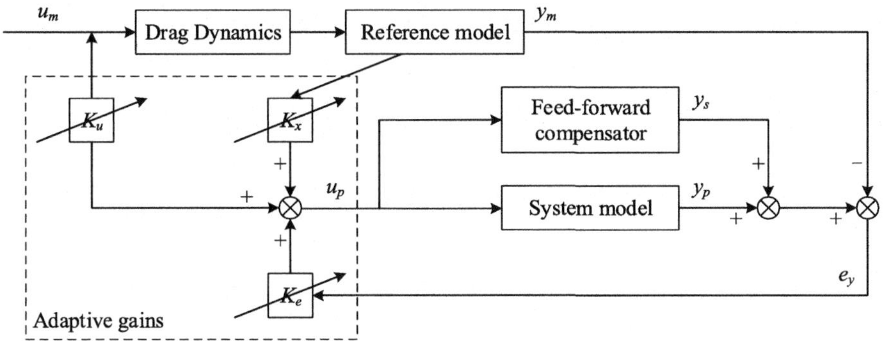

Fig. 11.1 Sketch of CGT-based direct adaptive tracking guidance

$$\begin{aligned} \dot{x}_2 &= f_1(x_1)x_1 + f_2(x_1)x_2 + b(x_1)u \\ y &= x_1 \end{aligned} \tag{11.26}$$

Re-writing Eq. (11.26) in matrix form

$$\begin{aligned} \dot{\boldsymbol{x}}_p &= \boldsymbol{A}_p \boldsymbol{x}_p + \boldsymbol{B}_p \boldsymbol{u}_p \\ \boldsymbol{y}_p &= \boldsymbol{C}_p \boldsymbol{x}_p \end{aligned} \tag{11.27}$$

where $\boldsymbol{x}_p = \begin{bmatrix} x_1 \ x_2 \end{bmatrix}^{\mathrm{T}}$, $\boldsymbol{A}_p = \begin{bmatrix} 0 & 1 \\ f_1(x_1) & f_2(x_1) \end{bmatrix}$, $\boldsymbol{B}_p = \begin{bmatrix} 0 \\ b(x_1) \end{bmatrix}$, $\boldsymbol{C}_p = \begin{bmatrix} 1 \ 0 \end{bmatrix}$.

Therefore, the drag dynamics system (11.6) can be considered as a single-input–single-output (SISO) secondorder nonlinear system with relative order of two. It can be easily concluded that this system is not ASP.

To assure that all states and gains in the adaptive system are bounded and the output error is asymptotically stable, it is necessary that a controlled system is ASP. This means, for practical purposes, that the controlled system can be stabilized by a constant or time-variable high-output gain and the controlled system is strictly minimum-phase. To apply CGT-based direct model reference adaptive control to the non-strictly passive (NSP) Mars entry system described in Eq. (11.6), various modifications have been developed. One feasible approach is to augment the non-strictly passive system with a so-called feed-forward compensator (FFC), such that the modified system becomes ASP and suited for the CGT-based direct model reference adaptive control algorithm mentioned above.

Parallel FFC is related to the controlled system, we assume that the parallel feed-forward compensation has the following state-space form

$$\begin{aligned} \dot{\boldsymbol{x}}_s &= \boldsymbol{A}_s \boldsymbol{x}_s + \boldsymbol{B}_s \boldsymbol{u}_p \\ \boldsymbol{y}_s &= \boldsymbol{C}_s \boldsymbol{x}_s \end{aligned} \tag{11.28}$$

and then the composite controlled system with parallel FFC can be represented as

$$\begin{aligned}\dot{\boldsymbol{x}}_a(t) &= \boldsymbol{A}_a(\boldsymbol{x}_a)\boldsymbol{x}_a + \boldsymbol{B}_a(\boldsymbol{x}_a)\boldsymbol{u}_p(t)\\ \boldsymbol{y}_a(t) &= \boldsymbol{C}_a(\boldsymbol{x}_a)\boldsymbol{x}_a\end{aligned} \tag{11.29}$$

where $\boldsymbol{x}_a = \begin{bmatrix}\boldsymbol{x}_p\\ \boldsymbol{x}_s\end{bmatrix}$, $\boldsymbol{y}_a = \begin{bmatrix}\boldsymbol{y}_p\\ \boldsymbol{y}_s\end{bmatrix}$, $\boldsymbol{A}_a = \begin{bmatrix}\boldsymbol{A}_p & 0\\ 0 & \boldsymbol{A}_s\end{bmatrix}$, $\boldsymbol{B}_a = \begin{bmatrix}\boldsymbol{B}_p\\ \boldsymbol{B}_s\end{bmatrix}$, $\boldsymbol{C}_a = \begin{bmatrix}\boldsymbol{C}_p & \boldsymbol{C}_s\end{bmatrix}$.

Now, it is assumed that the resulting closed-loop system with a constant output gain $\boldsymbol{K}_e$ (Fig. 11.1) is

$$\begin{aligned}\dot{\boldsymbol{x}}_a(t) &= \boldsymbol{A}_{ac}(\boldsymbol{x}_a)\boldsymbol{x}_a + \boldsymbol{B}_{ac}(\boldsymbol{x}_a)\boldsymbol{u}_{ac}(t)\\ \boldsymbol{y}_a(t) &= \boldsymbol{C}_{ac}(\boldsymbol{x}_a)\boldsymbol{x}_a\end{aligned} \tag{11.30}$$

According to the definition of ASP, if there exists a positive definite feedback matrix $\boldsymbol{K}_e$ (unknown and not needed for implementation) which makes the resulting closed-loop system formulated in Eq. (11.30) strictly passive, then the augmented controlled system in Eq. (11.29) is ASP. Since the relative degree of the controlled system is 2, the parallel FFC can be constructed in the frequency domain as follows

$$F(s) = \frac{k_c s}{(s + a_0)(s + a_1)} \tag{11.31}$$

where k_c, a_0 and a_1 are the adjustable parameters. The ASP of the proper composite system can be met by properly adjusting the parameters mentioned above. In general, the model following error will be bounded but not zero after the FFC is included into the controlled system. So parameter adjustment is necessary to keep the model following error as small as possible. Taking into account the fact that the dynamics equation in Eq. (11.6) is a second order nonlinear system, we select a second-order linear system as reference model to accurately track the nominal drag trajectory. Since the reference model is a representation of the desired system dynamics, it needs to have a smaller steady state error and better dynamic performance. It is assumed that the settling time is <0.2 s, the overshoot does not exceed 5% and steady-state error is zero. According to classical control theory, the reference model can be selected according to classic control theory as follows:

$$G(s) = \frac{1000}{s^2 + 45s + 1000} \tag{11.32}$$

Stability of the adaptive controller is a very important performance index and prerequisite for engineering application. Stability analysis of the longitudinal adaptive guidance developed in this paper can be found as follows.

Proof As an adaptive controller is desired to bring the control system to that idealistic situation when perfect output tracking is performed, we consider the ideal bounded trajectories $\boldsymbol{x}_a^*(t)$ and ideal control $\boldsymbol{u}_p^*(t)$ that satisfy

$$\dot{\boldsymbol{x}}_a^*(t) = \boldsymbol{A}_a^*(\boldsymbol{x}_a^*)\boldsymbol{x}_a^* + \boldsymbol{B}_a^*(\boldsymbol{x}_a^*)\boldsymbol{u}_p^*(t)$$

$$\boldsymbol{y}_a^*(t) = \boldsymbol{C}(\boldsymbol{x}_a)\boldsymbol{x}_a^* \tag{11.33}$$

where the matrices $\boldsymbol{A}_a^*(\boldsymbol{x}_a^*)$, $\boldsymbol{B}_a^*(\boldsymbol{x}_a^*)$ and $\boldsymbol{C}(\boldsymbol{x}_a)$ are uniformly bounded.

It is useful to represent $\boldsymbol{x}_a^*(t)$ and $\boldsymbol{u}_p^*(t)$ as linear combinations of $\boldsymbol{x}_m(t)$ and $\boldsymbol{u}_m(t)$

$$\begin{aligned} \boldsymbol{x}_a^*(t) &= \boldsymbol{S}_x\boldsymbol{x}_m(t) + \boldsymbol{S}_u\boldsymbol{u}_m(t) \\ \boldsymbol{u}_p^*(t) &= \boldsymbol{K}_{xm}^*\boldsymbol{x}_m(t) + \boldsymbol{K}_{um}^*\boldsymbol{u}_m(t) \end{aligned} \tag{11.34}$$

When the system states reach the ideal trajectory, it is required that

$$\boldsymbol{y}_a^*(t) = \boldsymbol{C}_a\left(\boldsymbol{x}_a^*\right)\boldsymbol{x}_a^* = \boldsymbol{y}_m(t) \tag{11.35}$$

Since the system states are required to reach the ideal states that allow prefect tracking, define the state error as

$$\boldsymbol{e}_x = \boldsymbol{x}_a^* - \boldsymbol{x}_a \tag{11.36}$$

and then the output error as

$$\begin{aligned} \boldsymbol{e}_y &= \boldsymbol{y}_m - \boldsymbol{y}_a = \boldsymbol{y}_a^* - \boldsymbol{C}_a\left(\boldsymbol{x}_a^*\right)\boldsymbol{x}_a^* + \boldsymbol{C}_a(\boldsymbol{x}_a)\boldsymbol{x}_a^* - \boldsymbol{y}_a \\ &= \boldsymbol{C}_a(\boldsymbol{x}_a)\boldsymbol{e}_x + \left(\boldsymbol{C}_a\left(\boldsymbol{x}_a^*\right) + \boldsymbol{C}_a(\boldsymbol{x}_a)\right)\boldsymbol{x}_a^* = \boldsymbol{C}_a(\boldsymbol{x}_a)\boldsymbol{e}_x + \boldsymbol{d}_1 \end{aligned} \tag{11.37}$$

where $\boldsymbol{d}_1 = \left(\boldsymbol{C}_a\left(\boldsymbol{x}_a^*\right) + \boldsymbol{C}_a(\boldsymbol{x}_a)\right)\boldsymbol{x}_a^*$.

Differentiating $\boldsymbol{e}_x$ in Eq. (11.36) gives

$$\begin{aligned} \dot{\boldsymbol{e}}_x &= \dot{\boldsymbol{x}}_a^* - \dot{\boldsymbol{x}}_a = \dot{\boldsymbol{x}}_a^* - \boldsymbol{A}_a(\boldsymbol{x}_a)\boldsymbol{x}_a^* + \boldsymbol{A}_a(\boldsymbol{x}_a)\boldsymbol{x}_a^* - \dot{\boldsymbol{x}}_a \\ &= \boldsymbol{A}_a^*\left(\boldsymbol{x}_a^*\right)\boldsymbol{x}_a^* + \boldsymbol{B}_a^*\boldsymbol{x}\left(\boldsymbol{x}_a^*\right)\boldsymbol{u}_p^*(t) - \boldsymbol{A}_a(\boldsymbol{x}_a)\boldsymbol{x}_a^* + \boldsymbol{A}_a(\boldsymbol{x}_a)\boldsymbol{x}_a^* - \boldsymbol{A}(\boldsymbol{x}_a)\boldsymbol{x}_a \\ &\quad - \boldsymbol{B}_a(\boldsymbol{x}_a)\boldsymbol{u}_a(t) \\ &= \boldsymbol{A}_a(\boldsymbol{x}_a)\boldsymbol{e}_x - \boldsymbol{B}_a(\boldsymbol{x}_a)\boldsymbol{K}_{ey}^*\boldsymbol{e}_y - \boldsymbol{B}_a(\boldsymbol{x}_a)\left(\boldsymbol{K} - \boldsymbol{K}^*\right)\boldsymbol{r} \\ &\quad + \left(\left(\boldsymbol{A}_a^*\left(\boldsymbol{x}_a^*\right) - \boldsymbol{A}_a(\boldsymbol{x}_a)\right)\boldsymbol{S}_x + \left(\boldsymbol{B}_a^*\left(\boldsymbol{x}_a^*\right) - \boldsymbol{B}_a(\boldsymbol{x}_a)\right)\boldsymbol{K}_{xm}^*\right)\boldsymbol{x}_m \\ &\quad + \left(\left(\boldsymbol{A}_a^*\left(\boldsymbol{x}_a^*\right) - \boldsymbol{A}_a(\boldsymbol{x}_a)\right)\boldsymbol{S}_u + \left(\boldsymbol{B}_a^*\left(\boldsymbol{x}_a^*\right) - \boldsymbol{B}_a(\boldsymbol{x}_a)\right)\boldsymbol{K}_{um}^*\right)\boldsymbol{u}_m \end{aligned} \tag{11.38}$$

Substituting $\boldsymbol{e}_y$ from Eq. (11.37) into Eq. (11.38) gives

$$\begin{aligned} \dot{\boldsymbol{e}}_x &= \boldsymbol{A}_a\,(\boldsymbol{x}_a)\boldsymbol{e}_x - \boldsymbol{B}_a\,(\boldsymbol{x}_a)\boldsymbol{K}_{ey}^*(\boldsymbol{C}_a(\boldsymbol{x}_a)\boldsymbol{e}_x + \boldsymbol{d}_1) - \boldsymbol{B}_a\,(\boldsymbol{x}_a)\left(\boldsymbol{K} - \boldsymbol{K}^*\right)\boldsymbol{r} \\ &\quad + \left(\left(\boldsymbol{A}_a^*\left(\boldsymbol{x}_a^*\right) - \boldsymbol{A}_a\,(\boldsymbol{x}_a)\right)\boldsymbol{S}_x + \left(\boldsymbol{B}_a^*\left(\boldsymbol{x}_a^*\right) - \boldsymbol{B}_a\,(\boldsymbol{x}_a)\right)\boldsymbol{K}_{xm}^*\right)\boldsymbol{x}_m \\ &\quad + \left(\left(\boldsymbol{A}_a^*\left(\boldsymbol{x}_a^*\right) - \boldsymbol{A}_a\,(\boldsymbol{x}_a)\right)\boldsymbol{S}_u + \left(\boldsymbol{B}_a^*\left(\boldsymbol{x}_a^*\right) - \boldsymbol{B}_a\,(\boldsymbol{x}_a)\right)\boldsymbol{K}_{um}^*\right)\boldsymbol{u}_m \\ &= \boldsymbol{A}_{ac}(\boldsymbol{x}_a)\boldsymbol{e}_x - \boldsymbol{B}_{ac}\,(\boldsymbol{x}_a)\left(\boldsymbol{K} - \boldsymbol{K}^*\right)\boldsymbol{r} + \boldsymbol{F} \end{aligned} \tag{11.39}$$

where

$$\begin{aligned} \boldsymbol{F} = & \left(\left(\boldsymbol{A}_a^*(\boldsymbol{x}_a^*)-\boldsymbol{A}_a(\boldsymbol{x}_a)\right)\boldsymbol{S}_x + \left(\boldsymbol{B}_a^*(\boldsymbol{x}_a^*)-\boldsymbol{B}_a(\boldsymbol{x}_a)\right)\boldsymbol{K}_{xm}^*\right)\boldsymbol{x}_m \\ & + \left(\left(\boldsymbol{A}_a^*(\boldsymbol{x}_a^*)-\boldsymbol{A}_a(\boldsymbol{x}_a)\right)\boldsymbol{S}_u + \left(\boldsymbol{B}_a^*(\boldsymbol{x}_a^*)-\boldsymbol{B}_a(\boldsymbol{x}_a)\right)\boldsymbol{K}_{um}^*\right)\boldsymbol{u}_m \\ & - \boldsymbol{B}_a(\boldsymbol{x}_a)\boldsymbol{K}_e\boldsymbol{d}_1 \end{aligned} \tag{11.40}$$

The augmented controlled system has been transformed into almost strictly passive system by adding into the parallel feed-forward compensator. According to Kalman–Yakubovich lemma, there exist two positive definite matrix $\boldsymbol{P}$ and $\boldsymbol{Q}$ such that

$$\dot{\boldsymbol{P}} + \boldsymbol{P}\boldsymbol{A}_{ac} + \boldsymbol{A}_{ac}^T\boldsymbol{P} = -\boldsymbol{Q} \tag{11.41}$$

The following quadratic Lyapunov function will be adopted for the stability analysis of longitudinal adaptive guidance control

$$V = \boldsymbol{e}_x^T\boldsymbol{P}\boldsymbol{e}_x + \mathrm{tr}\left(\left(\boldsymbol{K}_I - \boldsymbol{K}^*\right)\boldsymbol{\Gamma}^{-1}\left(\boldsymbol{K}_I - \boldsymbol{K}^*\right)^T\right) \tag{11.42}$$

where $\boldsymbol{P}$ is the positive definite matrix defined in Eq. (11.41) and $\boldsymbol{\Gamma}$ is the positive definite scaling factor. The derivative of $\boldsymbol{V}$ along the trajectory gives

$$\dot{V} = \boldsymbol{e}_x^T\dot{\boldsymbol{P}}\boldsymbol{e}_x + \dot{\boldsymbol{e}}_x^T\boldsymbol{P}\boldsymbol{e}_x + \boldsymbol{e}_x^T\boldsymbol{P}\dot{\boldsymbol{e}}_x + 2\mathrm{tr}\left(\left(\boldsymbol{K}_I - \boldsymbol{K}^*\right)\boldsymbol{\Gamma}^{-1}\left(\dot{\boldsymbol{K}}_I - \dot{\boldsymbol{K}}^*\right)^T\right) \tag{11.43}$$

Substituting $\dot{\boldsymbol{e}}_x$ from Eq. (11.39) and $\dot{\boldsymbol{K}}_I$ from Eq. (11.25) into Eq. (11.43) gives

$$\begin{aligned} \dot{V} = & \boldsymbol{e}_x^T\left(\dot{\boldsymbol{P}} + \boldsymbol{P}\boldsymbol{A}_{ac} + \boldsymbol{A}_{ac}^T\boldsymbol{P}\right)\boldsymbol{e}_x + \boldsymbol{e}_x^T\boldsymbol{P}\boldsymbol{B}_{ac}\left(\boldsymbol{K} - \boldsymbol{K}^*\right)\boldsymbol{r} \\ & - \boldsymbol{r}^T\left(\boldsymbol{K} - \boldsymbol{K}^*\right)^T\boldsymbol{B}_{ac}^T\boldsymbol{P}\boldsymbol{e}_x + 2\boldsymbol{e}_x^T\boldsymbol{P}\boldsymbol{F} - \boldsymbol{e}_y^T\left(\boldsymbol{K}_I - \boldsymbol{K}^*\right)\boldsymbol{r} \\ & + \boldsymbol{r}^T\left(\boldsymbol{K}_I - \boldsymbol{K}^*\right)^T\boldsymbol{e}_y - 2\sigma\mathrm{tr}\left(\left(\boldsymbol{K}_I - \boldsymbol{K}^*\right)\boldsymbol{\Gamma}^{-1}\boldsymbol{K}_I^T\right) \end{aligned} \tag{11.44}$$

Substituting $\boldsymbol{K}_I$ from Eq. (11.45) into Eq. (11.44)

$$\boldsymbol{K}_I = \boldsymbol{K} - \boldsymbol{K}_P = \boldsymbol{K} - \boldsymbol{e}_y\boldsymbol{r}^T \tag{11.45}$$

gives

$$\begin{aligned} \dot{V} = & -\boldsymbol{e}_x^T\boldsymbol{Q}\boldsymbol{e}_x - 2\boldsymbol{e}_y^T\boldsymbol{e}_y\boldsymbol{r}^T\bar{\boldsymbol{\Gamma}}\boldsymbol{r} - 2\sigma\mathrm{tr}\left(\left(\boldsymbol{K}_I - \boldsymbol{K}^*\right)\boldsymbol{\Gamma}^{-1}\left(\boldsymbol{K}_I - \boldsymbol{K}^*\right)^T\right) \\ & - 2\sigma\mathrm{tr}\left(\left(\boldsymbol{K}_I - \boldsymbol{K}^*\right)\boldsymbol{G}^{-1}\boldsymbol{K}^{*T}\right) - 2\boldsymbol{r}^T\left(\boldsymbol{K} - \boldsymbol{K}^*\right)^T\boldsymbol{d}_1 + 2\boldsymbol{e}_x^T\boldsymbol{P}\boldsymbol{F} \end{aligned} \tag{11.46}$$

As $\boldsymbol{x}_m$, $\boldsymbol{u}_m$ and $\boldsymbol{u}_p$ are all bounded, it is easy to conclude that there exist some positive coefficients $\alpha_1 \cdots \alpha_6 > 0$ such that

$$\begin{aligned}\dot{V} \le & -\alpha_1\|\boldsymbol{e}_x\|^2 - \alpha_2\left\|\boldsymbol{e}_y\right\|^2\|\boldsymbol{r}\|^2 - \alpha_3\|\boldsymbol{K}_I - \boldsymbol{K}^*\|^2 \\ & -\alpha_4\left\|\boldsymbol{r}(\boldsymbol{K}_I - \boldsymbol{K}^*)\right\| + \alpha_5\|\boldsymbol{e}_x\| + \alpha_6\left\|\boldsymbol{K}_I - \boldsymbol{K}^*\right\|\end{aligned} \tag{11.47}$$

Obviously, if either $\|\boldsymbol{e}_x\|$, $\left\|\boldsymbol{K}_I - \boldsymbol{K}^*\right\|$ or $\left\|\boldsymbol{r}(\boldsymbol{K}_I - \boldsymbol{K}^*)\right\|$ increase beyond some bound, the negative definite quadratic terms in Eq. (11.47) become dominant, and thus $\dot{V}(t)$ becomes negative

$$\dot{V}(t) < -\varepsilon V_1,\ \forall V(t) > V_1 \tag{11.48}$$

Therefore, $V(t)$ is uniformly bounded, and then the output tracking error of the controlled system is also uniformly bounded.

Q.E.D.

11.5 Heading Alignment Guidance Law

The role of the lateral guidance law is to command the sign of the bank angle, which couples into the lateral dynamics through Eq. (11.1) and allows one to control heading to meet the desired crossrange requirement. The logic for commanding the sign of the bank angle is based on a heading angle corridor. When the vertical plane of motion becomes too different from the vertical plane containing the vehicle and the target, a sign change is commanded. The attitude control system must then rotate the vehicle about the velocity vector for what is termed a bank reversal. In this section, we derive the lateral entry guidance using bank reversal to reduce the heading error and thus reduce the cross range error, which is similar to that used of Apollo and Mars Science Laboratory. The heading angle error is defined as the difference between the current heading angle ψ and the desired heading angle ψ_d that leads straight to the prescribed parachute deployment point

$$\Delta\psi = \psi - \psi_d \tag{11.49}$$

According to the physical meaning of the desired heading angle ψ_d, its range should be between 0 and π, namely, $0 \le \psi_{LOS} \le \pi$. The desired heading angle is

$$\begin{aligned}\psi_{LOS} = & \tan^{-1}\left(\frac{\sin(\theta_T - \theta)}{\cos\lambda\tan\lambda_T - \sin\lambda\cos(\theta_T - \theta)}\right) + 0.5\pi \\ & -0.5\pi sign\left(\frac{\sin(\theta_T - \theta)}{\cos\lambda\tan\lambda_T - \sin\lambda\cos(\theta_T - \theta)}\right)\end{aligned} \tag{11.50}$$

$$\begin{aligned}\psi_{LOS} = & \tan^{-1}\left(\frac{\sin(\theta_T - \theta)}{\cos\lambda\tan\lambda_T - \sin\lambda\cos(\theta_T - \theta)}\right) + 0.5\pi \\ & -0.5\pi sign\left(\frac{\sin(\theta_T - \theta)}{\cos\lambda\tan\lambda_T - \sin\lambda\cos(\theta_T - \theta)}\right)\end{aligned} \tag{11.51}$$

where, θ_T and λ_T are the longitude and latitude of the desired parachute deployment point. In order to assure the heading angle error $\Delta\psi$ lies in the heading angle error corridor, the entry vehicle should be flying towards the prescribed parachute deployment point in the initial stage of entry. Therefore, the symbol of bank angle in the initial phase can be determined

$$sign(\sigma_0) = -sign(\Delta\psi_0) \tag{11.52}$$

where σ_0 is the initial bank angle, and $\Delta\psi_0$ is the initial heading angle error. When the entry begins, the sign of bank angle is determined as follows:

$$sign(\sigma) = \begin{cases} sign(\sigma(t_{k-1})) & |\Delta\psi| < \Delta\psi_H(V) \\ -sign(\Delta\psi(V)) & |\Delta\psi| \geq \Delta\psi_H(V) \end{cases} \tag{11.53}$$

where $sign(\sigma(t_{k-1}))$ stands for the sign of the bank angle in the k-1 guidance loop, and $\Delta\psi_H(V)$ is the deviation threshold of the heading angle, which is usually designed as the piecewise linear function of velocity with the times of bank reversal and the magnitude of the final crossrange error taken into account. The heading angle error corridor adopted here is shown in Fig. 11.2.

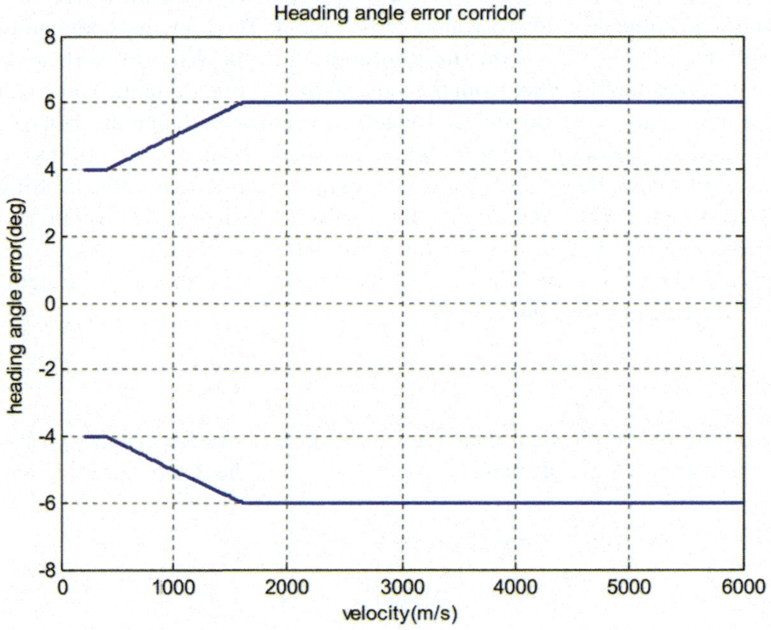

Fig. 11.2 Heading angle error corridor

11.6 Simulation and Results

In order to confirm the validity of the CGT-based direct model reference adaptive tracking guidance for Mars atmospheric entry developed in this paper, Monte Carlo simulation and analysis in the MATLAB/Simulink environment has been carried out. The simulation system can be divided into three parts, entry system dynamics model block, reference model block and adaptive guidance algorithm block, from the viewpoint of the structure and function. The entry system dynamics model block defines the detailed mathematical models (Eq. (11.1)) used in the simulation. The reference model block off-line plans the reference drag acceleration profile with the constraints (Eqs. (11.12)–(11.15),) taken into account. Adaptive gain coefficients are produced in the adaptive guidance algorithm block.

The configuration and physical parameters of Mars Science Laboratory vehicle are used as an example to perform simulation and analysis in this paper. The detailed parameters of a vehicle similar to the MSL entry vehicle are as follows: vehicle mass $m = 2200$ kg, reference area $S = 12.8825\ \text{m}^2$, drag coefficient $C_D = 1.450$, ballistic coefficient $\beta = m/(C_D S) = 117.78\ \text{kg/m}^2$, lift-to-drag ratio $L/D = 0.24$. The vehicle initial states at the entry interface are provided by the Deep Space Network (DSN), Table 11.1 presents the initial states and the corresponding errors used in our simulations. Other relevant simulation parameters are set as follows: the reference radius of Mars $r_0 = 3397$ km, the gravitational constant of Mars $\mu_{Mars} = 4.282829 \times 10^3\ \text{km}^3/\text{s}^2$, the adjustable parameters of the feed-forward compensator $a_0 = 1e - 5, a_1 = 0.2$, $k_c = -0.01$, initial adaptive gain coefficient matrix $\boldsymbol{K}_{I,0} = \left[0\,0\,0\,0\right]^{\mathrm{T}}$, initial integral gain matrix $T_{i,0} = \left[0.01\ 0.01\ 0.01\right]^{\mathrm{T}}$, positive definite weighting matrix $T_i = 15 I_{4\times4}$, positive semi-definite weighting matrix $T_p = diag\left(30\ 0.05\ 1\ 0.07\right)^{\mathrm{T}}$, and forgetting factor $\zeta = 6$. The adaptive gain coefficients are obtained according to Eqs. (11.18)–(11.21). A PD controller is adopted to track planned nominal bank angle, its parameters are set as $k_p = 6$ and $k_d = 10$. The simulation sample period/ step is set to one second due to the actuator limitation and the four-order Runge–Kutta algorithm is selected as numerical solver of differential equations (Eq. (11.1)). Due to the very thin atmosphere, Mars entry vehicles have low control authority. If the actual trajectories are far from the nominal trajectory,

Table 11.1 State initial values and errors of vehicle at entry interface (EI)

States	Initial guess	Error (3σ)
r (km)	3522	±1
v (m/s)	6000	±5
γ (°)	−11.2	±0.1
θ (°)	0	±0.0172
λ (°)	0	±0.0172
ψ (°)	90	±0.5

the entry vehicle may not have the capability to achieve precision landings. Therefore, the nominal trajectory must be carefully designed. The prescribed downrange is assumed to be 992 km, which satisfies the constraints above and makes the nominal trajectory not too hard to track. The nominal bank angle profile and corresponded drag trajectory used in our simulation, shown in Figs. 11.3 and 11.4, respectively, are planned using the fast online trajectory planning technology developed in [105]. The Mars atmosphere is very thin and the density is very small at the beginning of the entry, so the reference drag acceleration is very small too. With decreasing the altitude of the entry vehicle, the atmospheric density is becoming denser and denser, so the reference drag acceleration is correspondingly becoming larger and larger. At the same time, the velocity of the entry vehicle is gradually decreasing in the course of entry. Therefore, the reference drag acceleration is also gradually decreasing after a certain peak value, though the density is always increasing during entry. To access the performance of the guidance algorithm developed in this paper, three different initial simulation scenarios are defined with various disturbance sources and uncertainty range. In the baseline simulation scenario I, 10, 5 and 5% uncertainties are respectively added to the nominal atmospheric density, ballistic coefficient and lift-to-drag ratio. In the scenario II and III, the value of the uncertainty in the nominal atmospheric density and ballistic coefficient doubles and triples respectively, while the uncertainty of the lift-to-drag is assumed to be unchanged. There is little uncertainty for the lift-to-drag ratio in actual atmospheric flight, so, 5% uncertainty assumption is reasonable and adopted for all three simulation scenarios. The distribution type and mean value (3σ) of the disturbance are shown in Table 11.2. The 300-run Monte Carlo simulation results are shown in Figs. 11.5 and 11.7, which are conducted under the assumption of $\pm 10\%$ density uncertainty, $\pm 5\%$ ballistic coefficient uncertainty and $\pm 5\%$ lift-to-drag ratio uncertainty respectively. The time history of the actual bank angle is shown in Fig. 11.5, three or four time's bank reversals have been executed to track the reference profile and eliminate the effect of the uncertainty factors. Figure 11.5 shows that the maximum value of adaptive drag tracking error corresponds to the worst case is no larger than 3 m/s^2 and the final drag tracking error can be controlled to 0.025 m/s^2. It can be also seen from Fig. 11.5 that there is the existence of the phenomenon of saturation with bank angle between 60 and 110° due to the low control authority of aerodynamic lift. The altitude error, velocity error and flight path angle error are illustrated in Fig. 11.6. The maximum errors of altitude, velocity and flight path angle corresponding to the worst cases are <2 km, 30 m/s and 4° respectively. It can be easily concluded from Fig. 11.6 that the altitude and velocity error curves have better convergence than that of flight path angle. The reason why the altitude and velocity errors have better convergence performance lies in the fact that there is a closely relationship between atmospheric density, velocity and drag acceleration, and density is heavily related to altitude. Target miss distances at parachute deployment (end of the simulation) for the 300-run study are depicted in Fig. 11.7, the maximum downrange error and crossrange error corresponding to the worst case are <10 and 1 km respectively. Further statistical analysis shows that only 11 cases (3.67%) in all 300-run simulation end with downrange errors larger than 5 km, which meets the requirement of future Mars pinpoint landing exploration

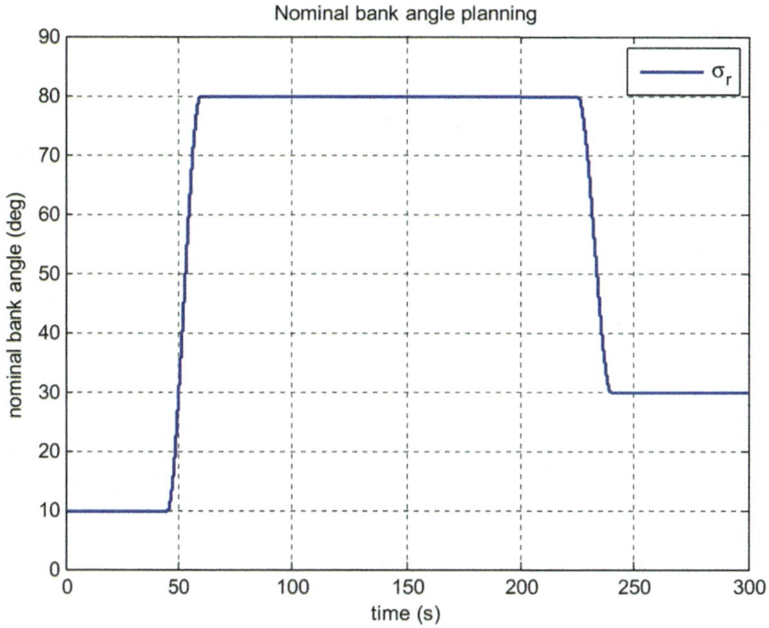

Fig. 11.3 Planned nominal bank angle

activities. Though larger hypothetical uncertainties are included into Mars environment parameters and entry vehicle parameters, the adaptive guidance errors are still satisfied with the requirement of precision Mars landing missions, which denotes the guidance algorithm developed here has a better robustness performance. For comparison purposes, the footprint dispersion of open-loop reference trajectory guidance with uncertainties range I is plotted in Fig. 11.8, it is seen that the maximum open loop tracking error reaches 50 km and is five-time magnitude of closed-loop adaptive guidance error (Table 11.3).

In order to further test the robustness of the adaptive guidance algorithm under the larger disturbance and uncertainty, Monte Carlo simulations for both closed-loop adaptive guidance and open-loop reference following guidance with scenario II and III have been carried out and the corresponding results are graphical displayed in Figs. 11.9 and 11.12. Figures 11.9 and 11.10 show that the maximum downrange errors of closed-loop adaptive guidance with simulation scenario II and III are not larger than 15 km and 30 km respectively. Further statistical analysis shows that only 4 cases (1.33%) in simulation scenario II end with downrange errors larger than 10 km and only 5 cases (1.67%) in simulation scenario III end with downrange errors larger than 15 km in all 300-run Monte Carlo simulations. The footprint dispersion of open-loop tracking guidance with uncertainties range II and III are plotted in Figs. 11.11 and 11.12. The open/close loop guidance error ratios with different uncertainty range

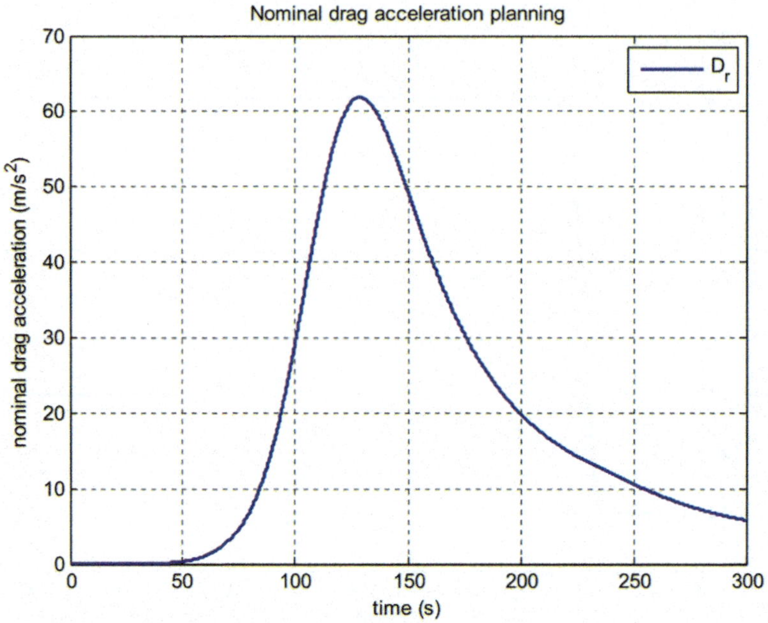

Fig. 11.4 Planned nominal drag acceleration

Table 11.2 Uncertainty in atmospheric density, ballistic coefficient and lift-to-drag ratio

Parameters with uncertainty	Distribution	3r Uncertainty		
		Range I (%)	Range II (%)	Range III (%)
Atmospheric density	Normal	±10	±20	±30
Ballistic coefficient	Normal	±5	±10	±15
Lift-to-drag ratio	Normal	±5	±5	±5

Table 11.3 Maximum downrange error ratio of open/close loop guidance with different uncertainty parameters

Uncertainty	Distribution	3σ Uncertainty		
		Range I (%)	Range II(%)	Range III (%)
Footprint dispersion error (km)	Open-loop	50	60	90
	Close-loop	10	15	30
Open/close loop error ratio	Normal	5	4	3

are given in Table 3, which are derived from Figs. 11.7, 11.8, 11.9, 11.10, 11.11 and 11.12 using the largest circle labeled value. It can be found that these ratios are gradually decreasing when the uncertainty ranges become larger and larger, which denotes the fact that the closed-loop adaptive guidance performance degrades with

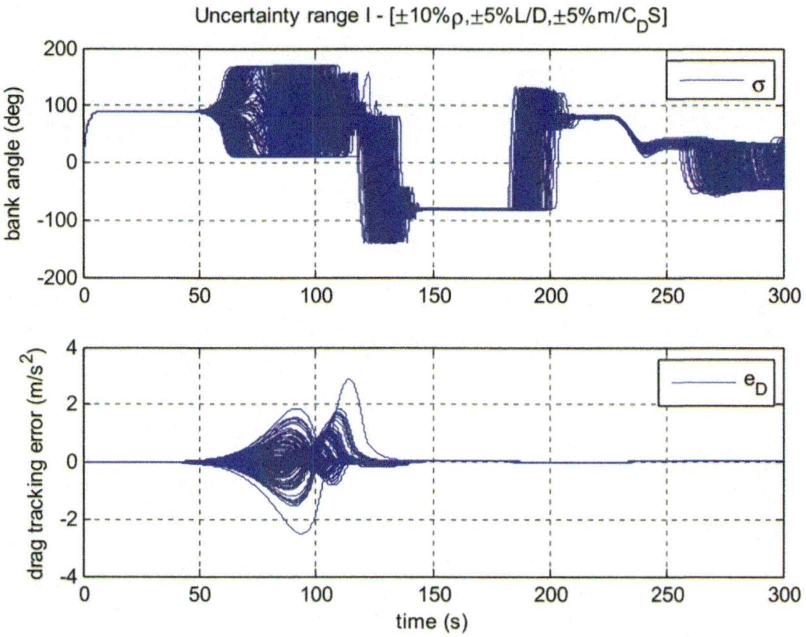

Fig. 11.5 Actual bank angle and corresponding drag tracking error of adaptive guidance with uncertainty range I

large uncertainty though the performance of the closed-loop adaptive guidance is always far superior to that of open loop tracking guidance. The reason why existence of the phenomenon mentioned above lies in the fact that the thin Martin atmospheric density and low lift-to-drag ratio lead to the lower control authority, and then it is difficult to accurately track the planned nominal trajectory when there is a large deviation. It is interesting to note that the closed-loop adaptive guidance usually leads to undershoot and open-loop reference tracking guidance usually leads to overshoot by comparing the target miss distance dispersions of adaptive guidance in Figs. 11.7, 11.9, and 11.10 with those of open-loop guidance in Figs. 11.8, 11.11, and 11.12. The reason why the closed-loop adaptive guidance leads to little overshoot compared with the open-loop reference tracking guidance maybe lies in the fact that the former has higher control authority than the latter.

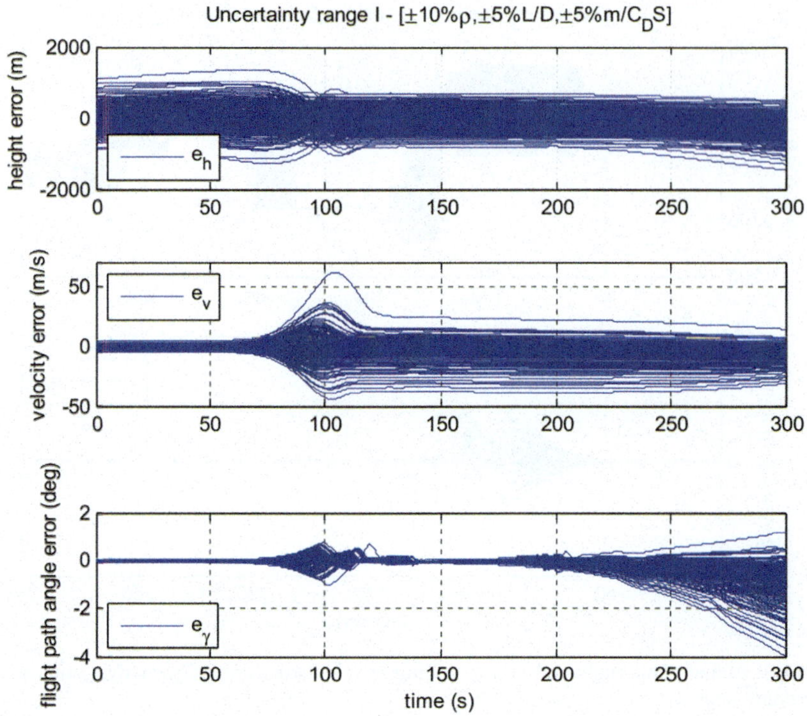

Fig. 11.6 Altitude/velocity/flight path angle error of adaptive guidance with uncertainty range I

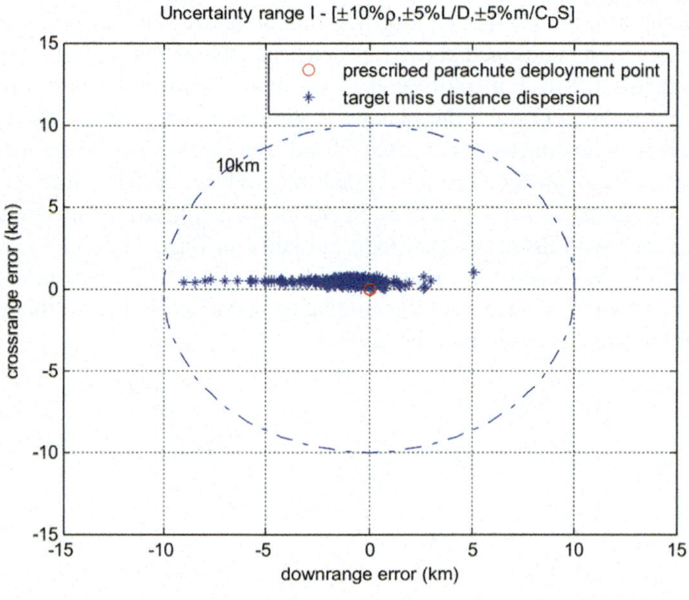

Fig. 11.7 Target miss distance dispersion of adaptive guidance with uncertainty range I

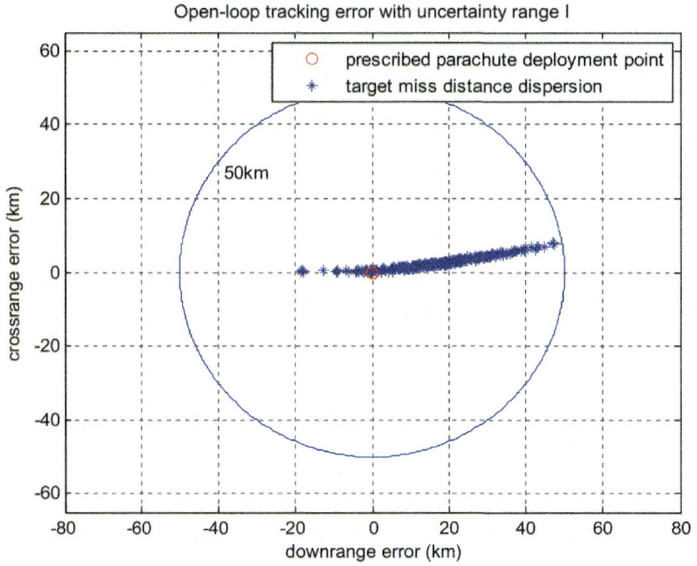

Fig. 11.8 Target miss distance dispersion of open-loop reference tracking guidance with uncertainty range I

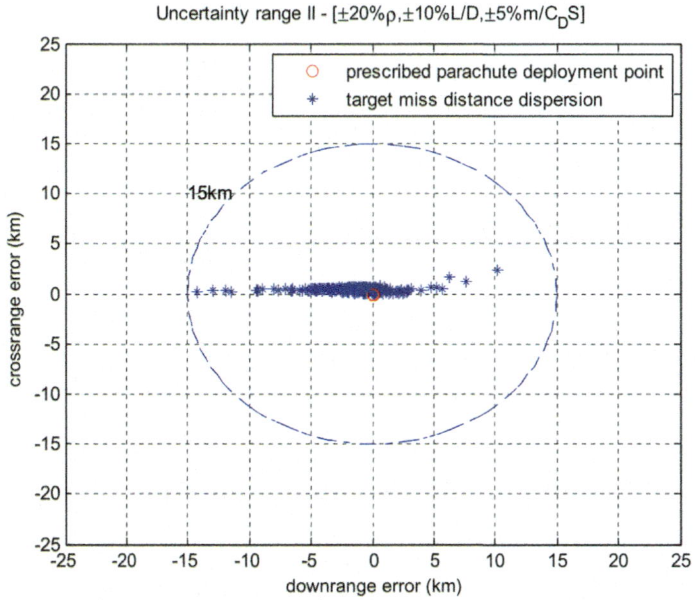

Fig. 11.9 Target miss distance dispersion of adaptive guidance with uncertainty range II

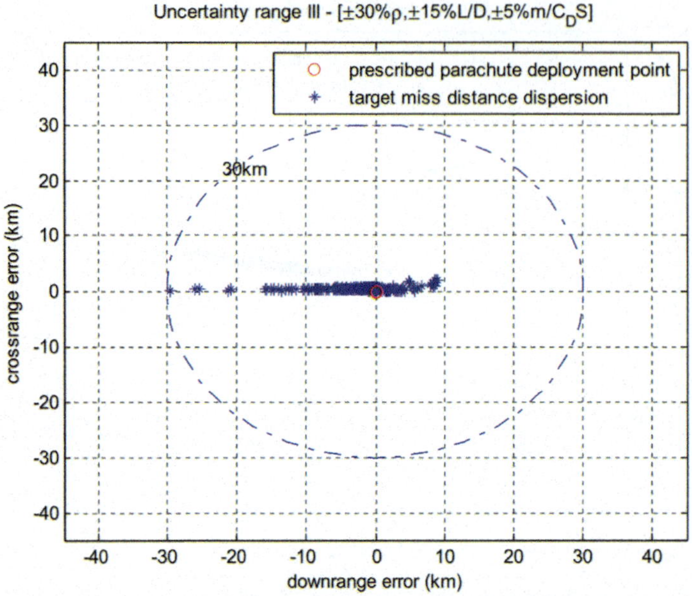

Fig. 11.10 Target miss distance dispersion of adaptive guidance with uncertainty range III

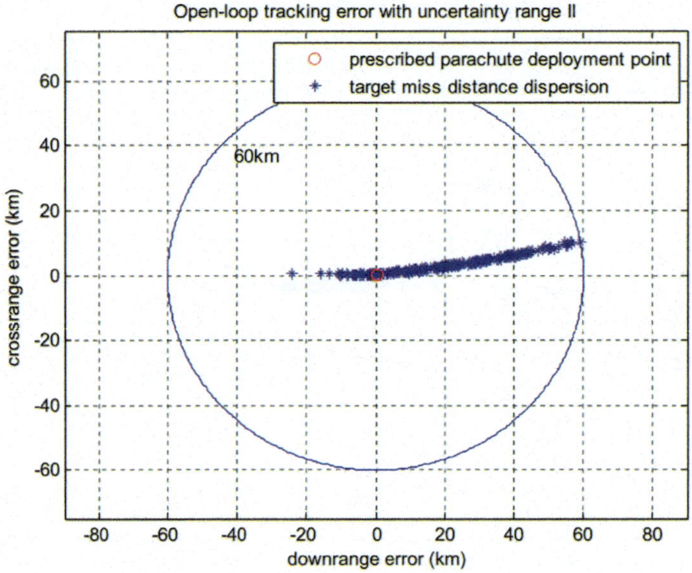

Fig. 11.11 Target miss distance dispersion of open-loop reference tracking guidance with uncertainty range II

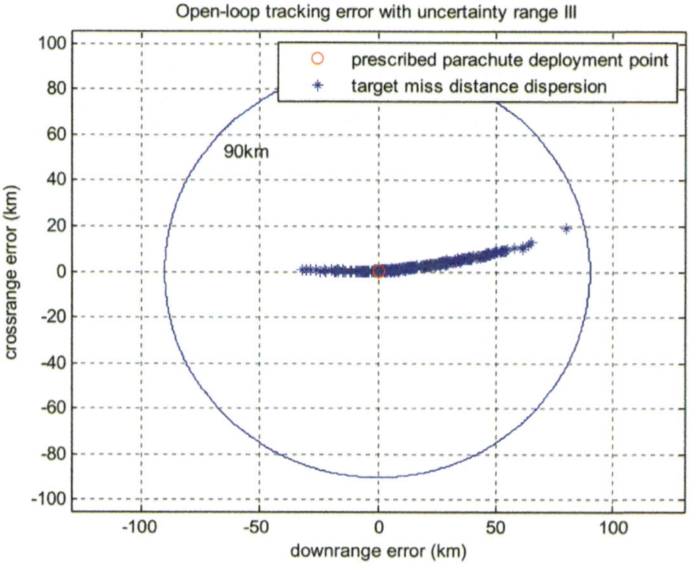

Fig. 11.12 Target miss distance dispersion of open-loop reference tracking guidance with uncertainty range III

11.7 Conclusions

This chapter addresses the Mars atmospheric entry guidance by use of command generator tracker based direct model reference adaptive controller to eliminate the effect of the bounded uncertainties on atmospheric density and aerodynamic coefficients. The proposed strategy is easily implemented on low-cost, low-weight entry vehicle from a practical point of view, because it requires only storing on board a pre-computed set of DMRAC control gains and a reference trajectory. The feasibility and robustness of the guidance algorithm developed in this paper is validated by Monte Carlo simulation and analysis. In order to obtain the better tracking performance in low-lift conditions using bank angle control only, the reference drag acceleration profile should be more carefully planned and optimized with various constraints taken into account. At the same time, the introduction of feed-forward compensator can guarantee the controlled system almost strictly passive, but the performance of steady-state tracking error maybe degrades. Therefore, the relative parameters in feed-forward compensator should be carefully selected to ensure almost strictly passive while reducing the steady-state tracking error as much as possible. In this paper, we only considered the uncertainties in the atmospheric density, ballistic coefficient and lift-to-drag ratio. The modeling errors in the dynamics (planetary rotation, winds, non-spherical gravity, etc.) should be also included to make the simulation more realistic. These three aspects will be analyzed in the further study.

Chapter 12
Computational Guidance Method for Mars Entry

The Entry Terminal Point Controller (ETPC) algorithm and its variants have been used to perform entry guidance for a variety Mars missions including the Mars Science Laboratory [23], Mars 2020 [26], and Tianwen-1 [21]. In essence, the ETPC guidance predicts the final downrange error at the reference final altitude and produces a feedback control correction for the guidance system to eliminate the downrange error rather than simply following the reference trajectory. However, path constraints such as heat flux and dynamic pressure are not explicitly considered in the guidance logic but are only implicitly considered in the process of reference trajectory design. That is, the reference trajectory is restricted within the feasible space defined by these path constraints. The computational complexity of guidance algorithms can be significantly reduced by ignoring path constraints; however, when the guidance system is faced with large uncertainties, the trajectory guided by the unconstrained ETPC algorithm may easily violate the path constraints and lead to infeasible flight trajectories.

The rapid development of high-performance computing platforms and highly efficient computational algorithms provides significant opportunities to generate guidance commands in real time [94, 141, 206, 207]. Among these algorithms, convex optimization has been widely used in the field of atmospheric entry guidance because of its superior characteristics of global optimization and polynomial complexity [54]. Although the feasibility of designing tracking guidance with the convex QCQP has been investigated, the uniform discretization strategy adopted in the literature, such as [93], typically requires a large number of discrete meshes to ensure acceptable solution accuracy, which causes large-scale optimization problems and high computational load. In this paper, we aim to fundamentally improve the QCQP-based tracking guidance by performing the discretization using the LGR pseudospectral method, which reduces the scale of the resulting optimization problem while preserving numerical accuracy. Additionally, the objective function has also been modified to include a quadratic term about the rate of change of the control

S. Li et al., *Trajectory Optimization and Guidance Methods for Mars Entry*,
https://doi.org/10.1007/978-981-99-6282-2_12

variable, enabling the possibility to generate a smooth control profile. Additionally, the effects of the quadratic coefficients on the proposed guidance law are discussed.

The remainder of this chapter is organized as follows: Sect. 12.1 introduces the problem formulation and reviews the relevant linear quadratic regulator (LQR) approach for Mars entry tracking guidance. The enhanced tracking guidance law based on QCQP and the pseudospectral method is described in Sects. 12.2 and 12.3. Section 12.4 presents the Monte-Carlo simulation results, and the conclusion is provided in Sect. 12.5.

12.1 Entry Guidance Problem

12.1.1 Mars Entry Dynamics

Under the assumption of a spherical, rotating Mars, the dimensionless, three-degrees-of-freedom entry dynamics is described as follows:

$$\begin{cases} \dot{r} = V\sin\gamma \\ \dot{\theta} = V\cos\gamma\sin\psi \big/ r\cos\phi \\ \dot{\phi} = V\cos\gamma\cos\psi \big/ r \\ \dot{V} = -D - \sin\gamma \big/ r^2 + \omega_0^2 r\cos\phi(\cos\phi\sin\gamma - \sin\phi\cos\gamma\cos\psi) \\ \dot{\gamma} = L\cos\sigma \big/ V + (V\big/ r - 1\big/ r^2 V)\cos\gamma + \\ 2\omega_0\cos\phi\sin\psi + \omega_0^2 r\cos\phi(\cos\phi\cos\gamma + \cos\psi\sin\phi\sin\gamma)\big/ V \\ \dot{\psi} = L\sin\sigma \big/ V\cos\gamma + V\cos\gamma\sin\psi\tan\phi\big/ r + \\ 2\omega_0(\sin\phi - \cos\phi\tan\gamma\cos\psi)\big/ V + \omega_0^2 r\sin\phi\cos\phi\sin\psi\big/ V\cos\gamma \end{cases} \tag{12.1}$$

where variables $\omega_0, r, \theta, \phi, V, \gamma, \psi$, and σ are the non-dimensional Mars self-rotation rate, radial distance, longitude, latitude, relative velocity, flight path angle, azimuth angle, and bank angle, respectively. In Eq. (12.1), the dimensionless factors of length, acceleration, speed, and time are R_0, g_0, $\sqrt{R_0 g_0}$, and $\sqrt{R_0\big/ g_0}$, respectively, where R_0 denotes the mean Mars radius and g_0 is the Mars gravitational acceleration at R_0. Additionally, the lift and drag accelerations in g_0 are defined as follows:

$$\begin{cases} L = R_0\rho V^2 S_r C_L \big/ (2m) \\ D = R_0\rho V^2 S_r C_D \big/ (2m) \end{cases} \tag{12.2}$$

where C_L, C_D, S_r, and m are the lift coefficient, drag coefficient, reference area, and mass of the entry vehicle, respectively. The variable $\rho = \rho_0\exp[-R_0(r-1)\big/ h_s]$ is the Martian atmospheric density, where ρ_0 is the density at R_0 and h_s is the scale height. It should be pointed out that S_r, m, ρ_0, and h_s are dimensional parameters and set to be constant in this paper.

For simplicity, Eq. (12.1) is expressed in the following state-space representation:

$$\dot{\boldsymbol{x}}(t) = \boldsymbol{f}(\boldsymbol{x}, u) + \boldsymbol{f}_{\omega}(\boldsymbol{x}) \tag{12.3}$$

where the state vector $\boldsymbol{x}(t) = [r, \theta, \phi, V, \gamma, \psi]^T$ and the control variable $u(t) = \sigma$. Functions $\boldsymbol{f}(\boldsymbol{x}, u)$ and $\boldsymbol{f}_{\omega}(\boldsymbol{x})$ compose the right-hand side of Eq. (12.1), and $\boldsymbol{f}_{\omega}(\boldsymbol{x})$ represents the Mars-rotation-dependent term.

12.1.2 Constraints

In addition to entry dynamics, there are three types of constraints that need to be considered during Mars entry: path constraints, boundary conditions, and control constraints. For path constraints, heat flux $\dot{Q}$, dynamic pressure q, and normal load a are typically considered as follows:

$$\begin{cases} \dot{Q} = k_Q\sqrt{\rho / R_n}(V_s V)^{3.15} \le \dot{Q}_{\max} \\ q = 0.5\rho(V_s V)^2 \le q_{\max} \\ a = \sqrt{L^2 + D^2} \le a_{\max} \end{cases} \tag{12.4}$$

where k_Q is a heating-related constant, R_n is the vehicle's nose radius, $V_s = \sqrt{R_0 g_0}$ represents the dimensionless factor of speed, and $\dot{Q}_{\max}$, $q_{\max}$, and $a_{\max}$ are the upper limits of the three path constraints.

Boundary conditions refer to the initial and final state conditions of the entry mission. Typically, the initial state is fully determined while the final state is partially constrained as follows:

$$\boldsymbol{x}(t_0) = \boldsymbol{x}_0, \quad \boldsymbol{x}(t_f) = \boldsymbol{x}_f \tag{12.5}$$

where $\boldsymbol{x}_0$ and $\boldsymbol{x}_f$ denote the given initial state and desired final state, respectively, t_0 and t_f are the initial and final times, respectively.

The control constraint is usually defined as a feasible bank angle margin and in some cases the bank angle rate as well. Nevertheless, only the constraint on bank angle magnitude is taken into consideration in this paper:

$$u_{\min} \le u(t) \le u_{\max}. \tag{12.6}$$

12.1.3 Tracking Guidance via LQR

Mars entry tracking guidance keeps the entry vehicle as close to the reference trajectory as possible. The reference trajectory $\{\boldsymbol{x}^*(t), u^*(t)\}$ is predetermined via offline or online trajectory optimization. Then, the tracking guidance starts with the linearization of the original nonlinear dynamics in Eq. (12.3). In this fashion, a linear time-varying (LTV) system for guidance is obtained as follows by approximating the Mars-rotation-dependent term as $\boldsymbol{f}_\omega(\boldsymbol{x}^*)$ evaluated at the reference trajectory:

$$\delta\dot{\boldsymbol{x}}(t) = \boldsymbol{A}^*(t)\delta\boldsymbol{x}(t) + \boldsymbol{B}^*(t)\delta u(t) \tag{12.7}$$

where $\delta\boldsymbol{x}(t) = \boldsymbol{x}(t) - \boldsymbol{x}^*(t)$ and $\delta u(t) = u(t) - u^*(t)$ are the state error and the control correction, respectively. The Jacobian matrices $\boldsymbol{A}^*(t)$ and $\boldsymbol{B}^*(t)$ are derived as follows:

$$\boldsymbol{A}^*(t) = \left.\frac{\partial \boldsymbol{f}}{\partial \boldsymbol{x}}\right|_{\boldsymbol{x}^*(t),\boldsymbol{u}^*(t)} = \begin{bmatrix} 0 & 0 & 0 & a_{14} & a_{15} & 0 \\ a_{21} & 0 & a_{23} & a_{24} & a_{25} & a_{26} \\ a_{31} & 0 & 0 & a_{34} & a_{35} & a_{36} \\ a_{41} & 0 & 0 & a_{44} & a_{45} & 0 \\ a_{51} & 0 & 0 & a_{54} & a_{55} & 0 \\ a_{61} & 0 & a_{63} & a_{64} & a_{65} & a_{66} \end{bmatrix}, \quad \boldsymbol{B}^*(t) = \left.\frac{\partial \boldsymbol{f}}{\partial \boldsymbol{u}}\right|_{\boldsymbol{x}^*(t),\boldsymbol{u}^*(t)} = \begin{bmatrix} 0 \\ 0 \\ 0 \\ 0 \\ b_5 \\ b_6 \end{bmatrix} \tag{12.8}$$

$a_{14} = \sin\gamma$, $a_{15} = V\cos\gamma$, $a_{21} = -\frac{V\cos\gamma\sin\psi}{r^2\cos\phi}$, $a_{23} = \frac{V\cos\gamma\sin\psi\sin\phi}{r\cos^2\phi}$, $a_{24} = \frac{\cos\gamma\sin\psi}{r\cos\phi}$, $a_{25} = -\frac{V\sin\gamma\sin\psi}{r\cos\phi}$, $a_{26} = \frac{V\cos\gamma\cos\psi}{r\cos\phi}$, $a_{31} = -\frac{V\cos\gamma\cos\psi}{r^2}$, $a_{34} = \frac{\cos\gamma\cos\psi}{r}$, $a_{35} = -\frac{V\sin\gamma\cos\psi}{r}$, $a_{36} = -\frac{V\cos\gamma\sin\psi}{r}$, $a_{41} = -D_r + \frac{2\sin\gamma}{r^3}$, $a_{44} = -D_V$, $a_{45} = -\frac{\cos\gamma}{r^2}$, $a_{51} = \frac{L_r\cos\sigma}{V} + \frac{2\cos\gamma}{r^3 V} - \frac{V\cos\gamma}{r^2}$, $a_{54} = \frac{L_V\cos\sigma}{V} - \frac{L\cos\sigma}{V^2} + \frac{\cos\gamma}{r^2V^2} + \frac{\cos\gamma}{r}$, $a_{55} = \frac{\sin\gamma}{r^2V} - \frac{V\sin\gamma}{r}$, $a_{61} = \frac{L_r\sin\sigma}{V\cos\gamma} - \frac{V\cos\gamma\sin\psi\tan\phi}{r^2}$, $a_{63} = \frac{V\cos\gamma\sin\psi}{r\cos^2\phi}$, $a_{64} = \frac{L_V\sin\sigma}{V\cos\gamma} - \frac{L\sin\sigma}{V^2\cos\gamma} + \frac{\cos\gamma\sin\psi\tan\phi}{r}$, $a_{65} = \frac{L\sin\sigma\sin\gamma}{V\cos^2\gamma} - \frac{V\sin\gamma\sin\psi\tan\phi}{r}$, $a_{66} = \frac{V\cos\gamma\cos\psi\tan\phi}{r}$, $b_5 = -\frac{L\sin\sigma}{V}$, $b_6 = \frac{L\cos\sigma}{V\cos\gamma}$, $D_r = -\frac{R_0}{h_s}D$, $D_V = \frac{R_0\rho_0 e^{-R_0(r-1)/h_s} V S_r C_D}{m}$, $L_r = -\frac{R_0}{h_s}L$, $L_V = \frac{R_0\rho_0 e^{-R_0(r-1)/h_s} V S_r C_L}{m}$.

To track the reference trajectory, a classic optimal guidance law is the LQR. In fact, the method proposed in this paper is similar to LQR. Therefore, this section reviews the LQR first, and subsequently, the next section introduces the QCQP-based tracking guidance algorithm.

In LQR, the quadratic objective function is defined in the time interval $[t_i, t_f]$:

$$J = \frac{1}{2}\delta\boldsymbol{x}^T(t_f)\boldsymbol{P}_f\delta\boldsymbol{x}(t_f) + \frac{1}{2}\int_{t_i}^{t_f}[\delta\boldsymbol{x}^T(t)\boldsymbol{Q}\delta\boldsymbol{x}(t) + \delta u^T(t)R\delta u(t)]dt \tag{12.9}$$

where $t_i \in [t_0, t_f)$ denotes the current time for guidance, $\boldsymbol{P}_f$ and $\boldsymbol{Q}$ are symmetric positive semidefinite matrices, while R is a positive real number. In Eq. (12.9), the first quadratic term minimizes the final state error, and the second integral term minimizes the state error and control correction over the time interval $[t_i, t_f]$. As such, the optimal control problem for Mars entry tracking guidance is written as follows:

P12-1: minimize Eq. (12.9)
subject to Eq. (12.7)

It should be noted that P12-1 does not take into account the constraints shown in Eqs. (12.4)–(12.6). As a result, the LQR essentially solves an unconstrained optimal control problem to generate the control correction for entry guidance. For the finite-horizon problem described in P12-1, the optimal solution can be derived from the Riccati matrix differential equation. However, obtaining the linear state feedback gain matrix is a time-consuming and laborious process. Nevertheless, the gain converges to a constant matrix as t_f approaches infinity in most real-world problems, and the convergence rate of the gain matrix is far beyond the actual response rate of the problem itself. Therefore, it is reasonable to solve an unconstrained infinite-horizon optimal control problem below with zero final state error:

$$J = \frac{1}{2}\int_{t_i}^{\infty} [\delta \boldsymbol{x}^T(t)\boldsymbol{Q}\delta \boldsymbol{x}(t) + \delta u^T(t) R \delta u(t)]dt \tag{12.10}$$

By solving the algebraic Riccati equation, the optimal solution can be obtained as follows:

$$\delta u(t) = -\boldsymbol{K}(t)\delta \boldsymbol{x} \tag{12.11}$$

where $\boldsymbol{K}(t)$ is the gain matrix determined by the algebraic Riccati equation.

12.2 QCQP-Based Tracking Guidance Law

12.2.1 Quadratically Constrained Quadratic Programming

QCQP is a special class of optimization problems with a quadratic objective function subject to quadratic inequality constraints and linear equality constraints if any. It has the following general form [51]:

$$\begin{aligned} &\underset{\boldsymbol{x}}{\text{minimize}}\ \frac{1}{2}\boldsymbol{x}\boldsymbol{P}_0^T\boldsymbol{x} + \boldsymbol{q}_0^T\boldsymbol{x} + r_0 \\ &\text{subject to}\ \frac{1}{2}\boldsymbol{x}\boldsymbol{P}_i^T\boldsymbol{x} + \boldsymbol{q}_i^T\boldsymbol{x} + r_i \le 0,\ i = 1, \ldots, m \\ &\boldsymbol{A}\boldsymbol{x} = \boldsymbol{b} \end{aligned} \tag{12.12}$$

where $\boldsymbol{x} \in \mathbb{R}^n$, $\boldsymbol{P}_i \in \mathbb{S}^n$, $\boldsymbol{q}_i \in \mathbb{R}^n$, $i = 0, \ldots, m$, $\boldsymbol{A} \in \mathbb{R}^{p \times n}$, $\boldsymbol{b} \in \mathbb{R}^p$, $\mathbb{R}$ and $\mathbb{S}$ are the sets of real and symmetric matrices, respectively. In general, QCQP is a non-convex, NP-hard problem; however, it degenerates into a convex form when $\boldsymbol{P}_i$ is positive semidefinite for $i = 0, \ldots, m$. In this case, the interior-point method can be used to solve the convex QCQP in polynomial time.

In contrast to LQR where a quadratic objective function is minimized without any constraints, QCQP-based tracking guidance takes constraints into account, and three non-convex path constraints are relaxed to linear ones as follows by the first-order Taylor approximation:

$$f_i \approx f_i(r^*, V^*) + f_i'(r^*, V^*) \ldots [r - r^*, V - V^*]^{\mathrm{T}} \leq f_{i,\max}, \quad i = 1, 2, 3 \tag{12.13}$$

where $f_1 \triangleq \dot{Q}$, $f_2 \triangleq q$, $f_3 \triangleq a$, $f_{1,\max} \triangleq \dot{Q}_{\max}$, $f_{2,\max} \triangleq q_{\max}$, $f_{3,\max} \triangleq a_{\max}$, and

$$\begin{cases} f_1'(r^*, V^*) = \left[\frac{\partial f_1}{\partial r}, \frac{\partial f_1}{\partial V}\right]\Big|_{r^*,V^*} \\ \quad = \left[-0.5R_0 k_Q (V V_s)^{3.15}\sqrt{\rho/R_n}\Big/ h_s, 3.15 k_Q V_s^{3.15} V^{2.15}\sqrt{\rho/R_n}\right]\Big|_{r^*,V^*} \\ f_2'(r^*, V^*) = \left[\frac{\partial f_2}{\partial r}, \frac{\partial f_2}{\partial V}\right]\Big|_{r^*,V^*} \\ \quad = \left[-0.5R_0\rho(V V_s)^2\Big/ h_s, V_s^2\rho V\right]\Big|_{r^*,V^*} \\ f_3'(r^*, V^*) = \left[\frac{\partial f_3}{\partial r}, \frac{\partial f_3}{\partial V}\right]\Big|_{r^*,V^*} \\ \quad = \left[-0.5R_0^2 S_r(\sqrt{C_L^2 + C_D^2})\rho V^2\Big/ (m h_s), R_0 S_r(\sqrt{C_L^2 + C_D^2})\rho V\Big/ m\right]\Big|_{r^*,V^*} \end{cases} \tag{12.14}$$

To rewrite the tracking guidance as a convex programming problem, another key step is to convert the LTV differential equality in Eq. (12.7) into a series of algebraic equalities. Typically, the trapezoidal rule is used to discretize the LTV at $N+1$ uniform time nodes such as $t_i = t_1 < \cdots < t_n < \cdots < t_{N+1} = t_f$ with $t_{n+1} = t_n + \Delta t$. We have

$$\begin{aligned} &\delta \boldsymbol{x}_{n+1} = \delta \boldsymbol{x}_n + \frac{\Delta t}{2}(\boldsymbol{A}_n^* \delta \boldsymbol{x}_n + \boldsymbol{B}_n^* \delta u_n + \boldsymbol{A}_{n+1}^* \delta \boldsymbol{x}_{n+1} + \boldsymbol{B}_{n+1}^* \delta u_{n+1}) \\ &\Rightarrow \delta \boldsymbol{x}_{n+1} = \left(\boldsymbol{I} - \frac{\Delta t}{2}\boldsymbol{A}_{n+1}^*\right)^{-1}\left[\left(\boldsymbol{I} + \frac{\Delta t}{2}\boldsymbol{A}_{n+1}^*\right)\delta \boldsymbol{x}_n + \frac{\Delta t}{2}(\boldsymbol{B}_n^* \delta u_n + \boldsymbol{B}_{n+1}^* \delta u_{n+1})\right] \end{aligned} \tag{12.15}$$

where $\delta \boldsymbol{x}_n \triangleq \delta \boldsymbol{x}(t_n)$, $\delta \boldsymbol{x}_{n+1} \triangleq \delta \boldsymbol{x}(t_{n+1})$, $\delta u_n \triangleq \delta u(t_n)$, $\delta u_{n+1} \triangleq \delta u(t_{n+1})$, $\delta \boldsymbol{A}_n^* \triangleq \delta \boldsymbol{A}^*(t_n)$, $\delta \boldsymbol{B}_{n+1}^* \triangleq \delta \boldsymbol{B}^*(t_{n+1})$, and $\boldsymbol{I}$ is a 6-by-6 identity matrix.

With the convexification of all the non-convex terms above, the optimal control problem for Mars entry tracking guidance becomes:

$$\text{P12 - 2: minimize } J = \frac{1}{2}\delta \boldsymbol{x}_{N+1}^T \boldsymbol{P}_f \delta \boldsymbol{x}_{N+1} + \frac{1}{2}\sum_{n=1}^{N}(\delta \boldsymbol{x}_n^T \boldsymbol{Q} \delta \boldsymbol{x}_n + \delta u_n^T R \delta u_n)$$

$$\text{subject to Eqs. (12.13) and (12.15)}$$

$$\delta \boldsymbol{x}(t_1) = \delta \boldsymbol{x}_1, \; u_{\min} \leq u_n \leq u_{\max},$$

$$\|\delta \boldsymbol{x}_n\| \leq \Delta_1, \; \|\delta u_n\| \leq \Delta_2$$

where $u_n \triangleq \delta u_n + u_n^*$. Note that two additional 2-norm constraints on state error $\delta \boldsymbol{x}_n$ and control correction δu_n are included in P12-2, except for the necessary path constraints, dynamics constraints, initial state conditions, and control constraints. Here, the two trust-region constraints are applied to restrain the search space of optimization in the neighborhood of the reference solution, so as to ensure the effectiveness of the first-order linear approximation to the original nonlinear system. Otherwise, the solution of an unbounded optimization problem is not an approximate solution to the original nonlinear tracking guidance problem.

By comparing the forms of problem P12-2 and QCQP, it is clear that P12-2 is not only a QCQP problem but also a convex problem because of the non-negative quadratic coefficients $\boldsymbol{P}_f$, $\boldsymbol{Q}$ and R. That is to say, the tracking guidance problem is approximated by a convex QCQP problem and can be solved in polynomial time via the interior-point method. However, the classic trapezoidal rule leads to a large-scale optimization problem because a large number of discrete nodes should be set to reduce the step size of integration to guarantee the numerical accuracy of Eq. (12.15). Otherwise, the accumulation of linearization and discretization errors may result in inaccurate linear approximation to the nonlinearity of the original tracking guidance problem.

12.2.2 Pseudospectral Discretization

To increase the computational efficiency and numerical accuracy, we introduce the Legendre–Gauss-Radau (LGR) pseudospectral method for discretization and quadrature, which can guarantee the numerical accuracy and reduce the scale of the transcribed numerical optimization problem [208]. The standard and flip types are two variants of the LGR pseudospectral method. In the standard LGR method, the N LGR collocation points $-1 = \tau_1, \tau_2, \ldots, \tau_N < 1$ are the zeros of the following equation:

$$P_N(\tau) + P_{N-1}(\tau) = 0, \quad P_N(\tau) = \frac{1}{2^N N!} \frac{d^N}{d\tau^N} \left[\left(\tau^2 - 1 \right)^N \right] \tag{12.16}$$

where $P_N(\tau)$ is Nth order Legendre polynomial. As for the flip LGR pseudospectral method, collocation points are the roots of the following equation:

$$P_N(\tau) - P_{N-1}(\tau) = 0 \tag{12.17}$$

The standard method is adopted in this paper. Therefore, the N-th order interpolating polynomials $\delta \boldsymbol{X}(\tau) \in \mathbb{R}^6$ and the $(N-1)$th order interpolating polynomials $\delta U(\tau) \in \mathbb{R}$ are used to approximate the state error $\delta \boldsymbol{x}(\tau)$ and control correction $\delta u(\tau)$ respectively as follow:

$$\delta \boldsymbol{x}(\tau) \approx \delta \boldsymbol{X}(\tau) = \sum_{i=1}^{N+1} \delta \boldsymbol{X}(\tau_i) L_i(\tau), \quad L_i(\tau) = \prod_{j=1, j \neq i}^{N+1} \frac{\tau - \tau_j}{\tau_i - \tau_j} \tag{12.18}$$

$$\delta u(\tau) \approx \delta U(\tau) = \sum_{i=1}^{N} \delta U(\tau_i) \tilde{L}_i(\tau), \quad \tilde{L}_i(\tau) = \prod_{j=1, j \neq i}^{N} \frac{\tau - \tau_j}{\tau_i - \tau_j} \tag{12.19}$$

where $L_i(\tau)$ and $\tilde{L}_i(\tau)$ are the base functions of the Lagrange interpolating polynomial $\delta \boldsymbol{X}(\tau)$ and $\delta U(\tau)$, respectively. These two functions are defined below:

$$L_i(\tau_n),\ \tilde{L}_i(\tau_n) = \begin{cases} 1 & n = i \\ 0 & n \neq i \end{cases} \tag{12.20}$$

In addition, the state error and control correction hold $\boldsymbol{X}(\tau_i) = \boldsymbol{x}(\tau_i)$ and $U(\tau_i) = u(\tau_i)$ at interpolation points $\{\tau_1, \ldots, \tau_N, \tau_{N+1}\}$ and $\{\tau_1, \tau_2, \ldots, \tau_N\}$, respectively.

Then, the derivative of state error are obtained by differentiating the state error in Eq. (12.18):

$$\delta \dot{\boldsymbol{x}}(\tau) \approx \delta \dot{\boldsymbol{X}}(\tau) = \sum_{i=1}^{N+1} \delta \boldsymbol{X}(\tau_i) \dot{L}_i(\tau), \quad i = 1, \ldots, N+1 \tag{12.21}$$

where the derivative of the Lagrange interpolating polynomials at the LGR collocation points are determined offline as follows:

$$D_{ni} \triangleq \dot{L}_i(\tau_n) = \sum_{l=1, l \neq i}^{N} \frac{\prod_{j=1, j \neq i, j \neq l}^{N+1} \tau_n - \tau_j}{\prod_{j=1, j \neq i}^{N+1} \tau_i - \tau_j}, \quad n = 1, 2, \ldots, N \tag{12.22}$$

where D_{ni} forms an $N \times (N+1)$ differentiation matrix $\boldsymbol{D}$.

By substituting Eq. (12.21) into Eq. (12.7) and performing the following time domain transformation, the LTV differential system Eq. (12.7) can be translated into a set of algebraic equalities as follows:

$$\tau = \frac{2t}{t_f - t_i} - \frac{t_f + t_i}{t_f - t_i} \tag{12.23}$$

$$\sum_{i=1}^{N+1} \delta \boldsymbol{X}(\tau_i) D_{ni} = \frac{t_f - t_i}{2} [\boldsymbol{A}_n^* \delta \boldsymbol{X}(\tau_n) + \boldsymbol{B}_n^* \delta \boldsymbol{U}(\tau_n)], \quad n = 1, \ldots, N \tag{12.24}$$

Similar to the discretization and integration via the trapezoidal rule, the resulting equalities are remained linear. Consequently, the convex QCQP problem can be defined as follows by introducing the Gauss-Radau quadrature:

$$\text{P12 - 3:}\ \text{minimize}\ \ J=\frac{1}{2}\delta\boldsymbol{X}_{N+1}^{T}\boldsymbol{P}_{f}\delta\boldsymbol{X}_{N+1}+\frac{1}{2}\sum_{n=1}^{N}\left[\omega_{n}\left(\delta\boldsymbol{X}_{n}^{T}\boldsymbol{Q}\delta\boldsymbol{X}_{n}+\delta U_{n}^{T}R\delta U_{n}\right)\right]$$

$$\text{subject to Eqs. (12.13) and (12.15)}$$

$$\delta\boldsymbol{X}(t_1)=\delta\boldsymbol{X}_1,\ u_{\min}\leq U_n\leq u_{\max},$$

$$\|\delta\boldsymbol{X}_n\|\leq\Delta_1,\ \|\delta U_n\|\leq\Delta_2$$

where $U_n=\delta U_n+u_n^*$, and ω_n is the Gauss-Radau quadrature weight at collocation points τ_n, which is given by:

$$w_n=\frac{1}{(N+1)^2}\frac{1-\tau_n}{\dot{P}_N^2(\tau_n)}. \tag{12.25}$$

12.3 Implementation of Algorithm

After discretization and linearization, the constrained nonlinear tracking guidance problem is cast as a convex QCQP problem. However, two key parametric issues should be emphasized and discussed before the implementation of the QCQP-based tracking guidance. In the following sections, the two QCQP-based tracking guidance laws described in Sects. 12.1.1 (P12-2) and 12.1.2 (P12-3) are referred to as U-QCQP and P-QCQP according to their uniform and pseudospectral discretization strategies, respectively.

12.3.1 Guidance Cycle

In the LQR method, the unconstrained optimal control problem is solved at each guidance time t_i. As the guidance process moves forward, t_i increases by a guidance cycle Δt in each step. Likewise, this strategy can also be used in both QCQP-based tracking guidance laws. However, solving the QCQP problem in each guidance cycle will significantly increase the computational effort, although the QCQP problem can be resolved in polynomial time. Therefore, model predictive control (MPC) is introduced to carry out the tracking guidance in order to reduce the computation of the QCQP-based algorithm.

In the generic framework of MPC shown in Fig. 12.1, there are three key elements, including model, optimizer, and plant. The model describing a real-world system is used to predict the system state from the current step to a future time, where the future timeslot is called the prediction horizon. Then, the optimizer generates the control actions in the prediction horizon according to the predicted state, the current state, and the optimization algorithm. These control actions are then fed into the plant, which

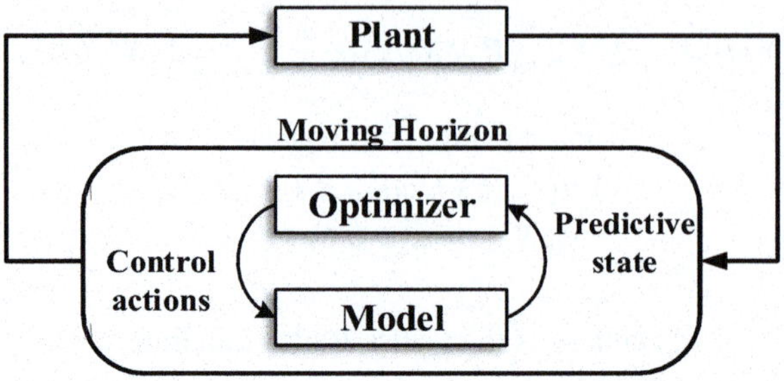

Fig. 12.1 Generic framework of MPC

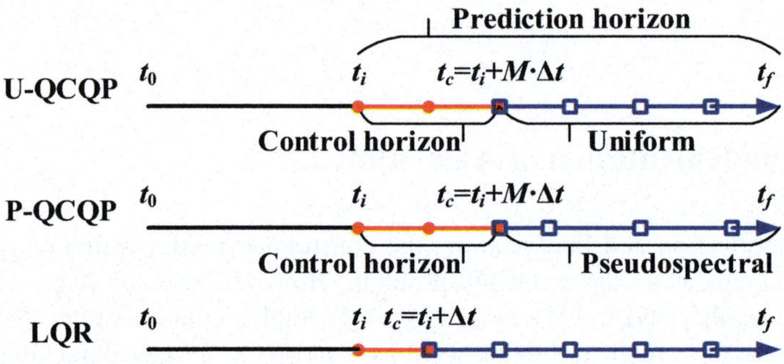

Fig. 12.2 Logic of guidance cycle for three guidance laws

will update the system state over the control horizon. At last, the aforementioned steps are implemented periodically with the shifting of the horizon.

For the QCQP-based guidance law, the prediction and control timeslots of the LQR, U-QCQP, and P-QCQP guidance laws are shown in Fig. 12.2. In both U-QCQP and P-QCQP, the prediction horizon is defined as the timeslots in the time interval $[t_i, t_f]$, and the control horizon $t_c - t_i$ is equal to M guidance cycles. Therefore, in every M guidance cycles, the convex QCQP problem is solved via IPM. Additionally, P-QCQP discretizes the system in the following prediction horizon using a non-uniform pseudospectral mesh as opposed to U-QCQP, which uses a uniform mesh. At the same time, the LQR is implemented once per guidance cycle.

In LQR and U-QCQP, the prediction horizon $t_f - t_i$ is discretized at a series of uniform points, and the discretization mesh number N is determined by the guidance cycle Δt as follows:

$$N = \left\lfloor \frac{t_f - t_i}{\Delta t} \right. \tag{12.26}$$

where the operator $\lfloor \cdot$ means rounding toward negative infinity. Under this circumstance, the number of discretization meshes is decreased by M after each optimization process. However, the number of the pseudospectral collocation points should match the prediction horizon for the P-QCQP method. In fact, too many collocation points will correspondingly increase the order of interpolation polynomials $\delta \boldsymbol{X}(\tau)$ and $\delta U(\tau)$, which in turn leads to the oscillation of the polynomial's value at the neighborhood interval of the two boundary points because the pseudospectral collocations are dense at both ends and sparse in the middle. As a result, the initial pseudospectral mesh number should be chosen carefully. Also, the mesh number should decrease by one after solving k QCQP problems to match the length of the prediction horizon. Otherwise, the numerical accuracy of the pseudospectral method cannot be guaranteed due to unnecessary oscillations. The values of parameters M and N in U-QCQP and P-QCQP will be given in Sect. 12.4.

12.3.2 Improvements of Objective Function

In addition to the guidance cycle, the objective function also has significant influence on the performance of the proposed tracking guidance algorithms. First, the quadratic coefficients $\boldsymbol{P}_f$, $\boldsymbol{Q}$, and R should be selected with care. In general, these coefficients are given by diagonal matrices, where the values of the diagonal elements represent the weights of state variables. For instance, $\boldsymbol{P}_f = 50\boldsymbol{I}$, $\boldsymbol{Q} = \boldsymbol{I}$, and $R = 1$ are adopted in an Earth entry tracking guidance scenario [93]. This combination of parameters allows the optimization problem to minimize the tracking error for each variable equally. However, the Mars entry mission focuses on the longitudinal and latitudinal errors and tends to minimize the horizontal position errors. Therefore, the four diagonal elements associated with the variables r, V, γ, and ψ are set to zero. The choice of these parameters has a significant effect on the guidance performance, which will be illustrated with simulation results in Sect. 12.4.

In addition, note that only the constraints on the magnitude and increment of the bank angle are considered in the formulated QCQP problem. However, if the perturbation is strong, the algorithm will generate a large control correction to reduce the reference trajectory tracking error, which results in a large difference between the two bank angles at the current and previous moments. In this case, the bank angle profile will show a sawtooth shape, and the magnitude of the bank-angle rate may exceed the tolerated maximum value allowed by the control system. To mitigate this issue, a penalty term on the control variable is added to the objective functions in problems P12-2 and P12-3:

P12-2:

$$J = \frac{1}{2}(u_1 - \overline{u})^T P(u_1 - \overline{u}) + \frac{1}{2}\delta \boldsymbol{x}_{N+1}^T \boldsymbol{P}_f \delta \boldsymbol{x}_{N+1} + \frac{1}{2}\sum_{n=1}^{N} (\delta \boldsymbol{x}_n^T \boldsymbol{Q} \delta \boldsymbol{x}_n + \delta u_n^T R \delta u_n) \tag{12.27}$$

P12-3:

$$J = \frac{1}{2}(U_1 - \overline{u})^T P(U_1 - \overline{u}) + \frac{1}{2}\delta \boldsymbol{X}_{N+1}^T \boldsymbol{P}_f \delta \boldsymbol{X}_{N+1} + \frac{1}{2}\sum_{n=1}^{N} \left[\omega_n \left(\delta \boldsymbol{X}_n^T \boldsymbol{Q} \delta \boldsymbol{X}_n + \delta U_n^T R \delta U_n\right)\right] \tag{12.28}$$

where $\overline{u}$ is the last control variable from the previous optimization step. The added term only suppresses the possible discontinuity between the QCQP problems solved in two consecutive steps, while the difference $u_{i+1} - u_i$ or $U_{i+1} - U_i, i = 1, \ldots, N$ is not considered because the Lagrange interpolation already ensures the smoothness of the control increment δu in the prediction horizon. This statement also indicates that the P-QCQP algorithm could have a smoother control profile than that of the U-QCQP. The reason is that the polynomials δX and δU obtained by the Lagrange interpolation are at least N-1 times continuously differentiable, while the state and control variables approximated by the trapezoidal rule are piecewise linear.

12.3.3 Synthesized Tracking Guidance Algorithm

Compared with the LQR, the two QCQP-based tracking guidance algorithms have the following advantages:

(1) Since the reference trajectory tracking guidance problem is approximated as a convex QCQP problem, the optimal control increment of the QCQP problem is an approximate solution of the original optimal control problem.
(2) Solving the QCQP problem is essentially solving a constrained optimal control problem, while the LQR method does not consider the path constraints and essentially solves an unconstrained optimal control problem.
(3) The QCQP-based guidance algorithm can be implemented online because it only solves the QCQP problem once every M guidance cycles, in contrast to the LQR method, which solves the optimization problem once at each guidance cycle.
(4) Different from the U-QCQP, the P-QCQP uses the LGR pseudospectral method instead of the trapezoidal rule for system discretization and integration, which

Table 12.1 Procedure of the QCQP-based tracking guidance laws

Step	Procedure
1	Initialize the algorithm with reference trajectory $(\boldsymbol{x}^*, u^*)$, guidance cycle Δt, control horizon length $M \cdot \Delta t$, discretization mesh number N in the prediction horizon $[t_i, t_f]$, and quadratic objective coefficients $\boldsymbol{P}_f$, $\boldsymbol{Q}$, and R
2	For P-QCQP, if $\text{mod}(t_i - t_0, k \cdot M) \equiv 0$ and $t_i \neq t_0$, then $N := N - 1$ For U-QCQP, if $t_i \neq t_0$, then $N := N - M$
3	Solve the QCQP problem P12-2 or P12-3 to obtain the feedback control corrections δu over the prediction horizon $[t_i, t_f]$
4	Input the control corrections δu over control horizon $[t_i, t_c]$ along with the reference control u^* to perform M guidance cycles under possible uncertainties and disturbances
5	If $t_c < t_f$, then let $t_i := t_c$, $t_c := t_c + M \cdot \Delta t$ (or $t_c := t_f$ when $t_c + M \cdot \Delta t > t_f$), and go to Step 2. Otherwise, the guidance process terminates

reduces the scale of the transcribed numerical optimization problem and improves the real-time performance of the algorithm.

To sum up, the QCQP-based optimal feedback tracking guidance laws consider the constraints, and the P-QCQP is superior to the U-QCQP in terms of computational efficiency. The procedure of the two QCQP-based guidance laws is summarized in Table 12.1.

12.4 Numerical Simulation

12.4.1 Simulation Set-Ups

To validate the performance of our developed QCQP-based tracking guidance law, we compare LQR, U-QCQP, P-QCQP, and ETPC algorithms through numerical simulation in this section. In U-QCQP and P-QCQP, the modeling toolbox YALMIP [130] and solver MOSEK [131] are utilized to solve the QCQP problem. As for the LQR, the MATLAB built-in function *lqr* is used. In the simulation, the parameters for Mars entry are given by: $R_0 = 3397.2$ km, $g_0 = 3.7114$ m/s^2, $\omega_0 = 7.0882 \times 10^{-5}$ rad/s, $\rho_0 = 0.0158$ kg/m^3, $h_s = 9354.5$ m, $C_D = 1.45$, $C_L = 0.36$, $m = 2804$ kg, $S_r = 15.9$ m^2, $k_Q = 1.9027 \times 10^{-4}$ kg$^{0.5}$s$^{0.15}$/m$^{1.15}$, $R_n = 6.476$ m, $\dot{Q}_{\max} = 70$ W/cm^2, $q_{\max} = 8.5$ kPa, $a_{\max} = 18.5\ g_0$, and the boundary conditions are shown in Table 12.2.

$$\tilde{\rho} = \rho(1 + \delta\rho) \quad \delta\rho = F(h)A(h) \tag{12.29}$$

Table 12.2 Boundary conditions for Mars entry simulation

Parameters	Initial value	Final value	Tolerated interval
h (km)	125	10	[0, 150]
θ (°)	−90	−70	[−180, 180]
ϕ (°)	−45	−41	[−180, 180]
V (m/s)	5500	Free	[0, 6000]
γ (°)	−13.5	Free	[−80, 80]
ψ (°)	85	Free	[−180, 180]
σ (°)	Free	Free	[−80, 80]

Note that the exponential model cannot accurately characterize the density of the Martian atmosphere, so the dispersion of the atmosphere density considered in the simulation is modeled as follows [82]

$$\begin{aligned} F(h) = a_1 + a_2 \sin[(h/1000 - a_3)/a_4] \\ + b_2 \cos^2[(h/1000 - b_3)/b_4] \\ + c_2 \cos^3[(h/1000 - c_3)/c_4] \end{aligned} \tag{12.30}$$

$$A(h) = d_2 \exp[(h/1000 - d_3)/40] \tag{12.31}$$

where parameters $a_1, a_2, b_2, c_2 \sim$ U(0, 1), $a_3, b_3, c_3 \sim$ U(30, 90), $a_4, b_4, c_4 \sim$ U(10, 30), $d_2 \sim$ N(0.36, 0.006^2), $d_3 \sim$ N(120, 5^2),h in meters represents the height, ρ represents the atmospheric density calculated according to the exponential density model in Sect. 12.1.1, $\delta\rho$ and $\tilde{\rho}$ are the dispersion and actual value of atmospheric density used in numerical simulation, respectively. The dispersion of Martian atmospheric density with respect to the height h is presented in Fig. 12.3.

The distribution types and values of other uncertainties, including the initial state and biases of aerodynamic coefficients, are shown in Table 12.3.

12.4.2 Monte-Carlo Campaign

In the simulation, the specific flow chart of QCQP-based tracking guidance is shown in Fig. 12.4. The guidance cycle Δt is set to 0.5 s in simulation. For the Mars entry mission, the trajectory control capability is limited due to the thin atmosphere at high altitude. Therefore, all algorithms start only when the drag acceleration exceeds 0.2 times the Earth's surface acceleration $g_E = 9.80665$ m/s^2.

In addition, the reference trajectory of a total of 360 s of flight for tacking guidance is optimized by using GPOPS [119]. To analyze the influence of quadratic term weight on guidance performance, cases 0 and 1 are set up and simulated in this section. In case 0, the coefficient matrices $\boldsymbol{Q} = diag([0, 10, 10, 0, 0, 0])$ and $R = 1$ are used in LQR. As for the QCQP-based tracking guidance law, $\boldsymbol{P}_f = diag([0, 10, 10, 0, 0, 0])$,

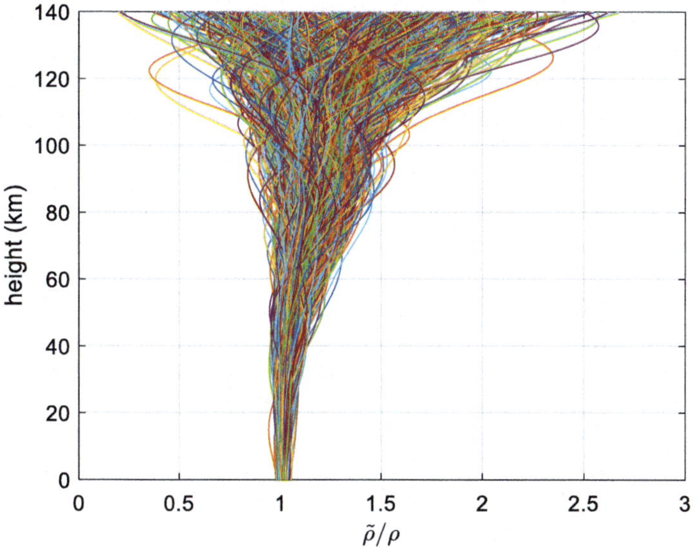

Fig. 12.3 The dispersion of Martian atmospheric density in simulation

Table 12.3 Uncertainties for Mars entry simulation

Initial state error	Distribution	3σ or min/max value
δh_0 (km)	Normal	0.2
$\delta\theta_0$ (°)	Normal	0.05
$\delta\phi_0$ (°)	Normal	0.05
δV_0 (m/s)	Normal	20
$\delta\gamma_0$ (°)	Normal	0.05
$\delta\psi_0$ (°)	Normal	0.05
$\delta C_L, \delta C_D$	Uniform	−0.5, 0.5

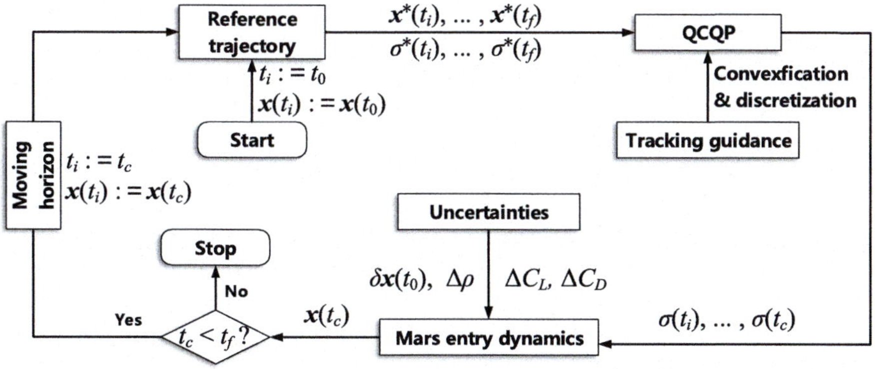

Fig. 12.4 The flow chart of QCQP-based tracking guidance

$\boldsymbol{Q} = diag([0, 1, 1, 0, 0, 0])$, $R = 1$, $M = 10$, $\Delta_1 = 0.6$, and $\Delta_2 = 0.12$ are used. Furthermore, we use $k = 8$ and initial $N = 15$ for the P-QCQP while initial $N = 360/\Delta t = 720$ for the U-QCQP. In other words, the QCQP problem is solved every 80 guidance cycles in P-QCQP and every 10 guidance cycles in U-QCQP, and the control horizon is 5 s. In case 1, we use $\boldsymbol{Q} = 10\boldsymbol{I}$ for LQR, $\boldsymbol{P}_f = 10\boldsymbol{I}$, $\boldsymbol{Q} = \boldsymbol{I}$ for both U-QCQP and P-QCQP. Other parameters in case 1 remain the same as those in case 0. With the aforementioned configurations and problem settings, 500 Monte-Carlo simulations are conducted in this section. All simulation results are shown in Figs. 12.5, 12.6, 12.7, 12.8, 12.9 and 12.10. However, it should be noted that the simulation results for both cases are very close, therefore, we show the full results for case 0 and partial results for case 1.

As shown in Fig. 12.5, the Monte-Carlo simulation results show that the profiles of the state variables obtained by all the four methods are close to the reference profile produced from GPOPS. Except for the LQR, all other methods in case 0 keep the final location, as shown in Fig. 12.6a, within a 5 km radius circle around the target location. Furthermore, the majority of the final locations of U-QCQP, P-QCQP, and ETPC are within a 2 km radius circle around the target location. This conclusion can also be observed from the histograms in Fig. 12.6-right. Similar results can be found in Fig. 12.7 as well, although the guidance accuracy in case 1 is somewhat degraded for both U-QCQP and P-QCQP. In conclusion, the guidance accuracy of QCQP is superior to that of LQR, and the U-QCQP and P-QCQP algorithms are almost identical in terms of the solution accuracy.

Both LQR and QCQP-based tracking guidance laws are established using the small disturbance assumption. In these two types of algorithms, the control correction is obtained by solving the algebraic Riccati equation and a QCQP problem, respectively. Therefore, the guided trajectory is always maintained near the reference trajectory, thus ensuring the effectiveness of the small disturbance assumption. However, the objective function of LQR mainly minimizes $\sum_{n=1}^{N} \delta\boldsymbol{X}_n^T \boldsymbol{Q} \delta\boldsymbol{X}_n$ rather than the final state error $\delta\boldsymbol{X}_{N+1}^T \boldsymbol{P}_f \delta\boldsymbol{X}_{N+1}$ as in the QCQP, so the final miss distance in LQR is larger. As for the ETPC algorithm, it eliminates the final downrange error by applying control correction, so the guidance accuracy is the best among the four algorithms.

Further, the final speed and final altitude are two critical parameters in guidance performance assessment and should be constrained to be within proper intervals to ensure the pressure required for the parachute deployment. As shown in Table 12.2, the Mars entry ends at an altitude of 10 km. However, from Fig. 12.10, it can be seen that the final altitude is influenced by the values of the quadratic coefficients $\boldsymbol{P}_f$ and $\boldsymbol{Q}$. In case 0, the objective function focuses on the latitudinal and longitudinal accuracy and does not consider the errors of the remaining variables. As a result, this strategy leads to a scattered final altitude profile, although the QCQP-based guidance law has high accuracy in the final location. On the contrary, case 1 has a lower location accuracy; however, the final altitude values are around the reference altitude of 10 km. By contrast, the ETPC minimizes the downrange error of the entry vehicle at the reference altitude, therefore, the final altitude values are close to 10 km. These discussions are consistent with the statistical results given in Table 12.4.

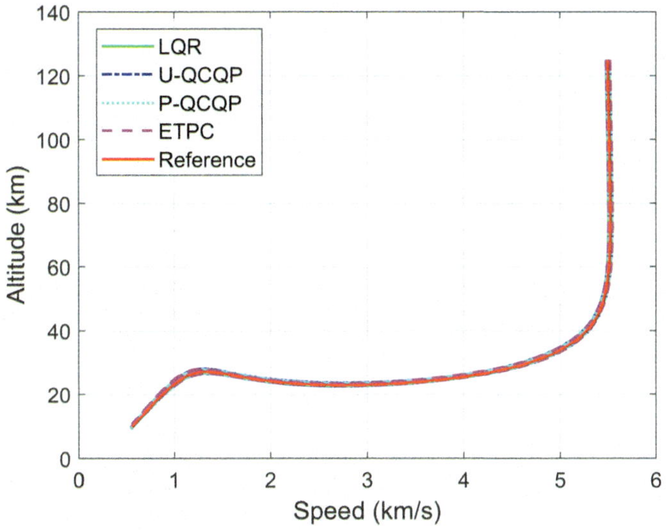

(a) The speed-altitude profiles

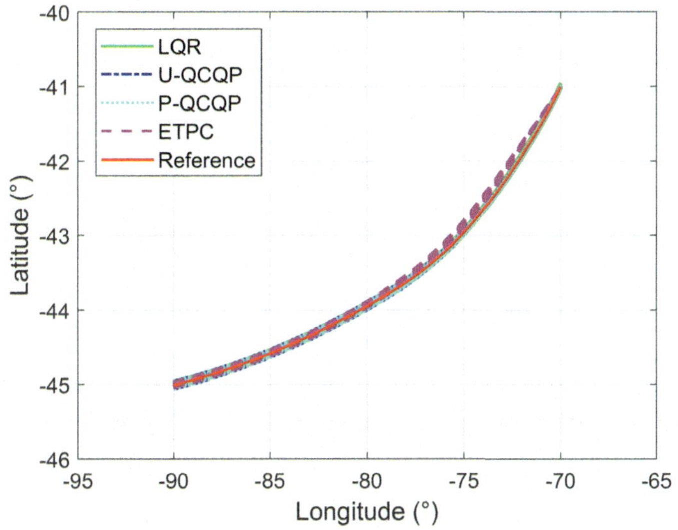

(b) The longitude-latitude profiles

Fig. 12.5 The state variables and path constraints profiles of four guidance laws (case 0)

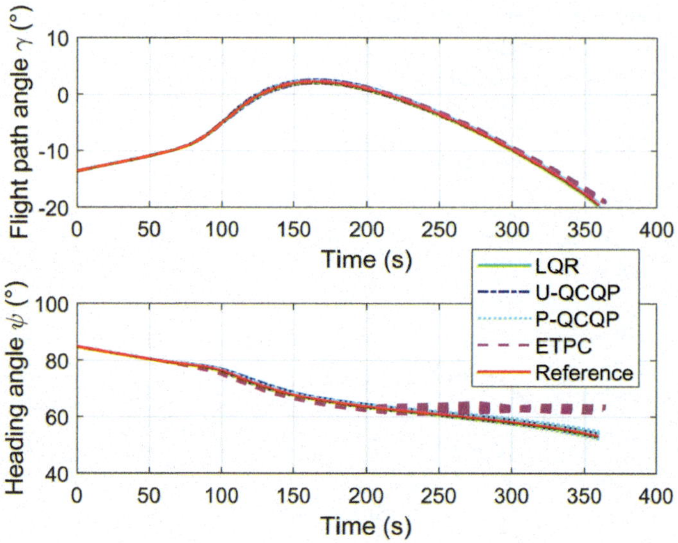

(c) The flight path angle and heading angle profiles

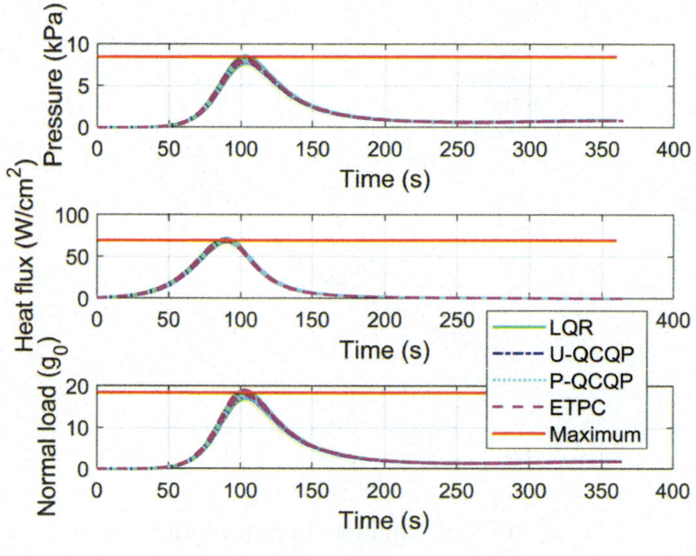

(d) The path constraints profiles

Fig. 12.5 (continued)

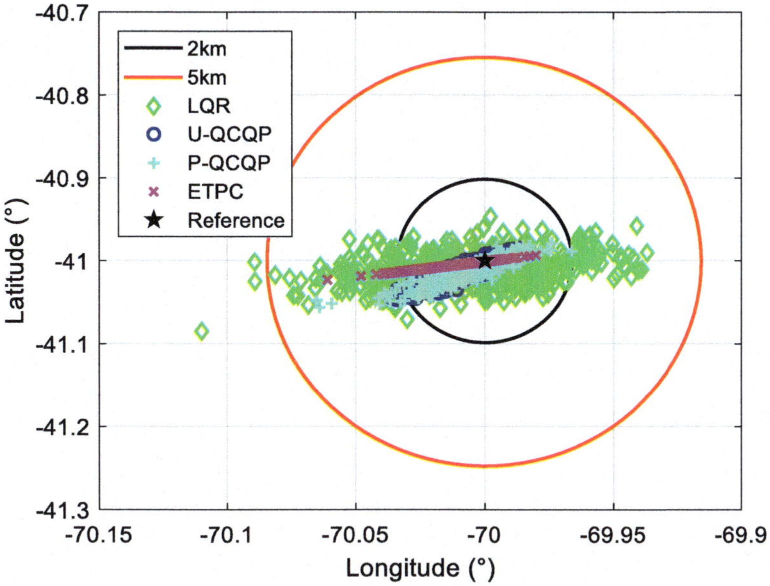

(a) The final location

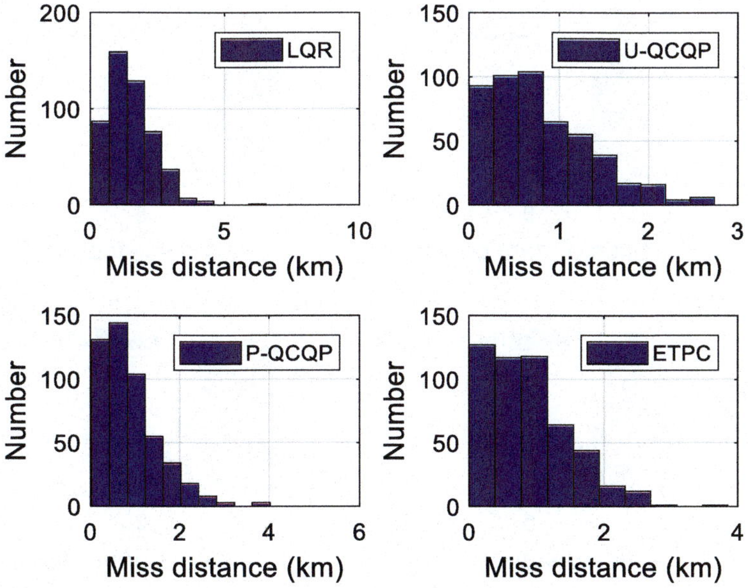

(b) The miss distance histogram

Fig. 12.6 The final accuracy of four guidance laws (case 0)

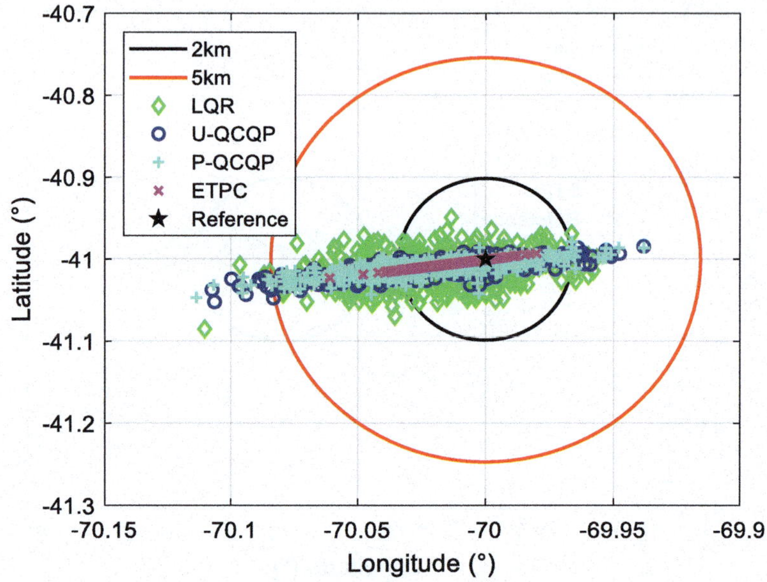

(a) The final locations

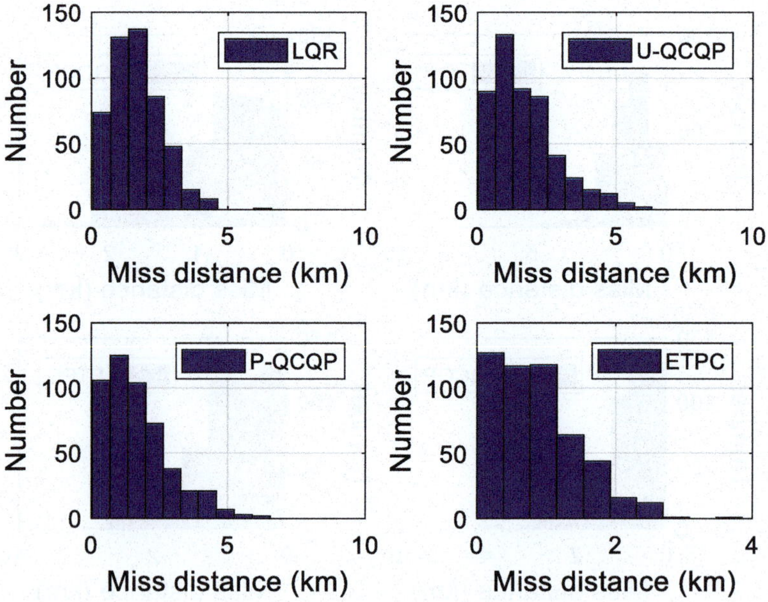

(b) The miss distance histogram

Fig. 12.7 The final accurayc of four guidance laws (case 1)

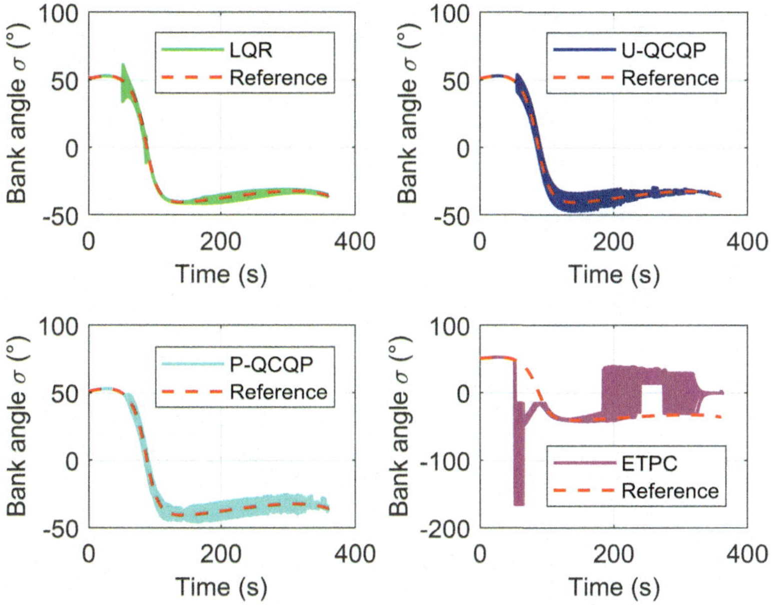

Fig. 12.8 The bank angle profiles of four guidance laws (case 0)

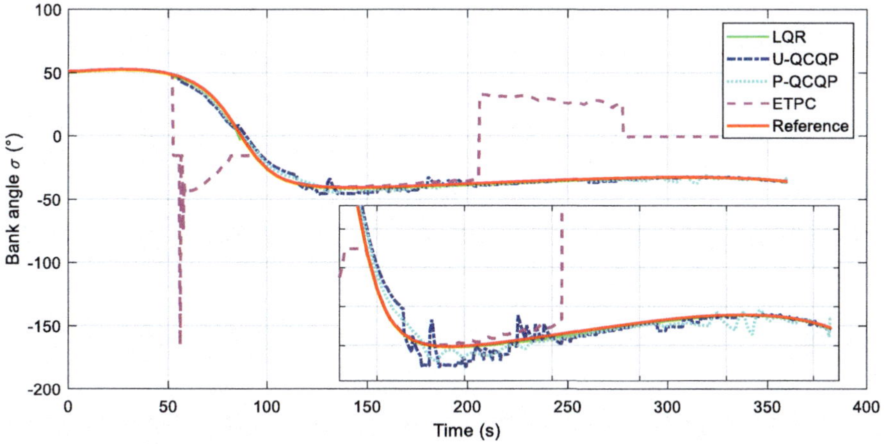

Fig. 12.9 The bank angle profiles of four guidance laws in one random simulation (case 0)

Table 12.4 not only indicates the accuracy of the four types of guidance methods but also gives the number of trajectories that violate the path constraints in the 500 Monte-Carlo simulations. Combining the results in Fig. 12.5d, it can be seen that the numbers of constraint violations of the QCQP-based and LQR guidance laws are smaller than that of the ETPC. In fact, by keeping the guided trajectory near

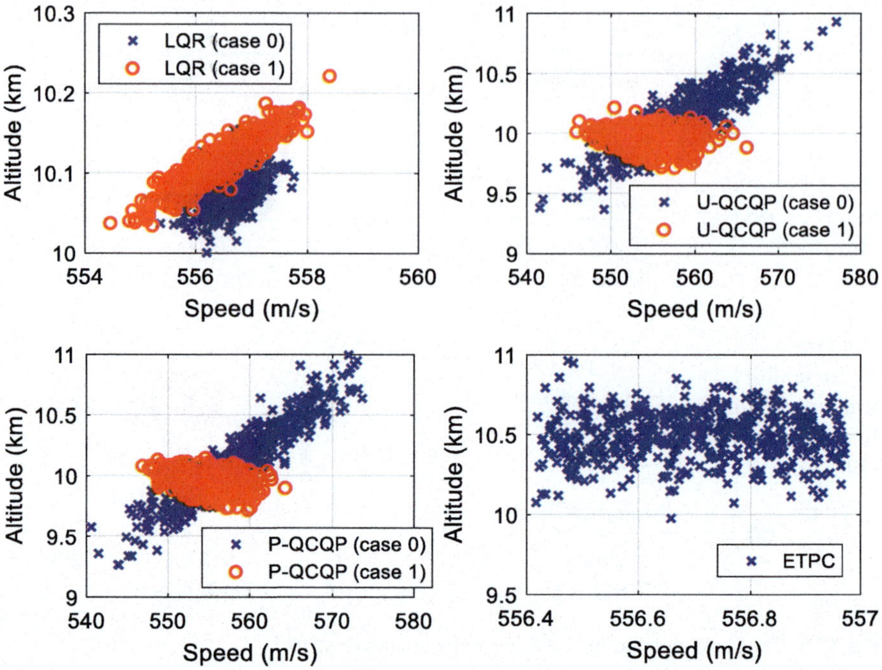

Fig. 12.10 The final speed-altitude profiles of four guidance laws in cases 0 and 1

Table 12.4 Statistical results of final locations

Parameters	LQR		U-QCQP		P-QCQP		ETPC
	Case 0	Case 1	Case 0	Case 1	Case 0	Case 1	
Mean (m)	1507.5	1669.9	808.5	1722.4	886.2	1698.0	875.6
Stand deviation (m)	840	922.9	562.1	1174.5	640.2	1190.5	613.6
Maximum (m)	6559.7	6588.7	2733.1	6308.9	4017.8	6543.7	3850.8
Minimum (m)	60.6	59.3	1.5	38.4	25.4	48.5	3.5
Violation	146	141	112	113	105	115	198

the reference trajectory, the LQR and QCQP algorithms have a higher probability of satisfying the path constraints. It should also be noted that although the QCQP-based method takes the path constraint into account, the guided trajectory may still violate the path constraints due to the large dispersion in atmospheric density and the biases of aerodynamic coefficients, although there is no notable reduction in the number of constraints violations, which is 146, 112, and 105 (or 141, 113, and 115) for LQR, U-QCQP, and P-QCQP, respectively. In addition, the ETPC based on an analytical prediction-correction process does not require maintaining the guided trajectory near the reference trajectory, and the path constraints are not considered in the ETPC guidance algorithm, resulting in 198 violations in our simulations.

In terms of the performance of the control profile, P-QCQP generates a smoother control profile and requires a smaller bank angle rate, as shown in Figs. 12.8 and 12.9. However, the result of U-QCQP has more jitters. In addition, Fig. 12.8 indicates that a smaller control increment is needed in the QCQP-based method than in LQR when the guidance system switches to the algorithm from the reference solution. In fact, although the ETPC guidance method has the best guidance accuracy, frequent bank reversals are performed to maintain the crossrange error within the allowable corridor. These reversals necessitate a high servo performance of the control actuator. Otherwise, the guidance accuracy will decrease. In summary, the QCQP-based method is better than other methods in terms of guidance accuracy and control profile smoothness.

In addition, the computational efficiency of the two types of QCQP-based tracking guidance laws is investigated for potential onboard applications. As stated in Sect. 12.2, the pseudospectral discretization reduces the number of optimization variables compared to the uniform discretization, and the model predictive control framework allows the algorithm to solve the QCQP problem only once every M guidance cycles. Therefore, the mean CPU times of 500 Monte-Carlo simulations consumed in U-QCQP and P-QCQP are shown in Fig. 12.11 to quantitatively compare their computation costs. It is obvious that the CPU times of P-QCQP are less than those of U-QCQP, although the computation cost of U-QCQP decreases when the prediction horizon is shrinking. Of course, there is room for further improvement in the computational efficiency of the algorithm, as the algorithm is currently executed in the time-consuming MATLAB environment.

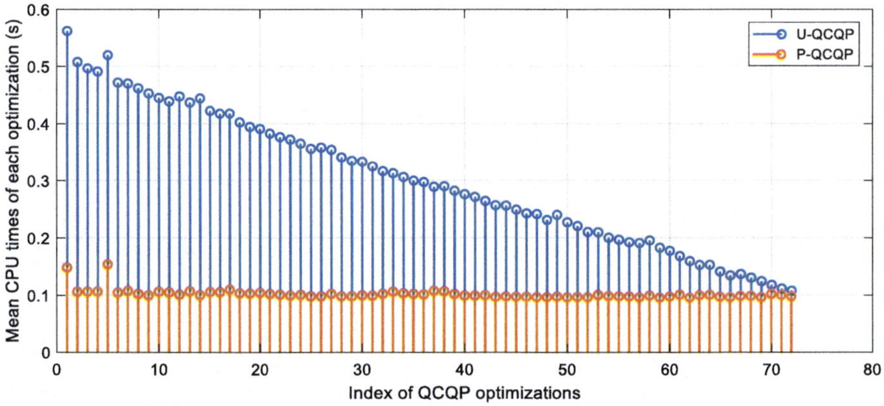

Fig. 12.11 Mean CPU times of U-QCQP and P-QCQP in Monte-Carlo simulations

12.5 Discussions

In this chapter, the pseudospectral discretization and modified objective function with penalty terms are used to develop a new P-QCQP entry tracking guidance algorithm to fundamentally reduce the scale of the problem, improve the computational efficiency, and smoothen the control profile based on the original U-QCQP algorithm. Compared with the existing LQR and ETPC algorithms, the QCQP-based method is less sensitive to the initial state error, atmospheric density perturbation, and aerodynamic coefficient bias because it takes path constraints into account. Furthermore, simulation results demonstrate that the quadratic coefficients in the objective function have obvious impacts on the guidance accuracy of the P-QCQP method. Therefore, these coefficients should be properly set according to the specific requirements of the entry mission. The results demonstrate the potential of the improved P-QCQP algorithm for real-time onboard applications.

References

1. https://en.wikipedia.org/wiki/List_of_missions_to_the_Mars#cite_note-Perminov1-8. Retrieved 10 Jan. 2023.
2. Soffen G A, Snyder C W. The first Viking mission to Mars[J]. Science, 1976, 193(4255): 759-766.
3. Spencer D A, Blanchard R C, Braun R D, et al. Mars Pathfinder entry, descent, and landing reconstruction[J]. Journal of Spacecraft and Rockets, 1999, 36(3): 357-366.
4. Desai P N, Knocke P C. Mars exploration rovers entry, descent, and landing trajectory analysis[J]. The Journal of the Astronautical Sciences, 2007, 55(3): 311-323.
5. Grover III M R, Cichy B D, Desai P N. Overview of the Phoenix entry, descent, and landing system architecture[J]. Journal of Spacecraft and Rockets, 2011, 48(5): 706-712.
6. Mendeck G F, Craig McGrew L. Entry guidance design and postflight performance for 2011 Mars Science Laboratory mission[J]. Journal of Spacecraft and Rockets, 2014, 51(4): 1094-1105.
7. Maddock R W, Cianciolo A M D, Korzun A M, et al. InSight Entry, Descent, and Landing Postflight Performance Assessment[J]. Journal of Spacecraft and Rockets, 2021, 58(5): 1530-1537.
8. Nelessen A, Sackier C, Clark I, et al. Mars 2020 entry, descent, and landing system overview[C]//2019 IEEE Aerospace Conference. IEEE, 2019: 1-20.
9. Chicarro A, Martin P, Trautner R. The Mars Express mission: an overview[J]. In: Mars Express: the scientific payload. Ed. by Andrew Wilson, scientific coordination: Agustin Chicarro. ESA SP-1240, Noordwijk, Netherlands: ESA Publications Division, ISBN 92-9092-556-6, 2004, p. 3–13, 2004, 1240: 3–13.
10. Albee A L, Arvidson R E, Palluconi F, et al. Overview of the Mars global surveyor mission[J]. Journal of Geophysical Research: Planets, 2001, 106(E10): 23291-23316.
11. Saunders R S, Arvidson R E, Badhwar G D, et al. 2001 Mars Odyssey mission summary[J]. Space Science Reviews, 2004, 110: 1-36.
12. Graf J E, Zurek R W, Eisen H J, et al. The Mars reconnaissance orbiter mission[J]. Acta Astronautica, 2005, 57(2-8): 566-578.
13. Sundararajan V. Mangalyaan-Overview and Technical Architecture of India's First Interplanetary Mission to Mars[C]//AIAA Space 2013 Conference and Exposition. 2013: 5503.
14. Bonetti D, De Zaiacomo G, Blanco G, et al. ExoMars 2016: Schiaparelli coasting, entry and descent post flight mission analysis[J]. Acta Astronautica, 2018, 149: 93-105.
15. Dutta S, Karlgaard C D, Kass D, et al. Postflight Analysis of Atmospheric Properties from Mars 2020 Entry, Descent, and Landing[J]. Journal of Spacecraft and Rockets, 2023, 60(3): 1022-1033.
16. Huang X, Li M, Wang X, et al. The Tianwen-1 guidance, navigation, and control for Mars entry, descent, and landing[J]. Space: Science & Technology, 2021, 9846185.

S. Li et al., *Trajectory Optimization and Guidance Methods for Mars Entry*,
https://doi.org/10.1007/978-981-99-6282-2

17. Beyer F, Duvet L, Karp A C, et al. Mars Sample Return Conceptual Mission Overview[R]. NASA Technical Reports Server, 20210005866.
18. Drake B G, Hoffman S J, Beaty D W. Human exploration of Mars, design reference architecture 5.0[C]//2010 IEEE Aerospace Conference. IEEE, 2010: 1–24.
19. Li S, Jiang X. Review and prospect of guidance and control for Mars atmospheric entry[J]. Progress in Aerospace Sciences, 2014, 69: 40-57.
20. Way D W, Dutta S, Zumwalt C, et al. Assessment of the Mars 2020 Entry, Descent, and Landing Simulation[C]//AIAA SciTech 2022 Forum. 2022: 0421.
21. Guo M, Huang X, Li M, et al. Adaptive entry guidance for the Tianwen-1 mission[J]. Astrodynamics, 2022, 6: 17-26.
22. Ives D, Geller D, Carman G. Apollo-derived Mars precision lander guidance[C]//23rd Atmospheric Flight Mechanics Conference. 1998: 4570.
23. Mendeck G, Carman G. Guidance design for Mars smart landers using the entry terminal point controller[C]//AIAA Atmospheric Flight Mechanics Conference and Exhibit. 2002: 4502.
24. Martin M S, Mendeck G F, Brugarolas P B, et al. In-flight experience of the Mars Science Laboratory guidance, navigation, and control system for entry, descent, and landing[J]. CEAS Space Journal, 2015, 7: 119-142.
25. Brugarolas P, San Martin A, Wong E. Attitude controller for the atmospheric entry of the Mars Science Laboratory[C]//AIAA Guidance, Navigation and Control Conference and Exhibit. 2008: 6812.
26. Brugarolas P. Guidance, navigation and control for the entry, descent, and landing of the mars 2020 mission[C]//Annual Guidance and Control Conference. 2017 (JPL-CL-CL# 17-0498).
27. Wei H, Rao W, Chen G, et al. Tianwen-1 Mars entry vehicle trajectory and atmosphere reconstruction preliminary analysis[J]. Astrodynamics, 2022, 6: 81-91.
28. Rao W, Sun Z, Dong J, et al. Design and implementation of the Mars entry, descent, and landing system for the Tianwen-1 mission[J]. SCIENTIA SINICA: Technologica, 2022, 52(8), 1162-1174. (In Chinese)
29. Li M, Huang X, Xu C, et al. Autonomous Navigation Technology of Tianwen-! Mars Probe During EDL Process [J]. Journal of Astronautics, 2022, 43(1): 11-20. (In Chinese)
30. Chen Z, Cui H, Tian Y, et al. Tianwen-1 Lander Entry, Descent and Landing Modeling and Simulation[J]. Journal of Astronautics, 2022, 43(1): 91-102. (In Chinese)
31. Zhao Y, Wang X, Huang X, et al. Tianwen-1 lander guidance navigation and control system for Mars soft landing[J]. Aerospace Control and Application, 2021, 47(5): 48-57. (In Chinese)
32. Zhao Y, Yuan L, Wang X, et al. Design and Implementation of GNC System for Entry Capsule of Tianwen-1 Probe[J]. Journal of Astronautics, 2022, 43(1): 1-10. (In Chinese)
33. Yang C, Qi Y, Rao W, et al. Design and implementation of the aerodynamic thermal protection system for the Tianwen-1 Mars probe[J]. SCIENTIA SINICA: Technologica, 2022, 52(2): 253-263. (In Chinese)
34. Bowles J V, Huynh L C, Hawke V M, et al. Mars sample return: Mars ascent vehicle mission and technology requirements[R]. NASA Technical Reports Server, 20140011316.
35. Rucker M. Emerging Technologies for Human Mars Exploration[J]. IEEE Aerospace 2018, 2018 (JSC-E-DAA-TN52796).
36. Drake B G, Hoffman S J, Beaty D W. Human exploration of Mars, design reference architecture 5.0[C]//2010 IEEE Aerospace Conference. IEEE, 2010: 1–24.
37. Craig D A, Herrmann N B, Troutman P A. The evolvable mars campaign-study status[C]// 2015 IEEE Aerospace Conference. IEEE, 2015: 1–14.
38. Polsgrove T, Dwyer-Cianciolo A M. Human mars entry, descent and landing architecture study overview[M]//AIAA SPACE 2016. 2016: 5494.
39. Percy T K, Polsgrove T, Sutherlin S, et al. Human Mars Entry, Descent, and Landing Architecture Study: Descent Systems[C]//2018 AIAA SPACE and Astronautics Forum and Exposition. 2018: 5193.
40. Cianciolo A D. Changing entry, descent, and landing paradigms for human mars lander[C]// 15th International Planetary Probe Workshop (IPPW-2018), 2018.

41. Polsgrove T, Dwyer-Cianciolo A M, Robertson E A, et al. Human Mars Entry, Descent, and Landing Architecture Study: Rigid Decelerators[C]//2018 AIAA SPACE and Astronautics Forum and Exposition. 2018: 5192.
42. Dwyer-Cianciolo A M, Polsgrove T. Human Mars entry, descent, and landing architecture study: phase 2 summary[C]//2018 AIAA SPACE and Astronautics Forum and Exposition. 2018: 5190.
43. Polsgrove T, Percy T K, Sutherlin S, et al. Human Mars Entry, Descent, and Landing Architecture Study: Deployable Decelerators[C]//2018 AIAA SPACE and Astronautics Forum and Exposition. 2018: 5191.
44. Calderon D, Garcia J A, Newton H, et al. Human Mars entry, Descent, and Landing Architecture Study: Phase 3 Summary[C]//AIAA SciTech Forum. 2020 (JSC-E-DAA-TN76588).
45. Jacob G L, Neeler G, Ramanan R V. Mars entry mission bank profile optimization[J]. Journal of Guidance, Control, and Dynamics, 2014, 37(4): 1305-1316.
46. Zheng Y, Cui H, Ai Y. Indirect trajectory optimization for mars entry with maximum terminal altitude[J]. Journal of Spacecraft and Rockets, 2017, 54(5): 1068-1080.
47. Long J, Gao A, Cui P, et al. Mars atmospheric entry guidance for optimal terminal altitude[J]. Acta Astronautica, 2019, 155: 274-286.
48. Mall K, Taheri E. Entry trajectory optimization for Mars science laboratory class missions using indirect uniform trigonometrization method[C]//2020 American Control Conference (ACC). IEEE, 2020: 4182–4187.
49. Kelly M. An introduction to trajectory optimization: How to do your own direct collocation[J]. SIAM Review, 2017, 59(4): 849-904.
50. Benson D A, Huntington G T, Thorvaldsen T P, et al. Direct trajectory optimization and costate estimation via an orthogonal collocation method[J]. Journal of Guidance, Control, and Dynamics, 2006, 29(6): 1435-1440.
51. Boyd S P, Vandenberghe L. Convex optimization[M]. Cambridge University Press, 2004.
52. Benson D. A Gauss pseudospectral transcription for optimal control[D]. Massachusetts Institute of Technology, 2005.
53. Acikmese B, Ploen S R. Convex programming approach to powered descent guidance for mars landing[J]. Journal of Guidance, Control, and Dynamics, 2007, 30(5): 1353-1366.
54. Liu X, Lu P, Pan B. Survey of convex optimization for aerospace applications[J]. Astrodynamics, 2017, 1: 23-40.
55. Liu X, Shen Z, Lu P. Entry trajectory optimization by second-order cone programming[J]. Journal of Guidance, Control, and Dynamics, 2016, 39(2): 227-241.
56. Liu X, Shen Z, Lu P. Solving the maximum-crossrange problem via successive second-order cone programming with a line search[J]. Aerospace Science and Technology, 2015, 47: 10-20.
57. Zhao D J, Song Z Y. Reentry trajectory optimization with waypoint and no-fly zone constraints using multiphase convex programming[J]. Acta Astronautica, 2017, 137: 60-69.
58. Wang Z, Grant M J. Constrained trajectory optimization for planetary entry via sequential convex programming[J]. Journal of Guidance, Control, and Dynamics, 2017, 40(10): 2603-2615.
59. Wang Z. Optimal trajectories and normal load analysis of hypersonic glide vehicles via convex optimization[J]. Aerospace Science and Technology, 2019, 87: 357-368.
60. Sagliano M, Mooij E. Optimal drag-energy entry guidance via pseudospectral convex optimization[C]//2018 AIAA Guidance, Navigation, and Control Conference. 2018: 1315.
61. Zhou X, He R, Zhang H, et al. Sequential convex programming method using adaptive mesh refinement for entry trajectory planning problem[J]. Aerospace Science and Technology, 2021, 109: 106374.
62. Wan C, Jing G, Dai R, et al. Fuel-optimal guidance for end-to-end human-mars entry, powered-descent, and landing mission[J]. IEEE Transactions on Aerospace and Electronic Systems, 2022, 58(4): 2837-2854.
63. Pei C, Wan C, Dai R, et al. A hybrid ADMM for six-degree-of-freedom entry trajectory optimization based on dual quaternions[J]. IEEE Transactions on Aerospace and Electronic Systems, 2022.

64. Grant M, Mendeck G. Mars science laboratory entry optimization using particle swarm methodology[C]//AIAA Atmospheric Flight Mechanics Conference and Exhibit. 2007: 6393.
65. Lafleur J, Cerimele C. Angle of attack modulation for Mars entry terminal state optimization[C]//AIAA Atmospheric Flight Mechanics Conference. 2009: 5611.
66. Lafleur J M, Cerimele C J. Mars entry bank profile design for terminal state optimization[J]. Journal of Spacecraft and Rockets, 2011, 48(6): 1012-1024.
67. Sorgenfrei M, Chester E. Exploration of the Mars entry, descent, and landing design space by means of a genetic algorithm[C]//AIAA Infotech@ Aerospace (I@ A) Conference. 2013: 4569.
68. Dutta S, Braun R D, Karlgaard C D. Uncertainty quantification for mars entry, descent, and landing reconstruction using adaptive filtering[J]. Journal of Spacecraft and Rockets, 2014, 51(3): 967-977.
69. Brune A J, West IV T K, Hosder S. Uncertainty quantification of planetary entry technologies[J]. Progress in Aerospace Sciences, 2019, 111: 100574.
70. Jiang X, Li S. Mars entry trajectory planning using robust optimization and uncertainty quantification[J]. Acta Astronautica, 2019, 161: 249-261.
71. Jiang X, Li S, Furfaro R, et al. High-dimensional uncertainty quantification for Mars atmospheric entry using adaptive generalized polynomial chaos[J]. Aerospace Science and Technology, 2020, 107: 106240.
72. Ye F, Jiang X, Luo Y, et al. Mixed uncertainty quantification for terminal flight state of Mars atmospheric entry[J]. Acta Astronautica, 2022, 199: 183-194.
73. Sarigul-Klijn N. Survey of planetary entry guidance algorithms[J]. Progress in Aerospace Sciences, 2014, 68: 64-74.
74. Li S, Jiang X. Review and prospect of guidance and control for Mars atmospheric entry[J]. Progress in Aerospace Sciences, 2014, 69: 40-57.
75. Quadrelli M B, Wood L J, Riedel J E, et al. Guidance, navigation, and control technology assessment for future planetary science missions[J]. Journal of Guidance, Control, and Dynamics, 2015, 38(7): 1165-1186.
76. Chai R, Tsourdos A, Savvaris A, et al. Review of advanced guidance and control algorithms for space/aerospace vehicles[J]. Progress in Aerospace Sciences, 2021, 122: 100696.
77. Tu K Y, Munir M S, Mease K D, et al. Drag-based predictive tracking guidance for Mars precision landing[J]. Journal of Guidance, Control, and Dynamics, 2000, 23(4): 620-628.
78. Leavitt J, Saraf A, Chen D, et al. Performance of evolved acceleration guidance logic for entry (EAGLE)[C]//AIAA Guidance, Navigation, and Control Conference and Exhibit. 2002: 4456.
79. Kluever C A. Entry guidance performance for Mars precision landing[J]. Journal of Guidance, Control, and Dynamics, 2008, 31(6): 1537-1544.
80. Xia Y, Chen R, Pu F, et al. Active disturbance rejection control for drag tracking in mars entry guidance[J]. Advances in Space Research, 2014, 53(5): 853-861.
81. Gong Y, Guo Y, Ma G, et al. Mars entry guidance for mid-lift-to-drag ratio vehicle with control constraints[J]. Aerospace Science and Technology, 2020, 107: 106361.
82. Zheng Y, Cui H, Ai Y. Constrained numerical predictor–corrector guidance for Mars precision landing[J]. Journal of Guidance, Control, and Dynamics, 2017, 40(1): 179-187.
83. Johnson B J, Cerimele C J, Stachowiak S, et al. Mid-lift-to-drag ratio rigid vehicle control system design and simulation for human Mars entry[C]//2018 AIAA Guidance, Navigation, and Control Conference. 2018: 0615.
84. Johnson B J, Lu P, Sostaric R R. Mid Lift-to-Drag Rigid Vehicle 6-DoF Performance for Human Mars Entry, Descent, and Landing: A Fractional Polynomial Powered Descent Guidance Approach[C]//AIAA SciTech 2020 Forum. 2020: 1513.
85. Lugo R A, Powell R, Dwyer-Cianciolo A M. Overview of a generalized numerical predictor-corrector targeting guidance with application to human-scale mars entry, descent, and landing[C]//AIAA Scitech 2020 Forum. 2020: 0846.
86. Ridderhof J, Tsiotras P, Johnson B J. Stochastic entry guidance[J]. Journal of Guidance, Control, and Dynamics, 2022, 45(2): 320-334.

87. Xue S, Lu P. Constrained predictor-corrector entry guidance[J]. Journal of Guidance, Control, and Dynamics, 2010, 33(4): 1273-1281.
88. Liang Z, Zhu S. Constrained predictor-corrector guidance via bank saturation avoidance for low L/D entry vehicles[J]. Aerospace Science and Technology, 2021, 109: 106448.
89. Lu P. Entry guidance: a unified method[J]. Journal of Guidance, Control, and Dynamics, 2014, 37(3): 713-728.
90. Introducing computational guidance and control[J]. Journal of Guidance, Control, and Dynamics, 2017, 40(2): 193–193.
91. Pan B, Ma Y, Yan R. Newton-type methods in computational guidance[J]. Journal of Guidance, Control, and Dynamics, 2019, 42(2): 377-383.
92. Reynolds T P. Computational guidance and control for aerospace systems[D]. University of Washington, 2020.
93. Wang Z, Grant M J. Autonomous entry guidance for hypersonic vehicles by convex optimization[J]. Journal of Spacecraft and Rockets, 2018, 55(4): 993-1006.
94. Sagliano M, Mooij E, Theil S. Onboard trajectory generation for entry vehicles via adaptive multivariate pseudospectral interpolation[J]. Journal of Guidance, Control, and Dynamics, 2016, 40(2): 466-476.
95. Bae J, Lee S D, Kim Y W, et al. Convex optimization-based entry guidance for spaceplane[J]. International Journal of Control, Automation and Systems, 2022, 20(5): 1652-1670.
96. Dominguez Calabuig G J, Mooij E. Optimal on-board abort guidance based on successive convexification for atmospheric re-entry[C]//AIAA Scitech 2021 Forum. 2021: 0860.
97. Tracy K, Manchester Z. CPEG: A Convex Predictor-corrector Entry Guidance Algorithm[C]//2022 IEEE Aerospace Conference (AERO). IEEE, 2022: 1–10.
98. Izzo D, Märtens M, Pan B. A survey on artificial intelligence trends in spacecraft guidance dynamics and control[J]. Astrodynamics, 2019, 3: 287-299.
99. Cheng L, Jiang F, Wang Z, et al. Multiconstrained real-time entry guidance using deep neural networks[J]. IEEE Transactions on Aerospace and Electronic Systems, 2020, 57(1): 325-340.
100. You S, Wan C, Dai R, et al. Learning-based optimal control for planetary entry, powered descent and landing guidance[C]//AIAA SciTech 2020 forum. 2020: 0849.
101. Shi Y, Wang Z. A deep learning-based approach to real-time trajectory optimization for hypersonic vehicles[C]//AIAA SciTech 2020 forum. 2020: 0023.
102. Jiang X, Li S, Furfaro R. Integrated guidance for Mars entry and powered descent using reinforcement learning and pseudospectral method[J]. Acta Astronautica, 2019, 163: 114-129.
103. Gaudet B, Drozd K, Furfaro R. Adaptive approach phase guidance for a hypersonic glider via reinforcement meta learning[C]//AIAA Scitech 2022 Forum. 2022: 2214.
104. Kazmierczak E J, Nakhjiri N. Improvements to modeling and trajectory simulation of mars aerocapture and aerobraking[C]//2018 Space Flight Mechanics Meeting. 2018: 2228.
105. Manrique J B. Advances in spacecraft atmospheric entry guidance[M]. University of California, Irvine, 2010.
106. Justh H L, Justus C G, Ramey H S. Mars-GRAM 2010: improving the precision of Mars-GRAM[C]//Fourth International Workshop on the Mars Atmosphere: Modelling and observations. 2011 (M11-0119).
107. Millour E, Forget F, Spiga A, et al. The Mars climate database (MCD version 5.2)[C]// European Planetary Science Congress. 2015, 10: 2015–2438.
108. Qin T, Wang S, Gao A, et al. Three-Dimensional Analytical Model for Mars Atmospheric Density[J]. Journal of Deep Space Exploration, 2022, 52(8), 1162-1174. (In Chinese)
109. Ushijima M, Hara H, Muramatsu A, et al. The operating characteristics of a ducted rocket in the Mars atmosphere[C]//46th AIAA/ASME/SAE/ASEE Joint Propulsion Conference & Exhibit. 2010: 6999.
110. Benito J, Mease K D. Reachable and controllable sets for planetary entry and landing[J]. Journal of Guidance, Control, and Dynamics, 2010, 33(3): 641-654.
111. Li Q, Rao W, Cheng X, et al. Aerodynamic design, analysis, and validation techniques for the Tianwen-1 entry module[J]. Astrodynamics, 2022, 6: 39-52.

112. Yao W, Chen X, Luo W, et al. Review of uncertainty-based multidisciplinary design optimization methods for aerospace vehicles[J]. Progress in Aerospace Sciences, 2011, 47(6): 450-479.
113. Kolmogorov A N, Bharucha-Reid A T. Foundations of the theory of probability: Second English Edition[M]. Courier Dover Publications, 2018.
114. Huang Y, Li H, Du X, et al. Mars entry trajectory robust optimization based on evidence under epistemic uncertainty[J]. Acta Astronautica, 2019, 163: 225-237.
115. Eberhart R, Kennedy J. Particle swarm optimization[C]//Proceedings of the IEEE International Conference on Neural Networks. 1995, 4: 1942–1948.
116. Shan J, Ren Y. Low-thrust trajectory design with constrained particle swarm optimization[J]. Aerospace Science and Technology, 2014, 36: 114-124.
117. Huntington G T. Advancement and analysis of a Gauss pseudospectral transcription for optimal control problems[D]. Massachusetts Institute of Technology, Department of Aeronautics and Astronautics, 2007.
118. Andrei N, Andrei N. A SQP algorithm for large-scale constrained optimization: SNOPT[J]. Continuous Nonlinear Optimization for Engineering Applications in GAMS Technology, 2017: 317–330.
119. Patterson M A, Rao A V. GPOPS-II: A MATLAB software for solving multiple-phase optimal control problems using hp-adaptive Gaussian quadrature collocation methods and sparse nonlinear programming[J]. ACM Transactions on Mathematical Software (TOMS), 2014, 41(1): 1-37.
120. Yang H, Bai X, Baoyin H. Rapid generation of time-optimal trajectories for asteroid landing via convex optimization[J]. Journal of Guidance, Control, and Dynamics, 2017, 40(3): 628-641.
121. Pinson R M, Lu P. Trajectory design employing convex optimization for landing on irregularly shaped asteroids[J]. Journal of Guidance, Control, and Dynamics, 2018, 41(6): 1243-1256.
122. Sagliano M, Mooij E. Optimal drag-energy entry guidance via pseudospectral convex optimization[J]. Aerospace Science and Technology, 2021, 117: 106946.
123. Wang J, Cui N, Wei C. Rapid trajectory optimization for hypersonic entry using convex optimization and pseudospectral method[J]. Aircraft Engineering and Aerospace Technology, 2019, 91(4): 669-679.
124. Szmuk M, Reynolds T P, Açıkmeşe B. Successive convexification for real-time six-degree-of-freedom powered descent guidance with state-triggered constraints[J]. Journal of Guidance, Control, and Dynamics, 2020, 43(8): 1399-1413.
125. Yang R, Liu X. Fuel-optimal powered descent guidance with free final-time and path constraints[J]. Acta Astronautica, 2020, 172: 70-81.
126. Wang Z, Grant M J. Optimization of minimum-time low-thrust transfers using convex programming[J]. Journal of Spacecraft and Rockets, 2018, 55(3): 586-598.
127. Mao Y, Szmuk M, Açıkmeşe B. Successive convexification of non-convex optimal control problems and its convergence properties[C]//2016 IEEE 55th Conference on Decision and Control (CDC). IEEE, 2016: 3636–3641.
128. Wang Z, Lu Y. Improved sequential convex programming algorithms for entry trajectory optimization[J]. Journal of Spacecraft and Rockets, 2020, 57(6): 1373-1386.
129. Wang Z, McDonald S T. Convex relaxation for optimal rendezvous of unmanned aerial and ground vehicles[J]. Aerospace Science and Technology, 2020, 99: 105756.
130. Lofberg J. YALMIP: A toolbox for modeling and optimization in MATLAB[C]//2004 IEEE International Conference on Robotics and Automation (IEEE Cat. No. 04CH37508). IEEE, 2004: 284–289.
131. ApS M. Mosek optimization toolbox for matlab[J]. User's Guide and Reference Manual, Version, 2019, 4: 1.
132. Hong H, Maity A, Holzapfel F, et al. Model predictive convex programming for constrained vehicle guidance[J]. IEEE Transactions on Aerospace and Electronic Systems, 2019, 55(5): 2487-2500.
133. Maity A, Oza H B, Padhi R. Generalized model predictive static programming and angle-constrained guidance of air-to-ground missiles[J]. Journal of Guidance, Control, and Dynamics, 2014, 37(6): 1897-1913.

134. Shen J, Tang T, Wang L L. Spectral methods: algorithms, analysis and applications[M]. Springer Science & Business Media, 2011.
135. Foust R, Chung S J, Hadaegh F Y. Optimal guidance and control with nonlinear dynamics using sequential convex programming[J]. Journal of Guidance, Control, and Dynamics, 2020, 43(4): 633-644.
136. Hong H, Piprek P, Gerdts M, et al. Computationally efficient trajectory generation for smooth aircraft flight level changes[J]. Journal of Guidance, Control, and Dynamics, 2021, 44(8): 1532-1540.
137. Hong H, Piprek P, Afonso R J M, et al. Trigonometric series-based smooth flight trajectory generation[J]. IEEE Transactions on Aerospace and Electronic Systems, 2020, 57(1): 721-728.
138. Oza H B, Padhi R. Impact-angle-constrained suboptimal model predictive static programming guidance of air-to-ground missiles[J]. Journal of Guidance, Control, and Dynamics, 2012, 35(1): 153-164.
139. Maity A, Oza H B, Padhi R. Generalized model predictive static programming and angle-constrained guidance of air-to-ground missiles[J]. Journal of Guidance, Control, and Dynamics, 2014, 37(6): 1897-1913.
140. Hong H, Piprek P, Afonso R J M, et al. Hierarchical parameterization for seamless transition trajectory generation[J]. IEEE Transactions on Aerospace and Electronic Systems, 2021, 57(5): 2878-2888.
141. Yang L, Liu X, Chen W, et al. Autonomous entry guidance using linear pseudospectral model predictive control[J]. Aerospace Science and Technology, 2018, 80: 38-55.
142. Li Y, Chen W, Yang L. Multistage linear Gauss pseudospectral method for piecewise continuous nonlinear optimal control problems[J]. IEEE Transactions on Aerospace and Electronic Systems, 2021, 57(4): 2298-2310.
143. Mathavaraj S, Padhi R. Quasi-spectral unscented MPSP guidance for robust soft-landing on asteroid[J]. Journal of Optimization Theory and Applications, 2021, 191: 823-845.
144. Mondal S, Padhi R. Constrained quasi-spectral MPSP with application to high-precision missile guidance with path constraints[J]. Journal of Dynamic Systems, Measurement, and Control, 2021, 143(3): 031001.
145. Liu X, Shen Z, Lu P. Exact convex relaxation for optimal flight of aerodynamically controlled missiles[J]. IEEE Transactions on Aerospace and Electronic Systems, 2016, 52(4): 1881-1892.
146. Guo X, Zhu M. Direct trajectory optimization based on a mapped Chebyshev pseudospectral method[J]. Chinese Journal of Aeronautics, 2013, 26(2): 401-412.
147. Cai W, Zhu Y, Yang L, et al. Optimal guidance for hypersonic reentry using inversion and receding horizon control[J]. IET Control Theory & Applications, 2015, 9(9): 1347-1355.
148. Berrut J P, Trefethen L N. Barycentric lagrange interpolation[J]. SIAM review, 2004, 46(3): 501-517.
149. Kosloff D, Tal-Ezer H. A modified Chebyshev pseudospectral method with an O (N-1) time step restriction[J]. Journal of Computational Physics, 1993, 104(2): 457-469.
150. Mall K, Taheri E. Entry trajectory optimization for Mars science laboratory class missions using indirect uniform trigonometrization method[C]//2020 American Control Conference (ACC). IEEE, 2020: 4182–4187.
151. Mall K, Grant M J, Taheri E. Uniform trigonometrization method for optimal control problems with control and state constraints[J]. Journal of Spacecraft and Rockets, 2020, 57(5): 995-1007.
152. Mall K, Taheri E. Unified trigonometrization method for solving optimal control problems in atmospheric flight mechanics[C]//AIAA Scitech 2020 Forum. 2020: 0022.
153. Li Y, Guan Y, Wei C, et al. Optimal control of ascent trajectory for launch vehicles: A convex approach[J]. IEEE Access, 2019, 7: 186491-186498.
154. Lu P. Propellant-optimal powered descent guidance[J]. Journal of Guidance, Control, and Dynamics, 2018, 41(4): 813-826.
155. Lu P. Augmented apollo powered descent guidance[J]. Journal of Guidance, Control, and Dynamics, 2019, 42(3): 447-457.

156. Darby C L. hp-Pseudospectral method for solving continuous-time nonlinear optimal control problems[D]. University of Florida, 2011.
157. Thrun S, Littman M L. Reinforcement learning: an introduction[J]. AI Magazine, 2000, 21(1): 103-103.
158. Gaudet B, Furfaro R. Adaptive pinpoint and fuel efficient mars landing using reinforcement learning[J]. IEEE/CAA Journal of Automatica Sinica, 2014, 1(4): 397-411.
159. Gaudet B, Linares R, Furfaro R. Deep reinforcement learning for six degree-of-freedom planetary landing[J]. Advances in Space Research, 2020, 65(7): 1723-1741.
160. Martinez W L, Martinez A R. Computational statistics handbook with MATLAB[M]. CRC press, 2015.
161. Bishop C M, Nasrabadi N M. Pattern recognition and machine learning[M]. New York: springer, 2006.
162. Zhao Y, Tsiotras P. A density-function based mesh refinement algorithm for solving optimal control problems[C]//Infotech and Aerospace Conference. 2009: 2009–2019.
163. Guimaraes D A, Floriano G H F, Chaves L S. A tutorial on the CVX system for modeling and solving convex optimization problems[J]. IEEE Latin America Transactions, 2015, 13(5): 1228-1257.
164. Seywald K, Seywald H. Desensitized optimal control[C]//AIAA SciTech 2019 forum. 2019: 0651.
165. Shen H, Seywald H, Powell R W. Desensitizing the minimum-fuel powered descent for Mars pinpoint landing[J]. Journal of guidance, control, and dynamics, 2010, 33(1): 108-115.
166. Makkapati V R, Maity D, Dor M, et al. C-DOC: Co-state desensitized optimal control[C]// 2020 American Control Conference (ACC). IEEE, 2020: 1761–1766.
167. Tu L, Yuan J. Reentry trajectory optimization using direct collocation method and nonlinear programming[C]//57th International Astronautical Congress. 2006: C1. 4.06.
168. Hargraves C R, Paris S W. Direct trajectory optimization using nonlinear programming and collocation[J]. Journal of Guidance, Control, and Dynamics, 1987, 10(4): 338-342.
169. Gill P E, Murray W, Saunders M A. SNOPT: An SQP algorithm for large-scale constrained optimization[J]. SIAM review, 2005, 47(1): 99-131.
170. Xiu D. Numerical methods for stochastic computations: a spectral method approach[M]. Princeton University Press, 2010.
171. Huang Y, Li H, Zhang J. Uncertainty analysis of reachable set for planetary entry using polynomial chaos[J]. Advances in Space Research, 2017, 60(11): 2491-2504.
172. Prabhakar A, Fisher J, Bhattacharya R. Polynomial chaos-based analysis of probabilistic uncertainty in hypersonic flight dynamics[J]. Journal of Guidance, Control, and Dynamics, 2010, 33(1): 222-234.
173. Cheng K, Lu Z. Adaptive sparse polynomial chaos expansions for global sensitivity analysis based on support vector regression[J]. Computers & Structures, 2018, 194: 86-96.
174. Wang F, Xiong F, Jiang H, et al. An enhanced data-driven polynomial chaos method for uncertainty propagation[J]. Engineering Optimization, 2018, 50(2): 273-292.
175. Nouy A, Le Maı^tre O P. Generalized spectral decomposition for stochastic nonlinear problems[J]. Journal of Computational Physics, 2009, 228(1): 202-235.
176. Le Maître O, Knio O M. Spectral methods for uncertainty quantification: with applications to computational fluid dynamics[M]. Springer Science & Business Media, 2010.
177. Jiang X, Li S. Analysis of uncertainty propagation in Mars entry using non-intrusive polynomial chaos[C]//AAS/AIAA Astrodynamics Specialist Conference, Snowbird, UT. 2018: 19-23.
178. Dutta S, Braun R D, Karlgaard C D. Uncertainty quantification for mars entry, descent, and landing reconstruction using adaptive filtering[J]. Journal of Spacecraft and Rockets, 2014, 51(3): 967-977.
179. Karniadakis G, Sherwin S. Spectral/hp element methods for computational fluid dynamics[M]. Oxford University Press, 2005.
180. Sepahvand K, Marburg S, Hardtke H J. Uncertainty quantification in stochastic systems using polynomial chaos expansion[J]. International Journal of Applied Mechanics, 2010, 2(02): 305-353.

181. Chi H, Mascagni M, Warnock T. On the optimal Halton sequence[J]. Mathematics and Computers in Simulation, 2005, 70(1): 9-21.
182. Ökten G, Göncü A. Generating low-discrepancy sequences from the normal distribution: Box–Muller or inverse transform?[J]. Mathematical and Computer Modelling, 2011, 53(5-6): 1268-1281.
183. Escribano T M V. Poincaré sections and resonant orbits in the restricted three-body problem[D]. Purdue University, 2010.
184. Luo Y, Yang Z. A review of uncertainty propagation in orbital mechanics[J]. Progress in Aerospace Sciences, 2017, 89: 23-39.
185. Jiang X. Uncertainty quantification for Mars atmospheric entry using polynomial chaos and spectral decomposition[C]//2018 AIAA Guidance, Navigation, and Control Conference. 2018: 1317.
186. Valdebenito M A, Schuëller G I. A survey on approaches for reliability-based optimization[J]. Structural and Multidisciplinary Optimization, 2010, 42: 645-663.
187. Huang Y, Li H. Reliability-based trajectory optimization using nonintrusive polynomial chaos for Mars entry mission[J]. Advances in Space Research, 2018, 61(11): 2854-2869.
188. Bäck J, Nobile F, Tamellini L, et al. Stochastic spectral Galerkin and collocation methods for PDEs with random coefficients: a numerical comparison[C]//Spectral and High Order Methods for Partial Differential Equations: Selected papers from the ICOSAHOM'09 conference, June 22–26, Trondheim, Norway. Springer Berlin Heidelberg, 2011: 43–62.
189. Bertsimas D, Brown D B, Caramanis C. Theory and applications of robust optimization[J]. SIAM Review, 2011, 53(3): 464-501.
190. Lee S H, Chen W. A comparative study of uncertainty propagation methods for black-box-type problems[J]. Structural and Multidisciplinary Optimization, 2009, 37: 239-253.
191. Klimke A, Wohlmuth B. Algorithm 847: spinterp: Piecewise multilinear hierarchical sparse grid interpolation in MATLAB[J]. ACM Transactions on Mathematical Software (TOMS), 2005, 31(4): 561-579.
192. Chai R, Savvaris A L, Tsourdos A. Violation learning differential evolution-based hp-adaptive pseudospectral method for trajectory optimization of space maneuver vehicle[J]. IEEE Transactions on Aerospace and Electronic Systems, 2017, 53(4): 2031-2044.
193. Ren Y, Shan J. Reliability-based soft landing trajectory optimization near asteroid with uncertain gravitational field[J]. Journal of Guidance, Control, and Dynamics, 2015, 38(9): 1810-1820.
194. Der Kiureghian A, Dakessian T. Multiple design points in first and second-order reliability[J]. Structural Safety, 1998, 20(1): 37-49.
195. Yang C, Kumar M. An adaptive Monte Carlo method for uncertainty forecasting in perturbed two-body dynamics[J]. Acta Astronautica, 2019, 155: 369-378.
196. Li S, Peng Y. Mars entry trajectory optimization using DOC and DCNLP[J]. Advances in Space Research, 2011, 47(3): 440-452.
197. Jiang X, Li S. Mars atmospheric entry trajectory optimization via particle swarm optimization and Gauss pseudo-spectral method[J]. Proceedings of the Institution of Mechanical Engineers, Part G: Journal of Aerospace Engineering, 2016, 230(12): 2320-2329.
198. D'Souza S, Sarigul-Klijn N. Investigation of trajectory generation for a mission adaptive planetary entry guidance algorithm[C]//AIAA Atmospheric Flight Mechanics Conference. 2012: 4508.
199. Patelli E, Broggi M, Angelis M, et al. OpenCossan: An efficient open tool for dealing with epistemic and aleatory uncertainties[M]//Vulnerability, Uncertainty, and Risk: Quantification, Mitigation, and Management. 2014: 2564–2573.
200. Cui P, Zhao Z, Yu Z, et al. Terminal altitude maximization for Mars entry considering uncertainties[J]. Acta Astronautica, 2018, 145: 446-455.
201. Kozynchenko A I. Analysis of predictive entry guidance for a Mars lander under high model uncertainties[J]. Acta Astronautica, 2011, 68(1-2): 121-132.
202. Dai J, Gao A, Xia Y. Mars atmospheric entry guidance for reference trajectory tracking based on robust nonlinear compound controller[J]. Acta Astronautica, 2017, 132: 221-229.

203. Benito J, Mease K. Nonlinear predictive controller for drag tracking in entry guidance[C]// AIAA/AAS Astrodynamics Specialist Conference and Exhibit. 2008: 7350.
204. Mease K D, Chen D T, Teufel P, et al. Reduced-order entry trajectory planning for acceleration guidance[J]. Journal of Guidance, Control, and Dynamics, 2002, 25(2): 257-266.
205. Mooij E. Model reference adaptive guidance for re-entry trajectory tracking[C]//AIAA Guidance, Navigation, and Control Conference and Exhibit. 2004: 4775.
206. Tian B, Zong Q. Optimal guidance for reentry vehicles based on indirect Legendre pseudospectral method[J]. Acta Astronautica, 2011, 68(7-8): 1176-1184.
207. Zhao J, Li S. Mars atmospheric entry trajectory optimization with maximum parachute deployment altitude using adaptive mesh refinement[J]. Acta Astronautica, 2019, 160: 401-413.
208. Liu X, Li S, Xin M. Pseudospectral Convex Optimization based Model Predictive Static Programming for Constrained Guidance[J]. IEEE Transactions on Aerospace and Electronic Systems, 2023, 59(3): 2232 - 2244.